# IRAM Conference "Multi-line Diagnostics of the Interstellar Medium"



EPJ Web of Conferences Volume 265 (2022)

Nice, France
4 - 6 April 2022

**Editors:**

**Laure Bouscasse**
**Carsten Kramer**
**Frederic Gueth**

ISBN: 978-1-7138-6059-4

**Printed from e-media with permission by:**

Curran Associates, Inc.
57 Morehouse Lane
Red Hook, NY 12571

**Some format issues inherent in the e-media version may also appear in this print version.**



No changes have been made to the content of these proceedings. There may be changes to pagination and minor adjustments for aesthetics.

Printed with permission by Curran Associates, Inc. (2022)

For additional information, please contact EDP Sciences – Web of Conferences at the address below.

EDP Sciences – Web of Conferences
17, Avenue du Hoggar
Parc d'Activité de Courtabœuf
BP 112
F-91944 Les Ulis Cedex A
France

Phone: +33 (0) 1 69 18 75 75
Fax: +33 (0) 1 69 28 84 91

contact-edps@webofconferences.org

**Additional copies of this publication are available from:**

Curran Associates, Inc.
57 Morehouse Lane
Red Hook, NY 12571 USA
Phone: 845-758-0400
Fax: 845-758-2633
Email: curran@proceedings.com
Web: www.proceedings.com

# TABLE OF CONTENTS

## Scientific Organizing Committee

Cecilia CECCARELLI, IPAG, France

José CERNICHARO, IFF-CSIC, Spain

Santiago GARCIA-BURILLO, OAN, Spain

Maryvonne GERIN, ENS, France

Stéphane GUILLOTEAU, LAB, France

Franck LE PETIT, Observatoire de Paris, Meudon, France

Roberto NERI, IRAM, France

Nami SAKAI, RIKEN, Japan

Eva SCHINNERER, MPIA, Germany

Karl SCHUSTER, IRAM, France

Linda J. TACCONI, MPE, Germany

Axel WEISS, MPIfR, Germany

## Local Organizing Committee

Arancha CASTRO-CARRIZO

Julien CHALAIN

Cécile DESCAMPS-PROTEAU

Jaimey DURAND

Frédéric GUETH

Carsten KRAMER

Sonja MOREAU

Jérôme PETY

## Acknowledgments

The Local Organizing Committee would like to thank the city of Nice for their help in the logistics.

All authors and anonymous referees should be thanked for their contribution to the editing process of the conference proceedings.

## List of Participants

Aghababaei Atefeh, U. of Cologne, Germany
Agúndez Marcelino, IFF-CSIC, Spain
Andreani Paola, ESO, Germany
Antonellini Stefano, IRAM, France
Arumugam Vinod, IRAM, France
Barcos-Muñoz Loreto, NRAO, USA
Barnes Ashley, AIfA, Germany
Beelen Alexandre, U. Aix-Marseille, France
Belete Asnakew, Chalmers U. of Technology, Sweden
Bergin Edwin, U. of Michigan, USA
Berta Stefano, IRAM, France
Beslic Ivana, AIfA, Germany
Beuther Henrik, MPIA, Germany
Bianchi Eleonora, IPAG, France
Bigiel Frank, AIfA, Germany
Bisbas Thomas, U. of Cologne, Germany
Bischetti Manuela, INAF, Italy
Black John, Chalmers U. of Technology, Sweden
Bonanomi Francesca, U. of Vienna, Austria
Bonfand Mélisse, LAB, France
Bouscasse Laure, IRAM, France
Bouvier Mathilde, IPAG, France
Bron Emeric, LERMA, Obs. de Paris, France
Brouillet Nathalie, LAB, France
Butler Kirsty, IRAM, France
Butterworth Joshua, Leiden U., The Netherlands
Carl Tadeus, Chalmers U. of Technology, Sweden
Carpenter John, JAO, Chile
Caselli Paola, MPE, Germany
Ceccarelli Cecilia, IPAG, France
Cernicharo José, IFF-CSIC, Spain
Chacón-Tanarro Ana, OAN-IGN, Spain
Chahine Layal, IRAM, France
Chapillon Edwige, IRAM, LAB, France
Chen Qingxiang, National U. of Singapore, Singapore
Christensen Ivalu, MPIfR, France
Ciocan Bianca-Iulia, U. of Vienna, Austria
Colzi Laura, CAB, CSIC-INTA, Spain
Combes Francoise, LERMA, Obs. de Paris, France
Cox Pierre, CNRS, IAP, France
De Simone Marta, IPAG, France, INAF, Italy
De Souza Magalhaes Victor, IRAM, France
Den Brok Jakob, AIfA, Germany
Duarte João, U. of Lisbon, Portugal
Dutrey Anne, LAB, France
Dzib Sergio, IRAM, France
Eibensteiner Cosima, AIfA, Germany
Esplugues Gisela, OAN, Spain

Falgarone Edith, LPENS, CNRS, ENS, France
Fehér Orsolya, IRAM, France
Fuente Asunción, OAN, Spain
Gao Yu, Xiamen U., China
García-Burillo Santiago, OAN, Spain
García-Rodríguez Axel, OAN-IGN, Spain
García Santa-Maria Miriam, IFF-CSIC, Spain
Gebek Andrea , Ghent U., Belgium
Genzel Reinhard, MPE, Germany
Gérin Maryvonne, LERMA, Obs. de Paris, France
Goicoechea Javier, IFF-CSIC, Spain
Gonzalez-Gaitan Santiago, U. of Lisbon, Portugal
Guélin Michel, IRAM, France
Gueth Frédéric, IRAM, France
Guilloteau Stéphane, LAB, France
Gururajan Gayathri, U. Aix-Marseille, France
Gusdorf Antoine, LERMA, Obs. de Paris, ENS, France
Hacar Alvaro, U. of Vienna, Austria
Hennebelle Patrick, CEA, France
Holguin Francisco , U. of Michigan, USA
Huang Ko-Yun, Leiden U., The Netherlands
Ismail Diana, LAM, France
Jeste Manali, MPIfR, Germany
Jiménez Donaire María Jesus, OAN, Spain
Jones Gareth, U. of Oxford, UK
Kaasinen Melanie, ESO, Germany
Kang Da Eun, U. of Heidelberg, Germany
Kapoor Anand Utsav, Ghent U., Belgium
Kauffmann Jens, Haystack Obs., MIT, USA
Kim Wonju, U. of Cologne, Germany
Klessen Ralf, U. of Heidelberg, Germany
Kramer Carsten, IRAM, France
Le Petit Franck, LERMA, Obs. de Paris, France
Lebouteiller Vianney, AIM, CEA, U. Paris-Saclay, France
Lefloch Bertrand, IPAG, France
Liszt Harvey, NRAO, USA
López-Sepulcre Ana, IRAM, IPAG, France
Madden Suzanne, AIM, CEA, U. Paris-Saclay, France
Maillard Vincent, LERMA, Obs. de Paris, France
Mancillas Vaquera Brisa, IRAM, Spain
Mangum Jeffrey, NRAO, USA
Marcelino Nuria, OAN-IGN, Spain
Mercimek Seyma, IRAM, France, INAF, Italy
Meshaka Raphaël, Obs. de Paris, France
Misquitta Persis, U. of Cologne, Germany
Montoya Arroyave Isabel, U. of Oslo, Norway
Muller Sebastien, Chalmers U. of Technology, Sweden
Muñoz-Elgueta Nahir, MPIfA, Germany
Navarro David, OAN, Spain
Neralwar Kartik Rajan, MPIfR, Germany

NERI Roberto, IRAM, France
NESVADBA Nicole, Obs. de la Cote d'Azur, France
NEUMANN Lukas, AIfA, Germany
OLIVARES Valeria, U. of Kentucky, USA
OMONT Alain, IAP, France
ORKISZ Jan, Chalmers U. of Technology, Sweden
PABST Cornelia, Leiden U., The Netherlands
PALUD Pierre, Centrale Lille Institut, France
PANTONI Lara, CEA-IRFU, DAP, France
PAUBERT Gabriel, IRAM, Spain
PENSABENE Antonio, U. of Milano-Bicocca, Italy
PETY Jérôme, IRAM, LERMA, France
PINEDA Jaime, MPE, Germany
PLUME Rene, U. of Calgary, Canada
PODIO Linda, INAF, Italy
PUNANOVA Anna, Ural Federal U., Russia
RAMAMBASON Lise, AIM, CEA, U. Paris-Saclay, France
RIECHERS Dominik, U. of Cologne, Germany
RIVIÈRE-MARICHALAR Pablo, OAN, Spain
RODRÍGUEZ BARAS Marina, OAN, Spain
RUIZ Dary, NRAO, USA
RYBAK Matus, Leiden U., The Netherlands
RYBARCZYK Daniel, U. of Wisconsin-Madison, USA
SALVESTRINI Francesco, INAF, Italy
SANCHEZ-PORTAL Miguel, IRAM, Spain
SATO Mamiko, Chalmers U. of Technology, Sweden
SCHINNERER Eva, MPIA, Germany
SCHLEMMER Stephan, U. of Cologne, Germany
SCHNEIDER Nicola, U. of Cologne, Germany
SCHUSTER Karl, IRAM, France
SEGURA-COX Dominique, U. of Texas at Austin, USA
SKRETAS Iason-Michail, MPIfR, Germany
SMART Brianna, U. of Hertfordshire, UK
SOCCI Andrea, U. of Vienna, Austria
SOLER Juan Diego, IAPS-INAF, Italy
SPEZZANO Silvia, MPE, Germany
STANIMIROVIC Snezana, U. of Wisconsin-Madison, USA
STANLEY Flora, IRAM, France
STUBER Sophia, MPIA, Germany
SURI Sümeyye, U. of Vienna, Austria
TABATABAEI MAZRAEH NO Farideh Sadat, MPE, Germany
TACCONI Linda, MPE, Germany
TAFALLA Mario, OAN-IGN, Spain
THEULE Patrice, U. Aix-Marseille, France
TRAFICANTE Alessio, IAPS-INAF, Italy
URQUHART Sheona, Open U., UK
USERO Antonio, OAN-IGN, Spain
VALDIVIA MENA Maria Teresa, MPE, Germany
VASTEL Charlotte, UPS-OMP, IRAP, France
VIDAL GARCIA Alba, LPENS, ENS, France

VITI Serena, Leiden U., The Netherlands
WEISS Axel , MPIfR, Germany
WINTERS Jan Martin, IRAM, France
WIRSTRÖM Eva, Chalmers U. of Technology, Sweden
WONG Ka Tat, IRAM, France
WYROWSKI Friedrich, MPIfR, Germany
YANG Chentao, Chalmers U. of Technology, Sweden
YTTERGREN Madeleine, Chalmers U. of Technology, Sweden

## List of Posters

First detection of extended SiO toward a filamentary hubcluster NGC6334 V
*Atefeh Aghababaei*

Mid-IR lines and continuum signatures of dust drift and accretion in Protoplanetary disks
*Stefano Antonellini*

Line emission and absorption dominating the optically thin spectrum in Arp 220 at 40 pc resolution with ALMA
*Loreto Barcos-Munoz*

The star forming galaxy HerBS-89a at z=2.95 and its gas inflow
*Stefano Berta*

Molecular spectroscopy across nearby star-forming disc galaxies
*Ivana Beslic*

The complex chemistry of young high-mass star forming regions
*Mélisse Bonfand*

Early warm-up phase chemistry towards a sample of emerging hot cores
*Laure Bouscasse*

The hunt for hot corinos and WCCC objects in the OMC–2/3 filament
*Mathilde Bouvier*

Characterisation of the best high-mass prestellar core candidate found so far and of the hot cores in W43–MM1
*Nathalie Brouillet*

Molecular Outflows in z>6 Unobscured Quasar Hosts - Driven by Star Formation
*Kirsty Butler*

Molecular Line Ratios as Tracers of Galactic Environments within NGC 1068
*Joshua Butterworth*

Deep Search for Glycine in Barnard 5
*Tadeus Carl*

The ALMA Wideband Sensitivity Upgrade
*John Carpenter*

Shocked gas and filamentary structures of OMC–2 FIR 4
*Layal Chahine*

Large Scale Deuteration in the Heart of the Swan, Cygnus–X
*Ivalu Christensen*

The VLT-MUSE and ALMA view of the MACS 1931.8–2635 brightest cluster galaxy
*Bianca-Iulia Ciocan*

CUBE: A new IRAM Science Software
*Victor De Souza Magalhaes*

A Sample of Dust Attenuation Laws for DES Galaxies
*João Duarte*

Molecular Fireworks: The high angular resolution 2–3 mm molecular line survey covering the central arc minute of the Fireworks Galaxy NGC 6946
*Cosima Eibensteiner*

A sulfur journey across star-forming regions: Study of thioformaldehyde emission
*Gisela Esplugues*

High resolution spectral imaging of CO(7–6), [CI](2–1), and continuum of three high-z lensed dusty star-forming galaxies using ALMA
*Gururajan Gayathri*

Investigating the relation between diffuse molecular gas and the dark neutral medium
*Maryvonne Gérin*

Extinction and interstellar lines in supernova spectra
*Santiago Gonzalez-Gaitan*

Imaging 100 000 channels with NOEMA  ALMA: The IMAGER program in GILDAS
*Stéphane Guilloteau*

Impact of cosmic rays on observational signatures in the circumgalactic medium
*Francisco Holguin*

The Nearby Evolved Stars Survey
*Manali Jeste*

First results from VERTICO: The Virgo Environment Traced in CO Survey
*Maria Jesus Jimenez Donaire*

Gas and Star Formation from HD and Dust Emission in a z~5.7 Strongly Lensed Starburst Galaxy
*Gareth Jones*

(Supra-)Thermalised CO in normal, z~2, star-forming galaxies
*Melanie Kaasinen*

Emission-line diagnostics of HII regions using conditional Invertible Neural Networks
*Kang Da Eun*

Probing the dust and gas evolution in starless cores
*Carsten Kramer*

Topological models to infer multi-phase ISM properties
*Vianney Lebouteiller*

Dynamical effects of the radiative stellar feedback on the HI-to-$H_2$ transition
*Vincent Maillard*

Formaldehyde deuteration from protostars to comets: Results from IRAM–30m and follow-up studies with ALMA
*Seyma Mercimek*

APEX at the QSO MUSEUM: Molecular gas reservoirs associated with z~3 quasars and their link to the extended Ly$\alpha$ emission
*Nahir Munoz-Elgueta*

The missing link: How do cloud-scale molecular gas properties connect to global dense gas fraction and dense gas star formation efficiency across nearby galaxies?
*Lukas Neumann*

Dealing with uncertainty in millimeter astronomy: Towards more realistic noise model and Bayesian approach
*Pierre Palud*

The crucial role of molecular gas to constrain galaxy evolution: The case of 4 Dusty Star-Forming Galaxies at the Cosmic Noon
*Lara Pantoni*

Cold methanol formation: Testing model predictions
*Anna Punanova*

Inferring the escape fractions of ionizing photons from HII regions in the Dwarf Galaxy Survey
*Lise Ramambason*

Gas phase Elemental abundances in Molecular cloudS (GEMS). Analysis of observational results and statistical trends
*Marina Rodríguez Baras*

Full of Orions? A 200 pc resolution mapping of ISM in a z~3 dusty galaxy
*Matus Rybak*

From atomic to molecular gas in the diffuse interstellar medium: The role of multi-phase neutral hydrogen
*Daniel Rybarczyk*

Unveiling the AGN impact on the ISM in local Seyfert 2 galaxies
*Francesco Salvestrini*

edp sciences

# Statement of Peer review

In submitting conference proceedings to *Web of Conferences*, the editors of the proceedings certify to the Publisher that

1. They adhere to its **Policy on Publishing Integrity** in order to safeguard good scientific practice in publishing.
2. All articles have been subjected to peer review administered by the proceedings editors.
3. Reviews have been conducted by expert referees, who have been requested to provide unbiased and constructive comments aimed, whenever possible, at improving the work.
4. Proceedings editors have taken all reasonable steps to ensure the quality of the materials they publish and their decision to accept or reject a paper for publication has been based only on the merits of the work and the relevance to the journal.

Title, date and place of the conference

Multi-line Diagnostics of the Interstellar Medium
4-6 April 2022
Hotel Le Saint Paul, Nice, France

Proceedings editor(s):

Laure Bouscasse, Carsten Kramer, Frédéric Gueth

Date and editor's signature

1-August-2022

# Foreword
# Multi-line Diagnostics of the Interstellar Medium

The IRAM conference "Multi-line Diagnostics of the Interstellar Medium" took place at the Côte d'Azur in Nice (France), April 4-6, 2022. The conference has brought together more than 150 scientists scientists from a wide interdisciplinary community to address recent advances, current challenges and future directions in millimeter and submillimeter astronomy. The program included 60 talks combined with dedicated sessions to discuss 50 posters covering a wide range of contemporary topics in mm-astrophysics and beyond. In the evening of the 2nd day, the participants met at the Negresco hotel at the Promenade des Anglais for a relaxed conference dinner with after-dinner speaker and Nobel prize winner Reinhard Genzel. After a long pause due to the pandemic, the participants really enjoyed meeting and exchanging again face-to-face in-person with their colleagues.

The conference program covered many aspects of millimeter and submillimeter astronomy, from the chemical complexity of the interstellar medium in the local and early Universe, to that of the environment of stars in the earliest stages of formation and the physical and chemical evolution of galaxies over cosmic time. Topics included in addition data analysis, theoretical modeling and simulations, and laboratory spectroscopy.

The meeting provided also a lively platform for discussing new approaches exploiting the unique capabilities of the world's best millimeter and submillimeter facilities. With a special focus on the expanded capabilities of the NOEMA interferometer with its 12 antennae and the unique capabilities of the IRAM 30-meter telescope, participants discussed the possibilities of spectral line measurements of molecular and atomic species with higher sensitivity, larger fields-of-view, better spectral and spatial resolutions than is currently possible.

Recent advances in broad band millimeter and submillimeter instrumentation, as well as modelling and laboratory measurements have changed the nature of research on interstellar matter. With modern observatories one can now measure a vastly increased number of lines from a great number of molecular and atomic species simultaneously. Spectroscopic data will also be available over larger areas by means of multi-beam instruments or interferometric mosaicking and on-the-fly imaging techniques. Current and upcoming modeling and analysis methods including statistical treatments of large data sets and source surveys will also open new opportunities.

These facts will form the basis for the next crucial steps towards a global description of cold and warm interstellar matter responsible for star and planet formation on local and cosmological scales. Modern multi-line observations have the power to grasp molecular complexity to its full extent and its dependence on local conditions, to determine physicochemical processes within and potentially also to reconstruct the history of the interstellar medium (ISM). This increased understanding will provide a quantitative foundation for sources where fewer detectable lines are available, such as in high-redshift galaxies. Comparisons to dust emission and multi-wavelength studies will pin down key parameters such as metallicity, temperature, density, and radiation fields. All these advances will lead to the possibility to understand structure formation from cosmological scales down to the circumstellar environment in greatly enhanced detail.

Laure Bouscasse, Carsten Kramer, Frédéric Gueth
Scientific editors of the conference on "Multi-line Diagnostics of the Interstellar Medium".

# Status of the IRAM observatories

*Roberto* Neri[1,*] and *Miguel* Sánchez-Portal[2,**]

[1]IRAM, 300 rue de la Piscine, 38406 Saint-Martin d'Hères, France
[2]IRAM, Avenida Divina Pastora 7, 18102 Granada, Spain

**Abstract.** The NOEMA interferometer and the IRAM 30-meter telescope offer a range of outstanding capabilities that will continue to be expanded as part of IRAM's long-term development strategy.

## 1 Introduction

NOEMA and the IRAM 30-meter telescope were designed and built to provide world-class observing capabilities for thousands of researchers across the world. The two observatories together, provide a unique combination of cutting-edge technology to help researchers address key astronomical questions such as those related to the molecular complexity of star- and planet-formation processes in the Milky Way, structure formation and galaxy evolution in the Universe, and along the way helping researchers to understand the link between the formation of complex organic molecules in space and the emergence of life on Earth.

## 2 NOEMA

NOEMA is in a constant cycle of expanding its capabilities and opening new avenues to exploring the Universe. Among the most notable improvements the observatory has seen in recent years are the doubling of the number of antennas from 6 to 12 and the implementation of the antenna phasing-up capability for very long baseline interferometry (VLBI) in support of the Event Horizon Telescope (EHT, 230 and 345 GHz) and Global mm-VLBI array (GMVA, 86 GHz) campaigns. Current capabilities and those expected in the near future are presented here.

### 2.1 Current Capabilities

The receivers are equipped with dual-polarization sideband-separating mixing blocks, covering a nominal bandpass of 7.744 GHz in each polarization and sideband, and operate in the four atmospheric windows 70-116 GHz (Band 1), 127-179 GHz (Band 2), 200-276 GHz (Band 3), and 275-373 GHz (Band 4). While NOEMA currently only allows observations in the lower three frequency bands, science operations in the highest frequency band are expected to resume in 2024.

*e-mail: neri@iram.fr
**e-mail: msanchez@iram.es

**Figure 1.** View of the NOEMA observatory as seen from the southeast. The antennas are arranged in the most compact configuration. The observatory is located in the French Alps at an altitude of 2550 m above sea level. NOEMA is operated by IRAM, an international research institute supported by the CNRS (France), MPG (Germany) and IGN (Spain). Photo by J. Boissier.

The PolyFiX correlator is able to process a 7.744 GHz bandwidth for each receiver in each sideband and polarization, for a total of 31 GHz bandwidth. The correlator, which incorporates FPGA technology to take advantage of flexible correlator configurations provides NOEMA with three basic operating modes. One mode allows observations with a channel spacing of 2 MHz over the entire 31 GHz bandwidth, with up to 128 high-resolution windows in parallel, each 64 MHz wide and with a channel spacing of 62.5 kHz. Another mode, recently commissioned (Q2/2022) and offered in the 2022 summer semester as a shared-risk opportunity, allows observations with a channel spacing of 250 kHz over the entire correlation bandwidth (Fig. 2). A third correlator mode is available that phases up the NOEMA antennas for VLBI observations.

### 2.2 Future upgrades

To ensure that NOEMA remains a major pathfinder in the development of instrumentation for millimeter- and submillimeter-wave research, developments are underway that will further expand its capabilities. These are presented below in the order in which they will be deployed in the near future.

#### *2.2.1 New Configurations*

By the end of the year (Q4/2022), the interferometer is expected to offer a very extended configuration (A) with an east-west baseline of 1.7 km for observations up to 276 GHz, along with a new set of intermediate and compact configurations for optimal coverage of the uv-plane. With an angular resolution as high as 0.19", 12 antennas, and sidelobe levels of 13 dB below the synthesized beam, NOEMA will offer much improved imaging capabilities. At the lowest available observing frequency (70 GHz), NOEMA will enable the mapping of large-scale emission from regions up to 25" in extent and resolve details down to an angular resolution of 0.6".

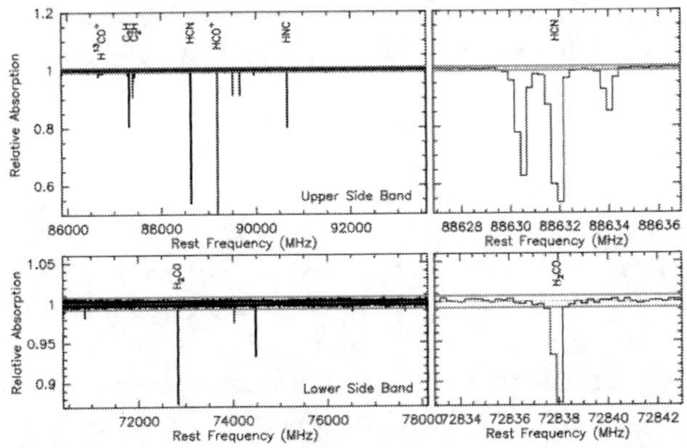


**Figure 2.** First science observations with the PolyFiX correlator operating in the 250 kHz mode. Narrow absorption lines towards the continuum of the BL Lac quasar reveal the presence of foreground Galactic clouds. The upper and lower left panels show the upper and lower sidebands, , the right panels zoom in on the HCN and $H_2CO$ absorption lines. The red lines are at 5× the noise level. Courtesy of H.Liszt, M.Gerin and J.Pety.

### 2.2.2 Spectral Scan Capabilities

Since about 20% of the observing time is invested in spectral surveys, work is underway to offer contiguous spectral coverage over frequency ranges four times larger than the current 7.744 GHz wide IF bandwidth. Alternating between two LO frequencies separated by exactly one IF bandwidth after each calibration cycle will open the possibility of performing spectral sweeps over bandwidths of 31 GHz, under virtually identical observing conditions. This mode will be best suited to conduct NOEMA zGAL-like redshift searches [1] and ALMA PILS-like spectral surveys [2] with uniform uv-plane sampling and homogeneous sensitivity across the covered frequency window. In addition, it will help achieve higher calibration accuracy and a much improved ability to leverage spectral indices. While not yet fully commissioned, work is progressing towards offering this new capability in Q2/2023 for Band 1 observations.

### 2.2.3 Dual Band Receivers

In the framework of the Dual Band Receiver (DBR) project, it is planned to equip all antennas with DBRs for simultaneous observations in Band 1 and Band 3 by the end of 2024. Concurrent operation in the two bands, coupled with the use of a second PolyFiX correlator to cover the entire 62 GHz bandwidth, will open a new era of unprecedented spectral coverage for NOEMA. Not only will this improve the observatory's scientific productivity by accelerating the completion of projects requiring observations in the two bands, but the DBRs will also allow for an overall improvement in data quality. They will pave the way for band-to-band phase referencing to improve the dynamic range of astronomical images, significantly reduce systematic calibration uncertainties, facilitate the completion of multiple transition line surveys, and last but not least, open up the possibility of observing transient and time-critical phenomena that evolve on short time scales.

### 2.2.4 Polarimetry and Band 4

Work is underway to enable full-Stokes polarization observations with NOEMA. Current plans call for initial Band 1 science demonstration programs in Q3/2022, first polarimetric observations in 2023 as part of ECOGAL, an ERC Synergy 2020 funded project, and a possible opening to the general user community for regular Band 1 observations in 2024, and successively in other bands.

Last but not least, work is also underway to reinstate the ability to observe in NOEMA's highest frequency band. Under current plans, all antennas would need to be equipped with

Band 4 receivers and be operational by Q4/2024. Observations in Band 4 will open up new opportunities, including observations with an angular resolution down to 0.12", and provide new momentum for dust studies.

## 3 IRAM 30-meter

The IRAM 30-meter millimetre radio telescope in the Observatorio de Pico Veleta near Granada, Spain, is one of the most sensitive and powerful single-dish telescopes worldwide in the 73-350 GHz frequency range. In operation since 1984, it continues to be a leading facility thanks to: *(a)* a large primary reflector size (30 m), ensuring a high sensitivity and resolving power (11"/7" HPBW at 230/340 GHz); *(b)* a high mirror surface accuracy (around 60 $\mu$m r.m.s. at an elevation of 50°) that makes the telescope very efficient; *(c)* a panoply of state-of-the-art instruments (see section 3.1); *(d)* a high-altitude (2850 m above sea level), low-latitude (37°N) site that allows observations at frequencies up to 350 GHz and the best possible access to the Galactic Center from continental Europe. In addition, the excellent thermal control of the back-up structure (BUS), yoke and quadrupod, along with an efficient de-icing system, ensures continuous operation around the clock in all seasons, and rapid recovery of operations after snow storms. The IRAM 30-meter has been recognized as one of the facilities in the Spanish Map of Unique Scientific and Technical Infrastructures (ICTS).

### 3.1 Current Capabilities

The IRAM 30-meter is currently offering two instruments to the community: the Eight-MIxer Receiver (EMIR), a very efficient single-pixel heterodyne receiver with four bands at 90, 150, 230 and 330 GHz and two orthogonal polarizations, and the New IRAM KID Array 2 (NIKA2), a novel 1.2 and 2 mm continuum camera based on Kinetic Inductance Detector (KID) technology, with polarimetric capabilities at 1.2 mm. The status of the instruments is kept up to date in the documentation available for each call for proposals[1] [3]. The main innovation in EMIR is the change of the folding oscillator for the outer bands, which allows observation of the $DCO^+$ line at 72.039 GHz with the FTS backend at 50 kHz resolution and continuous coverage of a 32 GHz bandwidth at 200 kHz resolution with two tunings. For NIKA2, the commissioning of the polarimetric mode at 1.2 mm was completed by the end of 2021, and first observations of B-FUN, a Large Program aimed at probing the magnetic field in star-forming regions, were started in the winter semester of 2021/2022. The next updates will include a new, improved dichroic filter, planned for late 2022, and the adaptation of the PIIC data reduction software for polarimetry. In the longer term, new KID arrays with higher sensitivity and higher pixel yield will be installed.

The pointing and tracking capabilities of the IRAM 30-meter are outstanding [3], although improvements are underway (see section 3.2). In particular, the maximum tracking speed, which is 150"/s due to limitations in the control system, limits the maximum observable elevation to 83°. The tracking accuracy is excellent at low wind speeds (~0.2"), but decreases at wind speeds above ~10 m/s.

### 3.2 Telescope Upgrades

IRAM is currently carrying out an ambitious upgrade of the telescope to keep it at the forefront of research for the years to come. The planned refurbishment includes a new servo control system for the mount and subreflector, as well as improvements to the surface of the primary reflector. These measures are partially funded by the European Regional Development Fund (ERDF) program and will be briefly described in this section.

[1] https://www.iram-institute.org/EN/content-page-57-7-57-0-0-0.html

### *3.2.1 Servo Control System*

The new servo control system aims to: *(i)* prolong the operational telescope lifetime by replacing obsolete components; *(ii)* improve the tracking accuracy at high wind speeds; and, *(iii)* allow faster slews and maximum tracking speeds, which will reduce overheads and extend the operating range near the zenith. In addition, it will be possible to design new scanning patterns. The new control system is being implemented by OHB Digital Connect GmbH in Mainz, Germany. It features state-of-the-art electronics, including ACU, amplifiers, motors and most of the safety sensors, and a control loop that runs at much higher rates than the current system. The bandwidth of the control loop will be extended to improve the tracking performance by 30% in high wind conditions. The system also implements a fully renewed servo control of the subreflector, including the six spindle drives of the hexapod and the wobbler control amplifiers. In addition, we are taking the opportunity to update and streamline parts of the control software. An overhaul of the telescope gear boxes is also planned in parallel. The Factory Acceptance Test (FAT) was recently performed in the contractor's premises, but the on-site deployment of the system, originally scheduled for summer 2022, has been postponed due to the delay in the provision of some critical hardware components. The updated plan is to start on-site activities in late February or early March 2023.

### *3.2.2 Surface Quality*

The 30-meter telescope has an outstanding, although yet improvable, surface quality. The telescope primary mirror was initially covered with a $50\,\mu$m $TiO_2$ paint layer that degraded over the years. In some places, the paint layer is as thin as $\sim 10\,\mu$m, exposing the dark primer. This leads to a deterioration of the thermal behaviour in day time that manifests itself in mirror deformations and buckling of panels, which is especially critical for the imaging instrument NIKA2. In addition, the uneven paint layer contributes to the RMS error budget of the surface by at least $30\,\mu$m. Hence, the renewal of the paint layer is likely the most important and urgent action to be taken to improve the imaging quality of the telescope. In addition, the point source sensitivity is strongly dependent on elevation, decreasing by about 25% above 80° and below 20°. Therefore, the following targets were defined: *(i)* improve the thermal behaviour of the surface and reduce the surface RMS due to optical path differences; *(ii)* improve the surface RMS error to $50\,\mu$m at any elevation, with a goal of $40\,\mu$m which would correspond to a 1.5 to 2-fold improvement in efficiency at high frequencies; and, *(iii)* reduce the dependence of mirror efficiency on elevation, so that the contribution to the RMS error due to gravity-only effects should be less than $30\,\mu$m over the entire elevation range from 20° to 80°. On-going studies are considering a number of strategies, with "conservative" and "aggressive" approaches. In the "conservative" strategy, all actions are performed in-situ without dismounting the panels from the BUS. The main action would be to clean and repaint the panels with very stringent accuracy requirements to greatly improve the thermal behaviour of the telescope and reduce the error induced by the optical path. This measure will be likely implemented in summer 2023 if the tests currently on-going are successful. In addition, it would be possible to reinforce some structures to improve the dish gravitational deformation at low and high elevations. The "aggressive" approach implies dismounting the panels with their sub-frames from the BUS, repaint and realign the panels within the sub-frames at a carefully chosen elevation, install a number of actuators at the corners of the sub-frames using predefined lookup tables to improve the gain-elevation curve, and finally remount the panels in the BUS. The final study report is due in fall 2022, and consolidated strategies will emerge from its outcome.

### 3.3 Future Instruments

Future instrumental developments are geared towards multi-beam (MB) heterodyne receivers and their associated backends. The pioneering HERA instrument, a $3 \times 3$ MB SSB receiver at 1 mm, with 1 GHz bandwidth per pixel and polarization, will have two successors, both using SIS 2SB mixers and dual polarizations: the SHERA instrument with $7 \times 7$ pixels at 1 mm and the 3 mm MB, with $5 \times 5$ pixels. Both instruments will produce 32 GHz of bandwith per pixel. Hence, new backends will be required to handle the huge amount of data. These instruments are in the preliminary design phase, but a prototype MB instrument at 3 mm, using HEMT LNA technology, with $3 \times 3$ pixels and dual polarization is being developed in collaboration with the IAF, INAF and MPIfR institutes in the framework of the AETHRA/RadioNet project. This receiver, equipped with a test array of $1 \times 3$ pixels was tested at the telescope in April 2022 (Figure 3).

**Figure 3.** The 3 mm AETHRA prototype installed in the receiver cabin of the 30-meter telescope in April 2022. The cryostat (currently housing a $1 \times 3$ pixel array), warm optics and calibration unit can be easily identified in the optical bench. Picture courtesy S. Navarro.

## References

[1] Cox P. et al., Multi-line Diagnostics of the Interstellar Medium, Nice Proceedings (2022)

[2] Jørgensen J.K. et al., A&A, 595, A117 (2016)

[3] Kramer, C. and Sánchez-Portal, M., IRAM 30-meter Telescope: Observing Capabilities and Organisation (2022). http://www.iram.fr/GENERAL/calls/s22/30mCapabilities-delta-s22.pdf

# Multi-line Observations, Models, and Data Needed to Understand the Nature of UV-irradiated Interstellar Matter

*Javier R.* Goicoechea[1,*], *Sara* Cuadrado[1], and *Franck* Le Petit[2]

[1]Instituto de Física Fundamental (IFF), CSIC. Calle Serrano 121-123, 28006, Madrid, Spain
[2]LERMA, Observatoire de Paris, PSL University, CNRS, Sorbonne Université, 92190 Meudon, France

**Abstract.** Far-ultraviolet photons from OB-type massive stars regulate the heating, ionization, and chemistry of much of the neutral interstellar gas in star-forming galaxies. The interaction of FUV radiation and interstellar matter takes place in environments broadly known as *photodissociation regions* (PDRs). PDR line diagnostics are the smoking gun of the radiative feedback from massive stars. Improving our understanding of stellar feedback in the ISM requires quantifying the energy budget, gas dynamics, and chemical composition of PDR environments. This goal demands astronomical instrumentation able to deliver multi-line spectroscopic images of the ISM (of the Milky Way and nearby galaxies). It also requires interdisciplinary collaborations to obtain the rate coefficients and cross sections of the many microphysical processes that occur in the ISM and that are included in models such as the *Meudon* PDR code.

## 1 Introduction

Far-UV (FUV: $E < 13.6\,\mathrm{eV}$) photons emitted by massive O- and B-stars govern, or at least greatly influence, the heating, ionization, and chemistry of the neutral interstellar gas: everywhere hydrogen atoms are in predominantly neutral form. The interaction of stellar FUV radiation and interstellar baryonic matter (atoms, molecules, and dust grains) takes place in so-called *photodissociation regions* (PDRs; [1]). This interaction occurs for different doses of FUV radiation ($G_0$) and at very different spatial scales: from the illuminated rims of star-forming clouds to kpc scales in starburst galaxies. The emission from PDRs reflects the radiative feedback from massive stars, a collection of processes that establish the phases and pressures of the ISM, and also regulate star formation and molecular cloud destruction [2].

The physical state of the interstellar gas depends on a plethora of detailed microphysical processes that determine its chemical composition and how the gas is heated and cooled. Therefore, in addition to understanding the (macro) astrophysical processes driving the dynamics and evolution of the ISM, it is also mandatory to understand the subtle microprocesses that form, destroy, and excite its basic constituents. Accompanied by new developments in astronomical instrumentation, last years on PDR research have reinforced closer collaborations with molecular physicists and laboratory experimentalists able to determine the rate coefficients and cross sections of the relevant microprocesses.

Bright molecular cloud rims, such as the iconic Orion Bar or the Horsehead, have always been excellent laboratories to study the molecular content of FUV-irradiated interstellar gas

*e-mail: javier.r.goicoechea@csic.es

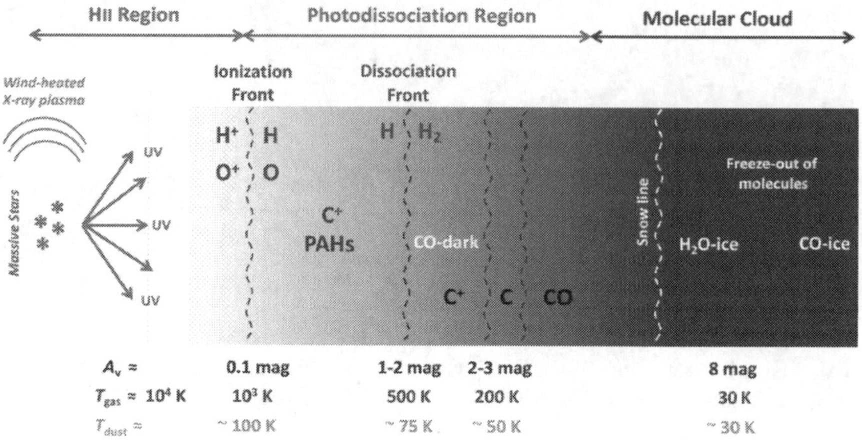


**Figure 1.** Structure of a strongly irradiated molecular cloud near massive stars (e.g., the Orion Bar).

and the wealth of physical and chemical processes that occur at microscopic (molecular) level. The number of molecules, atoms, and ions observed in these dense PDRs are not so easily detected toward fainter environments also influenced by stellar FUV radiation (e.g, the surface layers of protoplanetary disks and distant star-forming galaxies).

In this invited paper we summarize some of the latest developments in observations and modeling of PDRs. Due to the lack of space, we mostly limit to our own recent work. Readers are referred to Wolfire *et al.* [3] for a modern review on PDRs. Here we strengthen the interdisciplinary aspects of PDR research, and we emphasize the need of precise atomic and molecular data to correctly interpret future observations of the ISM in galaxies.

## 2 PDRs everywhere

PDRs host the critical conversion from atomic to molecular gas (the $H/H_2$ and $C^+/C/CO$ transition zones, Fig. 1). This stratification occurs as the column density of gas and dust increases deeper inside molecular clouds, and as the flux of FUV photons is attenuated by dust (for high $G_0/n_H$ ratios) and by $H_2$ absorption lines (for low $G_0/n_H$ ratios) [4]. The penetration of FUV radiation depends on metallicity and grain properties. FUV photons reach larger depths in lower metallicity environments (lower dust-to-mass ratios) and in regions affected by grain growth (flatter extinction curves). Both effects shift the $C^+/C/CO$ transition to higher $A_V$ and leave a larger mass fraction of $H_2$ gas that is "CO-dark" (or too faint).

PDRs emit most of the IR radiation arising from the ISM of star-forming galaxies: FUV-pumped IR bands from polycyclic aromatic hydrocarbons (PAHs), $H_2$ ro-vibrational, [C II]158 $\mu$m and [O I]63, 145 $\mu$m fine-structure, and mid-$J$ CO lines, as well as warm dust continuum – grains are heated by FUV photons and reemit FIR contimuum. Photoelectrons ejected from small grains and PAHs heat the gas. In dense PDRs, collisional de-excitation of FUV-pumped $H_2$ is an important gas heating mechanism too. The presence of vibrationally excited $H_2(v \geq 1)$ overcomes the endoergicity of the initiating reactions $H_2 + (C^+, S^+, O, N, ...)$ [5] and leads to the formation of $CH^+$, $SH^+$, OH, and NH [6].

In their most general definition – *neutral gas regulated by FUV radiation* –, PDRs represent the dominant fraction of the neutral atomic and molecular gas in the ISM of star-forming galaxies [1, 3]. Indeed, except for the cold and dense molecular cores associated with the first stages of star formation, most of the ISM (in volume and mass) is effectively at $A_V < 8$ mag. Hence, permeated by stellar FUV photons.

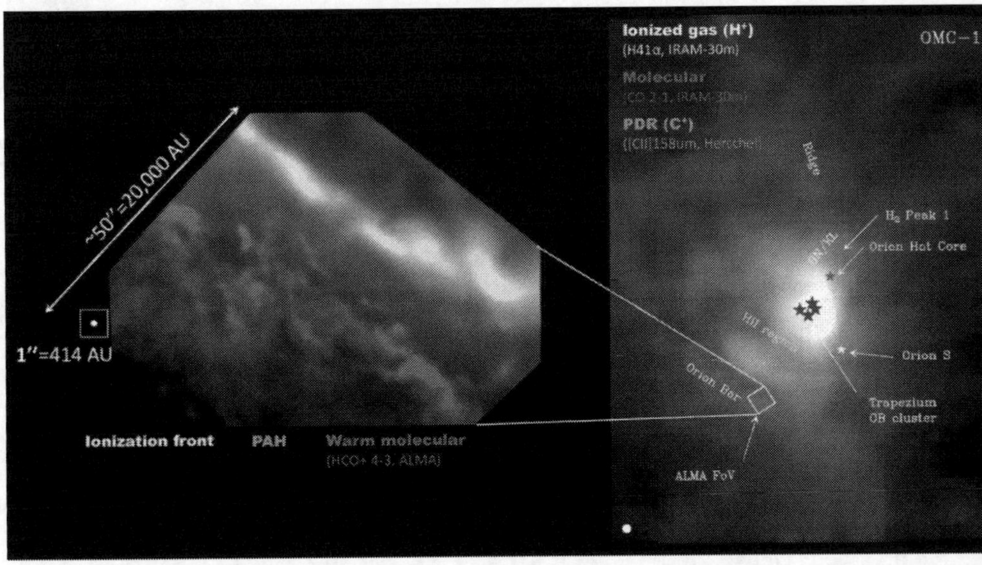


**Figure 2.** Central parsec ($\sim 7.5' \times 11.5'$) of the Orion molecular core OMC-1 around the Trapezium cluster (*right*, [7]) and $1''$ resolution zoom ($\sim 50'' \times 50''$) resolving a high pressure ($P_{\rm th}/k_{\rm B} \sim 10^8$ K cm$^{-3}$) and nearly edge-on H II/PDR/molecular cloud interface in the Orion Bar PDR (*left*, [8]).

## 2.1 New-generation observations of the ISM and PDR environments

State-of-the-art observations allow us to access a plethora of multi-line diagnostics at increasingly higher sensitivity, angular resolution, and field-of-view. Recent observations challenge some of our previous views of PDRs, for example, their fundamental small-scale structure(s), the origin of the observed rich chemistry, and their gas dynamics: propagation of ionization/dissociation fronts, photoevaporation, gas compression, and radiation pressure on grains. Last years have seen rapid progress in heterodyne receiver technology and specific techniques involving (sub)mm spectral-imaging of the ISM:

*i)* Increased sensitivity and angular resolution of interferometric mosaics (with ALMA and NOEMA), providing astonishing sub-arcsecond resolution images of the fundamental structure of the ISM (e.g., see Fig. 2 for the Orion Bar PDR [8]).

*ii)* Broader instantaneous bandwidth, allowing us to *a)* map large portions of star-forming clouds in several molecular lines simultaneously (CO, HCN, HNC, $HCO^+$, $N_2H^+$, ... [9]), and *b)* to obtain deep line surveys of the molecular content (through the detection of hundreds of molecular lines [10, 11]) and physical conditions of representative environments. IRAM, its forefront instrumentation, and the variety of available observational techniques (from on-the-fly mapping to line surveys) have played a pivotal role in improving our understanding of the ISM and its underlying PDR processes.

*iii)* Development of airborne multi-pixel arrays up to the THz domain, able to map the main gas cooling lines (i.e., the cloud energetics) at high-spectral resolution and providing, for example, velocity-resolved square-degree maps of the [C II]158 $\mu$m emission [2].

As a consequence of these developments, current research on PDRs and associated processes is not only about the detailed study of small fields in bright cloud rims such as the Orion Bar – from which we learn so much – but more generally about the role of stellar feedback and the evolution of the FUV-illuminated ISM at all relevant spatial scales: from cores and giant molecular clouds (GMCs) [2, 7] to distant star-forming galaxies.

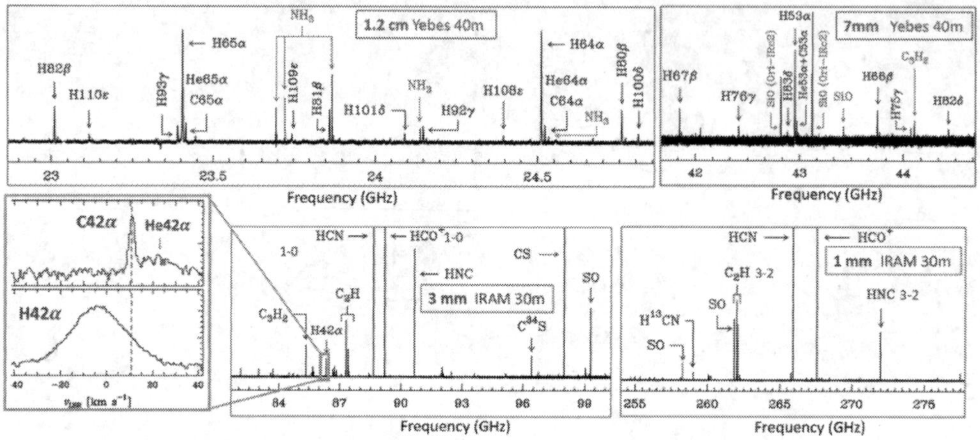


**Figure 3.** Multi-line diagnostics (molecular and recombination lines) observed toward the Orion Bar PDR and adjacent H II region using the IRAM 30m [10–12] and Yebes 40m [18] radiotelescopes.

## 2.2 Molecular content of PDRs and peculiarities

PDRs were once thought to be too harsh to develop a rich chemistry. However, our IRAM 30 m and Yebes 40 m line surveys of the Orion Bar, a strongly irradiated PDR ($G_0 \simeq 10^4$), exemplify that PDRs do host multiple lines and unique chemical features that are not seen (or are too faint to be detected) in regions shielded from FUV radiation. These specific lines (Fig. 3) have a great diagnostic power as tracers of FUV radiation from short-lived massive stars, and thus, of star formation: narrow carbon radio recombination lines (with line intensities proportional to $n_e^2 T_e^{-1.5}$) tracing the presence of $C^+$ ions and of high ionization fractions ($n_e/n_H \approx C^+/H$; [12]), reactive molecular ions ($CH^+$, $SH^+$, $HOC^+$, and $CO^+$; [13, 14]), as well as enhanced abundances of certain hydrocarbons ($C_3H^+$, $l$-$C_3H$; [10]), radicals (NH [6], OH [11], and HCO [15]), and unstable isomers of organic molecules (*cis*–HCOOH; [16]). In addition, specific sulfur-bearing molecules such as $H_2S$ show bright emission lines and enhanced abundances (as revealed by the detection of the rare $H_2^{33}S$ isotopologue; [17]). Indeed, the detection of sulfur radio recombination lines [18] implies a gas-phase $S^+/H$ abundance of $(1.4 \pm 0.4) \cdot 10^{-5}$ at edge of the Orion Bar. This is the elemental gas-phase sulfur abundance before an undetermined fraction goes into S-bearing molecules and deplete as grain ice mantles in the cloud interior. The inferred sulfur abundance in the PDR edge matches the solar abundance [17]. It thus leaves little room for large depletions of sulfur onto rocky grains in molecular clouds.

Some of the above "PDR species" are readily detected in diffuse clouds [19] as well as in starburst galaxies [20]. These detections prove the interaction of stellar radiation with interstellar matter in different regimes of FUV flux, gas density, and spatial scale.

# 3 Models, need of precise atomic and molecular data, and outlook

The theoretical study and first thermo-chemical models of dense PDRs started to develop more than 50 years ago. PDR models solve the penetration of FUV radiation into molecular clouds. This includes treating the absorption of FUV photons to electronically excited levels of H, $H_2$, and CO, as well as the absorption and anisotropic scattering by dust grains [21]. Among other things, the wavelength-dependent FUV ($\lambda > 911$Å) radiation field at different cloud depths determines the rate at which low-ionization-potential atoms are ionized

**Table 1.** Microphysical processes relevant to the study of interstellar PDRs.

| Process | Relevance | Required parameter | Methodology |
|---|---|---|---|
| Inelastic collisions with $o$-$H_2$, $p$-$H_2$, H, and $e^-$ | Non-LTE excitation. Precise comparison with observed line intensities | Inelastic collisional rate coefficient $\gamma_{ij}(T) \sim \sigma_{ij} \cdot v$ ($cm^3\,s^{-1}$) | Scattering calculations. Laboratory. |
| Chemical reactions | Formation and destruction of molecules and PAHs | Reaction rate coefficient $k(T)$ ($cm^3\,s^{-1}$) | Reaction dynamics calculations. Laboratory. |
| Reactions with vibrationally excited $H_2$ ($v{\geq}1, J$) | Reactions of FUV-pumped $H_2$ with atoms and molecules overcomes reaction endoergicities | State specific rate coefficient $k_{v,J}(T)$ ($cm^3\,s^{-1}$) | Reaction dynamics calculations. Laboratory. |
| Photo-ionization and dissociation of atoms, molecules, and PAHs | Photochemistry induced by stellar FUV photons | Photo cross sections $\sigma_{ion}(\lambda)$, $\sigma_{diss}(\lambda)$ ($cm^{-2}$) | Experiments (e.g., synchrotron). Calculations. |
| Low-$T_e$ radiative and dielectronic recombination of $S^+$ ($Si^+$, $Fe^+$, ...) | Defines $S^+$/S transition. Determines recombination line spectrum | Recombination rate coefficients $k_{RR,DR}(T_e)$ ($cm^3\,s^{-1}$) | Calculations. |
| Adsorption/desorption of atoms and molecules onto/from dust grains | Freeze-out, grain surface chemistry and sublimation of atoms/molecules from dust grains | Binding energies $E_b/k_B$ (K) | Temperature-programmed desorption experiments. Calculations. |
| Non-thermal desorption from grains | Photodesorption and cosmic-ray induced desorption of ice mantles | Desorption yields $Y$ (molecule photon$^{-1}$) $k_{CR}$ ($s^{-1}$) | FUV, X-ray, and particle irradiation experiments. |
| FUV extinction Photoelectric effect | FUV penetration. Gas heating | Grain/PAH optical props. $Q_{abs}(\lambda)$, $Q_{sca}(\lambda)$, $g(\lambda)$ | Laboratory. |

(C, S, Si, Fe, ...) and at which molecules are ionized and dissociated. The FUV field, gas density, and dust/PAH properties set the main heating mechanisms, whereas cooling from [C II], [O I], $H_2$, and mid-$J$ CO emission lines [22] determines the cloud-depth variation of the gas temperature. This temperature sets the rates of a large network of gas-phase ion-neutral and neutral-neutral chemical reactions forming and destroying molecules. At deeper depths into the molecular cloud, typically $A_V > 8$ mag, the flux of stellar FUV photons is almost completely attenuated and gas-phase molecules and atoms freeze-out on dust grains as temperatures drop. Cosmic-rays drive the ionization of atoms and molecules, but at much lower rates than in gas directly exposed to stellar FUV radiation. This leads to low ionization fractions and a slower chemistry. The catalytic formation of molecules on grain surfaces and subsequent thermal (sublimation) and nonthermal (photodesorption, cosmic-ray induced, or chemical) desorption adds a new layer of chemical complexity to model.

In order to correctly interpret the amount of information and the many details provided by state-of-the-art multi-line observations, astrochemical models need to accurately calculate the rates of the above microphysical processes (interstellar gas is rarely in thermal or chemical equilibrium). This is the main goal of the *Meudon* PDR code [23], a publicly available and open source model (https://ism.obspm.fr/) that is continuously upgraded toward more realistic descriptions of these processes [24] and as new rate coefficients and cross sections are published in the literature. Hence, it is mandatory to strengthen and foster interdisciplinary collaborations aimed to better characterize these processes, and to determine the precise atomic and molecular data needed in astrochemical models. These data can be obtained in sophisticated laboratory experiments or through high-level *ab initio* calculations. Table 1 summarizes some of the most relevant microphysical processes taking place in the ISM (with emphasis on PDRs) together with the physical parameters and rate coefficients typically needed in accurate thermo-chemical and non-LTE excitation models.

PDR models play a central role to correctly interpret spectroscopic observations of a very significant fraction of the ISM in star-forming galaxies (all neutral gas at $A_V < 8$ mag). New instrumentation will allow mapping several square-degree areas of GMCs (reaching the scales that dominate the extragalactic emission) in multiple rotational lines, critical to derive accurate physical conditions and precise chemical abundances. More automatic statistical data analysis tools – designed to link large grids of astrochemical models and hydrodynamic PDR simulations with observations – will be used to extract all the information contained in the million spectra these velocity-resolved cubes will generate [9]. These cubes provide access to the gas kinematics, thus to the driving forces in the ISM, and reveal the dynamical and non-stationary aspects of stellar feedback in the ISM (e.g., [25]). In parallel, interferometric mosaics will allow us to locally zoom into particular fields/templates at very high angular resolution. These images reveal the spatial scales at which chemical abundances and physical conditions abruptly change in clouds (Fig. 2). Interferometric observations also spatially resolve GMCs in nearby galaxies, with emission features often dominated by gas at low $A_V$. JWST will soon reveal the evolution of PAHs, warm dust grains, and $H_2$ emission (e.g., [26]) in all kind of FUV-irradiated environments. The IR to cm emission from these "PDRs" is determined by many microphysical processes that form, excite, and destroy the gas and dust in the ISM. Knowing the rates of these processes is equally important to correctly understand the many faces of stellar feedback in the ISM.

## References

[1] Hollenbach, D. J. & Tielens, A. G. G. M. 1999, Reviews of Modern Physics, 71, 173
[2] Pabst, C. H. M., Goicoechea, J. R., Teyssier, D., et al. 2020, A&A, 639, A2
[3] Wolfire, M. G., Vallini, L., & Chevance, M. 2022, ARAA, 60, arXiv:2202.05867
[4] Sternberg, A., Le Petit, F., Roueff, E., et al. 2014, ApJ, 790, 10
[5] Agúndez, M., Goicoechea, J. R., Cernicharo, J., et al. 2010, ApJ, 713, 662
[6] Goicoechea, J. R. & Roncero, O. 2022, A&A, arXiv:2206.10441
[7] Goicoechea, J. R., Teyssier, D., Etxaluze, M., et al. 2015, ApJ, 812, 75
[8] Goicoechea, J. R., Pety, J., Cuadrado, S., et al. 2016, Nature, 537, 207
[9] Pety, J., Guzmán, V. V., Orkisz, J. H., et al. 2017, A&A, 599, A98
[10] Cuadrado, S., Goicoechea, J. R., Pilleri, P., et al. 2015, A&A, 575, A82
[11] Cuadrado, S., Goicoechea, J. R., Cernicharo, J., et al. 2017, A&A, 603, A124
[12] Cuadrado, S., Salas, P., Goicoechea, J. R., et al. 2019, A&A, 625, L3
[13] Nagy, Z., Van der Tak, F. F. S., Ossenkopf, V., et al. 2013, A&A, 550, A96
[14] Goicoechea, J. R., Cuadrado, S., Pety, J., et al. 2017, A&A, 601, L9
[15] Goicoechea, J. R., Joblin, C., Contursi, A., et al. 2011, A&A, 530, L16
[16] Cuadrado, S., Goicoechea, J. R., Roncero, O., et al. 2016, A&A, 596, L1
[17] Goicoechea, J. R., Aguado, A., Cuadrado, S., et al. 2021, A&A, 647, A10
[18] Goicoechea, J. R. & Cuadrado, S. 2021, A&A, 647, L7
[19] Gerin, M., Liszt, H., Neufeld, D., et al. 2019, A&A, 622, A26
[20] Fuente, A., García-Burillo, S., Usero, A., et al. 2008, A&A, 492, 675
[21] Goicoechea, J. R. & Le Bourlot, J. 2007, A&A, 467, 1
[22] Joblin, C., Bron, E., Pinto, C., et al. 2018, A&A, 615, A129
[23] Le Petit, F., Nehmé, C., Le Bourlot, J., et al. 2006, *ApJS*, 164, 506
[24] Bron, E., Le Bourlot, J., & Le Petit, F. 2014, A&A, 569, A100
[25] Maillard, V., Bron, E., & Le Petit, F. 2021, A&A, 656, A65
[26] Berné, O., Habart, É., Peeters, E., et al. 2022, PASP, 134, 054301

# HCN/HNC ratio: A new chemical thermometer at 3 mm

*Alvaro* Hacar[1,*] and *Sümeyye* Suri[1,**]
[1]Department of Astrophysics, University of Vienna, Türkenschanzstrasse 17, 1180 Vienna, Austria

**Abstract.** The gas kinetic temperature ($T_K$) determines the physical and chemical evolution of the interstellar medium (ISM). This work explores the use of HCN/HNC (J=1–0) line ratio as new probe of the gas kinetic temperature in the molecular ISM. Based on a new set of IRAM-30m observations at 30 arcsec resolution towards Orion Nebula Cluster, we find a two-part linear correlation of the observed HCN/HNC line ratio and $T_K$. This empirical calibration allows to obtain direct estimates of the $T_K$ values across a wide range of column densities ($A_V \gtrsim 5$ mag) and up to scales of $\sim 10$ pc. Comparisons with additional studies highlight the potential use of the HCN/HNC line ratio as novel chemical thermometer at 3mm.

## 1 Introduction

The gas kinetic temperature ($T_K$) plays a key role in the most important gas properties in the Interstellar Medium (ISM). $T_K$ determines the gas pressure ($P/k = n\ T_K$), the sound speed ($c_s = \sqrt{kT_K/\mu}$), as well as fragmentation scale ($\lambda_{frag} \propto T_K^{3/2}$) in molecular clouds. Similarly, it influences the chemical structure and molecular emission properties of the molecular gas in the ISM. The description of the thermodynamic state of the ISM gas therfore requires a detailed study of $T_K$.

Multiple observational techniques have been developed to investigate the gas kinetic temperature in the ISM. In dedicated molecular line studies, the gas kinetic temperature has been traditionally inferred from the study of $NH_3$ rotational temperures ($T_{rot}$, [1]) and the combination of the study of excitaton temperatures ($T_{ex}$) and radiative transfer calculations in multi-line observations (e.g. [10]). The wide coverage of FIR observations in the ISM has also popularized the use of dust temperatures measurements ($T_{dust}$) as proxies of the gas temperatures in nearby clouds ([9]). However, the above approaches require of additional assumptions such as Local Thermodynamical Equilibrium (LTE; $T_K = T_{ex} = T_{rot}$) or an effective gas-to-dust coupling ($T_K = T_{dust}$) which are difficult to justify in a wide range of physical conditions. In practical terms, these studies are typically limited in dynamic range (e.g. $NH_3$ at $A_V \gtrsim 10$ mag), require observations of multiple transitions as well as expensive radiative transfer modeling (e.g. CO-ladders), or become only accesible at certain frequency ranges (e.g. $H_2CO$ at 1 mm or $NH_3$ at cm wavelengths). During the last years, most of the large scale observations of the local ISM, in which IRAM has been playing a leading role (e.g. [11]), are carried our at 3 mm (i.e. ALMA Band 3). Finding a versatile, reliable, easy-to-use tracer of the gas kinetic temperature at 3 mm is therefore of paramount importance for future ISM studies both in local and extragalactic contents.

*e-mail: alvaro.hacar@univie.ac.at
**e-mail: suemeyye.suri@univie.ac.at

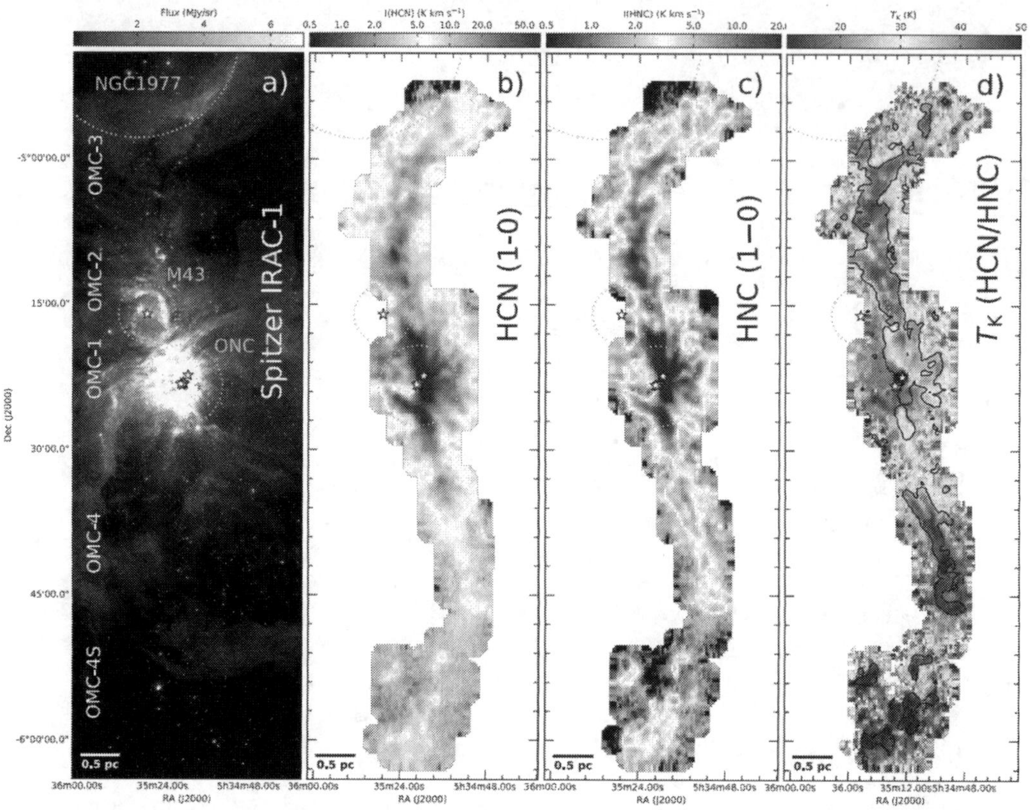


**Figure 1.** IRAM-30m observations along the Orion ISF. From left to right: (a) Spitzer IRAC-1 emission map; (b) HCN (J = 1–0), (c) HNC (J = 1–0), and (d) gas kinetic temperature derived using the HCN/HNC ratio (see [6]). The black contours in panel d indicate the regions with significant $N_2H^+$ (J=1–0) [4], denoting those filamentary regions at high densities (> $10^5$ cm$^{-3}$).

A large variety of astrochemical studies have investigated the variation of the HCN and HNC emission properties in the ISM (e.g. [7, 12]). This work explores the use of the HCN/HNC line ratio as a novel chemical thermometer of the molecular ISM. A detailed discussion of these results can be found in [6].

## 2 HCN & HNC IRAM-30m observations in Orion

We investigated the molecular emission of both HCN (88.6 GHz) and HNC (90.6 GHz) (J=1–0) lines along the Integral Shape Filament (ISF) in Orion (D~ 415 pc; [8]). The ISF contains the most active star-forming region within solar neighbourhood (i.e., Orion Nebula Cluster; ONC) and the only one forming massive stars (Trapezium & Orion BN/KL), becoming a testbed for star-formation studies. As illustrated in Figure 1a, ISF extends over ~ 8 pc in length, including the OMC 1/2/3/4/4S regions, and is shaped by different feedback events (e.g. ONC, M43, and NGC1977 bubbles; see Suri et al in prep.). The ISF is the most massive filament within 1 kpc ($M/L \sim 300 - 500$ $M_\odot$ pc$^{-1}$) with high emission levels in multiple molecular tracers widely studied in the literature (e.g. [5]).

We obtained On-the-Fly, Position-Switching maps of the HCN and HNC (J=1–0) emission along the ISF at 3 mm using the IRAM-30m telescope (see also [6]). Both transitions

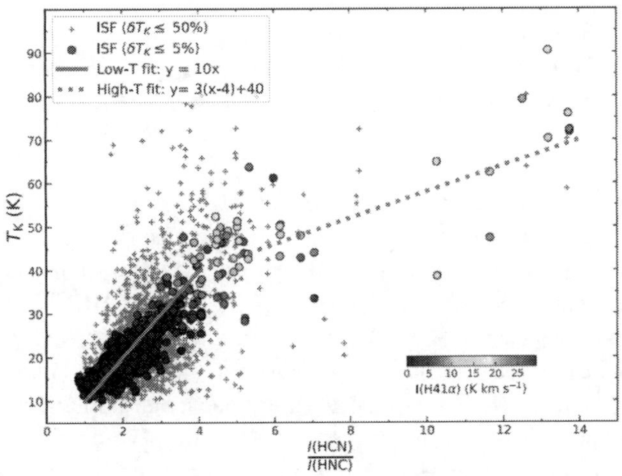


**Figure 2.** Empirical comparison between the I(HCN)/I(HNC) line intensity ratio and previous $NH_3$-based gas temperatures ($T_K(NH_3)$) estimates along the ISF. Colour filled symbols indicate those points with accurante $T_K(NH_3)$ measurements. The colour scale indicates the intensity of H41$\alpha$ detected in this region.

were observed simultaneously with the EMIR090 receiver in combination of the FTS back-end at low spectral resolution (0.6 km s$^{-1}$). Figures 1b and 1c show the integrated emission of the HCN and HNC (J=1–0) lines, respectively, at a final resolution of 30 arcsec, both showing bright emission along the entire ISF down to $A_V \sim 5$ mag (limited by our map boundaries).

## 3 HCN/HNC: A new chemical thermometer for the ISM

We compare the observed I(HCN)/I(HNC) intensity ratio with previous $T_K$ estimates previously obtained using $NH_3$ observations in the ISF ([3]) in Figure 2. Our results show a positive and strong correlation between these quantities in different parts of the cloud. Outside the Orion Nebula, indicated by low intensities of the H41$\alpha$ emission (see color scale), the I(HCN)/I(HNC) shows a tight and steep dependence on $T_K$. A shallower correlation tentatively continues inside the nebula showing higher H41$\alpha$ values. A two-part linear fit to the data in Fig.2 allows to empirically calibrate the observed I(HCN)/I(HNC) line ratio into a direct proxy of the $T_K$ following:

$$T_K[K] = 10 \times \left(\frac{I(HCN)}{I(HNC)}\right) \quad \text{when} \quad \frac{I(HCN)}{I(HNC)} \leq 4 \tag{1}$$

$$T_K[K] = 3 \times \left(\frac{I(HCN)}{I(HNC)} - 4\right) + 40 \quad \text{when} \quad \frac{I(HCN)}{I(HNC)} > 4 \tag{2}$$

These equations describe two physically distinguished regimes: (a) Eq.1 describes a low-temperature regime between $T_K < 40$ K characteristic of the typical cloud conditions, while (b) Eq.2 can be applied of the more extreme gas conditions found in the Orion Nebula.

The observed dependence of the HCN/HNC line ratio with temperature appears to be controlled by a combination of chemical and emission properties of both HCN and HNC tracers. Similar temperature-dependent abundance variations are produced by the combination of the selective destruction of $HNC + O \rightarrow CO + NH$ (reaction 1) at low-temperatures and the effective isomerization of $HNC + H \rightarrow HCN + H$ (reaction 2) at higher temperatures. Previous studies assumed high activation energies (> 1200 K) inhibiting reaction 1. Tests using different chemical models, however, indicate that the observed dependence can be reproduced using a low-energy barrier ($\Delta E \sim$ 15-25 K) in addition to the temperature dependence excitation and opacity changes of HCN and HNC (see [6] for a full discussion).

Independent of its origin, the I(HCN)/I(HNC) line ratio provides an empirical measurement of the gas kinetic temperature in the ISM. The parameterization obtained in Eqs.1 and 2 allows to obtain estimates of the $T_K$ based on two direct observables, namely, the integrated intensities of I(HCN) and I(HNC) (J=1–0) lines at 3 mm. This method simplifies previous estimates based on more time-consuming calculations (radiative transfer calculations) and observationally expensive techniques (e.g. CO-ladders that require multiple observations).

We have tested our new temperature estimates along the ISF. Figure 1d shows the gas kinetic temperature map obtained after converting the integrated intensity maps of HCN and HNC (Figs.1b & 1c) into $T_K$(HCN/HNC) according to Eqs.1 and 2. Despite its complexity, our temperature estimates reproduce the expected thermal gas behaviour in this region. Overall, the coldest part of the cloud down to 15-20 K correspond with the densents filamentary regions identified in $N_2H^+$ (black contours). Outside, the Interstellar Radiation Field (ISRF) increases the gas tempeartures up to lukewarm temperatures of 30-40 K. Additionally, the effects of the different feedback mechanisms (e.g. Orion Nebula) are clearly seen in our maps by their enhanced gas temperatures above 40-50 K. Particularly within the optimal range of this method (i.e. $T_K \sim [15, 40]$ K, see Eq.1), the HCN/HNC line ratio therefore acts as an effective observational probe of the gas kinetic temperature in this cluod.

Additional comparisons with previous results shows the a similar behaviour of this HCN/HNC line ratio tracing the thermal properties of molecular clouds. The use of Herschel column density maps [9] indicate that the HCN/HNC line ratio provides information of the gas temperatures over a large range of column densities down to $A_V \sim 5$ mag compared to $NH_3$-based temperature measurements limited to $A_V \sim 10$mag. Temperature variations of the HCN/HNC ratio are seen in gas surveys across the galactic plane (e.g. MOPRA survey, [2]). Additional application of this technique show positive results in other star-forming regions in Orion (Hacar et al, in prep). As primary advantage for (extra-)galactic studies, both HCN and HNC (J=1–0) integrated intensities can be easily measured in 3mm-band surveys using broadband receivers such as EMIR. With a robust behaviour, the observed HCN/HNC line ratio could in principle be used as chemical thermometer of the molecular ISM.

*Acknowledgements:* This project has received funding from the European Research Council (ERC) under the European Union's Horizon 2020 research and innovation programme (Grant agreement No. 851435).

## References

[1] Ho, P. T. P., & Townes, C. H. 1983, ARA&A, 21, 239
[2] Foster, J. B., Jackson, J. M., Barnes, P. J., et al. 2011, ApJS, 197, 25
[3] Friesen,R.K., Pineda,J.E., Rosolowsky,E., etal. 2017, ApJ, 843, 63
[4] Hacar, A., Alves, J., Tafalla, M., & Goicoechea, J. R. 2017, A&A, 602, L2
[5] Hacar, A., Tafalla, M., Forbrich, J., et al. 2018, A&A, 610, A77
[6] Hacar, A., Bosman, A. D., & van Dishoeck, E. F. 2020, A&A, 635, A4
[7] Herbst, E., Terzieva, R., & Talbi, D. 2000, MNRAS, 311, 869
[8] Johnstone, D., & Bally, J. 1999, ApJ, 510, L49
[9] Lombardi, M., Bouy, H., Alves, J., & Lada, C. J. 2014, A&A, 566, A45
[10] Nishimura, A., Tokuda, K., Kimura, K., et al. 2015, ApJS, 216, 18
[11] Pety, J., Guzmán, V. V., Orkisz, J. H., et al. 2017, A&A, 599, A98
[12] Schilke, P., Walmsley, C. M., Pineau Des Forets, G., et al. 1992, A&A, 256, 595

# Gas phase Elemental abundances in Molecular cloudS (GEMS)

*Asunción* Fuente[1,*] and the GEMS IRAM 30m Large Program team

[1]Observatorio Astronómico Nacional (OAN), Alfonso XII, 3, 28014 Madrid, Spain

**Abstract.** Gas phase Elemental abundances in Molecular CloudS (GEMS) is an IRAM 30m Large Program aimed at determining the elemental abundances of carbon (C), oxygen (O), nitrogen (N), and sulphur (S) in a selected set of prototypical star forming filaments. Elemental abundances have a key role in the star formation process by driving the gas cooling rate and the gas ionization degree. Defining elemental abundance as the amount of a given atom in volatiles, the elemental abundances of C, O, and N in the interstellar medium are currently known within a factor of a few. However, the elemental abundance of S remains uncertain by several orders of magnitude, and its determination is one of the most challenging goals of this program. This is a review of the observational, theoretical, and modeling effort that GEMS team is doing to improve our knowledge of the sulphur chemistry. Our results suggest that the sulphur elemental abundance depends on the local physical conditions and the star formation activity in the environment.

## 1 Introduction

In recent years, space telescopes such as Spitzer and Herschel have revolutionized our view of star-forming regions. Now, we know that interstellar filaments are almost everywhere in the Milky Way and are the preferred site for star formation. Herschel FIR continuum maps probed the dust to visual extinctions of $A_V \sim 1$ mag, thus establishing the physical conditions of pre-stellar cores and those of the diffuse clouds out of which they are formed. These maps provide an unprecedented observational database allowing for the first time to have a good description of the different phases of collapsing molecular clouds. However, we need additional ingredients to complete our understanding of the star formation process, in particular we need to include chemistry. Gas chemistry regulates fundamental processes to form a star, such as the gas temperature and the coupling of the gas with the magnetic field. Molecular filaments can fragment to form pre-stellar cores to a large extent because molecules cool the gas, thus diminishing the thermal support relative to self-gravity. Gas chemistry is also regulating the abundance of electrons in the gas. The ionization fraction controls the coupling of magnetic fields with the gas, driving the dissipation of turbulence and angular momentum transfer, and therefore it plays a crucial role in the cloud collapse and the dynamics of accretion discs. The gas ionization fraction as well as the molecular abundances depend on the elemental depletion factors [2]. In particular, Carbon (C) is the main donor of electrons in the cloud surface ($A_v$<4 mag) and, because of its lower ionization potential, Sulfur (S) might

*e-mail: a.fuente@oan.es

 https://doi.org/10.1051/epjconf/202226500005

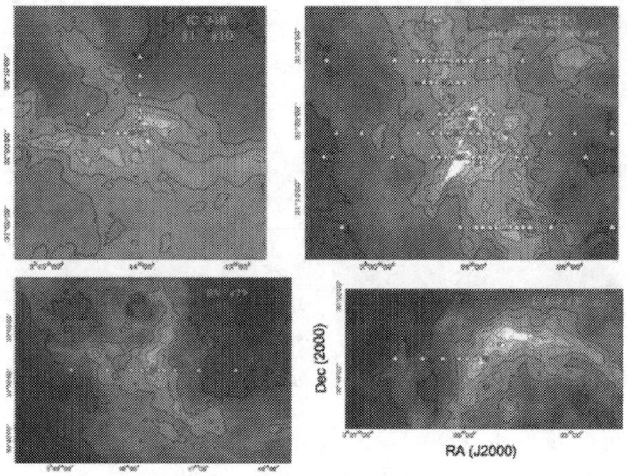


Table 1.- Molecular tracers

| | $A_V$ < 10 mag | $A_V$ > 10 mag |
|---|---|---|
| X(e⁻) | $^{13}$CO, $HCO^+$, $H^{13}CO^+$ | $C^{18}O$, $HC^{18}O^+$, $N_2H^+$ |
| n($H_2$) | CS, $C^{34}S$ | $C^{34}S$, $^{13}$CS, SO |
| C/H | $^{13}$CO, HCN, CS | $C^{18}O$, $H^{13}CN$ |
| O/H | $^{13}$CO, SO | $C^{18}O$, SO, $^{34}$SO |
| N/H | HCN | $H^{13}CN$, $N_2H^+$ |
| S/H | CS, $C^{34}S$, SO, $HCS^+$ | $C^{34}S$, $^{13}$CS, SO, $^{34}$SO |

**Figure 1.** Maps of the dust opacity at 850 $\mu$m towards IC 348, NGC 1333, B5, and L1448 [13]. Positions observed with the IRAM 30 m telescope are indicated with triangles. Red circles indicate the positions of the starless cores. Table 1 shows the species used as tracers of the physical and chemical conditions in the different cloud layers.

be the main donor in the ~3.7−7 magnitudes range that encompasses a large fraction of the molecular cloud mass. Depletions of C and O determine the cooling gas rate since CO and CII are main coolants in molecular clouds. The determination of the elemental abundances at different depths into prototypical molecular clouds is the goal of the "Gas phase Elemental abundances in Molecular CloudS" (GEMS) IRAM 30m Large program.

## 2 Observational Database

GEMS methodology consists of deriving the elemental abundances and the gas ionization fraction by fitting the fractional abundances of a set of selected molecules using a state-of-the-art chemical model. These species were selected for being well-known tracers of the gas density and for being the most abundant compounds containing a given element that can be observed in the (sub-)mm range (see Table 1). The observations were carried out with the IRAM 30m telescope from July, 2017 to December 2019, and the details of the observational strategy were described by [6]. These observations are now public and can be downloaded from IRAM Data Archive.

The GEMS sample is composed of 29 starless cores located in Taurus, Perseus and Orion. In each cloud, we observed sets of typically ~9-20 positions, separated by uniform steps of visual extinction, and aligned forming a cut intersecting the filament at the location of a visual extinction peak. Examples of these cuts in different regions of Perseus are shown in Fig. 1. This strategy allows us to probe the different layers of the cloud in the vicinity of the selected starless cores. The cuts were selected to avoid protostars and molecular outflows that could affect the chemical composition of the molecular cloud, hindering the study of molecular depletion. In total, we observed 12 cuts (99 positions) in Taurus, 14 cuts (153 positions) in Perseus, and 3 cuts (53 positions) in Orion.

Our sample includes cuts in regions with different physical conditions and star formation activity. The Taurus molecular cloud is considered as the prototype of isolated formation of low mass stars. In contrast to Taurus, the molecular clouds in Perseus are known to be forming star clusters. The group of cores in IC348 and NGC 1333 are close to the optical

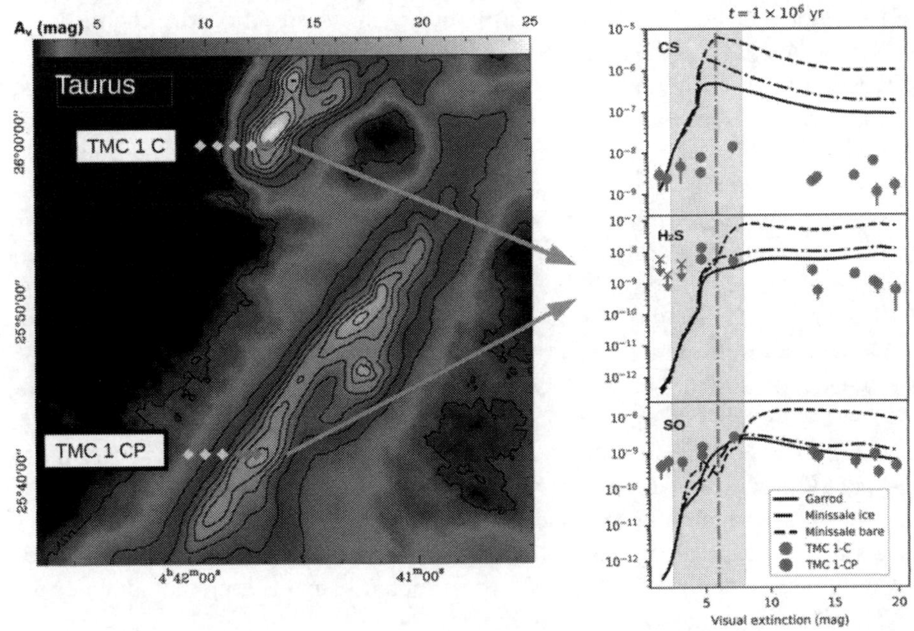


**Figure 2.** *left* Visual extinction map of the filament TMC 1. The positions observed are indicated with diamonds. The cuts labeled TMC 1 C and TMC 1 CP are modeled assuming that the associated cores are are well described with the physical structure of a Bonnor-Ebert sphere. Furthermore, we use the chemical model Nautilus [11] to predict the N(CS), N($H_2S$), and N(SO) column densities at different impact parameters. In the *right* panel, the obtained values normalized with $N_H$ are compared with model predictions with [S/H] = $1.5\times10^{-5}$. It should be noticed that while the results of $H_2S$ and SO are well described by the model, the column density of CS is over-predicted by more than an order of magnitude (see [7]).

clusters and might be the future generation of clustered stars. However, Barnard 1 and L1448 are located in a more quiescent region. Orion is the closest massive star formation region and the only one in which this kind of study can be done with single-dish telescopes. A detailed description of the sample was reported by [10].

We were able to detect several lines of CS to derive molecular hydrogen densities, only in 244 positions out of the 305 positions observed. Using these densities, we calculated the chemical abundances of 14 species: $^{13}CO$, $C^{18}O$, $HCO^+$, $H^{13}CO^+$, $HC^{18}O^+$, HCN, $H^{13}CN$, HNC, $HCS^+$, CS, SO, $^{34}SO$, $H_2S$, and OCS [10]. These abundances constitute the GEMS molecular database that is used here to derive the sulphur elemental abundance. In addition to these species, a few complex compounds such as $CH_3OH$ and $H_2CS$ were detected towards the visual extinction peaks. The chemistry of $CH_3OH$ and $H_2CS$ has been studied by [12] and [3].

## 3 Testing chemical models: TMC1 and Barnard 1b

Chemical models are a key stone for deriving elemental abundances. In practice, we estimated the elemental abundances through chemical modeling of the 14 species included in the GEMS molecular database. Unfortunately, the sulphur chemistry was poorly known, which introduces large uncertainties in the estimate of the sulphur elemental abundance ([S/H]). For this reason, the GEMS team have carried out a large experimental and theoretical effort to improve the sulphur chemical network. In particular we focused on the chemistry of SO, $H_2S$, and CS, which are ubiquitous species in the interstellar medium.

The reaction rates of the main formation and destruction paths for SO were estimated using *ab initio* calculations by [4] and [6]. It is well-known that $H_2S$ is formed on the grain surfaces under the physical conditions prevailing in dark clouds. Then, the molecules detected in gas phase should have been ejected from the grain surfaces through thermal and non-thermal desorption processes. In the cold and dense gas, only non-thermal desorption processes are efficient. [5] estimated the photo-desorption yield of $H_2S$. Photo-desorption is an efficient desorption mechanism in the cloud border where UV photons are not attenuated. [7] investigated the efficiency of the reactive chemical desorption of $H_2S$ as a function of the optical depth into the cloud. They concluded that the efficiency of chemical desorption decreases when the grains are covered with a thick ice mantle. [1] estimated the reaction rate of CS + O $\rightarrow$ CO + S that was proposed to be an important destruction mechanism of CS.

The updated sulphur chemical network was used by [7] to fit the observations of CS, SO, and $H_2S$, in TMC1 and Barnard 1b. They concluded that it was not possible to accurately fit the three molecules using the same value of [S/H]. While the abundances of $H_2S$ were better fitted assuming the solar value ([S/H]=$1.5\times10^{-5}$), the abundance of CS could only be fitted assuming a sulphur depletion of >10 [1, 8]. This problem remains even including the new CS + O $\rightarrow$ CO + S reaction rate as calculated by [1].

## 4 Sulphur elemental abundance

Our goal is to determine the sulphur elemental abundance, i.e., the amount of sulphur atoms in volatiles ([S/H]). Our large molecular database allows us to carry out a statistic study to explore possible trends of [S/H] with the local physical parameters (n, $T_g$) and environment ($\zeta(H_2)$, $\chi_{UV}$). We predicted the chemical abundance using NAUTILUS 1.1 [11], which is a three-phase model that takes into account gas, ice surface, and ice mantle chemistries. The chemical network updated by [7] was adopted for these calculations. We run the three grids of models shown in Table 2 in order to fit the abundances of CO, $HCO^+$, HCN, HNC, CS, $HCS^+$, $H_2S$, SO, OCS towards the 244 positions of the molecular database. In order to avoid degeneracies, we imposed some restrictions to the values of $A_V$, n($H_2$) and $T_g$: i) the visual extinction is assumed to be in the range $A_V \times 0.5$ and $A_V \times 0.5 + 2$, where $A_V$ is the value obtained from Herschel data, and the factor 0.5 accounts for the fact the the cloud is expected to be illuminated from the back and the front; ii) the density is assumed to be in the range 0.5×n($H_2$) and 5 × n($H_2$) where n($H_2$) is the value derived from the fitting of CS lines [10]; and iii) the gas kinetic temperature is allowed to vary between $T_d$ and $T_d$+ 5 K, where $T_d$ is the dust temperature obtained from Herschel data. During the fitting process, [S/H], $\zeta(H_2)$ and *Age* are allowed to vary among the values: *Age* = 0.1,1,10 Myr, [S/H] = $8\times10^{-8}$, $1.5\times10^{-6}$, $1.5\times10^{-5}$, $\zeta(H_2) = (1, 5, 10, 50) \times 10^{-17}$ s$^{-1}$. In order to find the *best-fit* model, we minimize the parameter $D_{iff}$ that is defined by,

$$D_{iff} = 1/n_{obs} \times \Sigma_i [log_{10}(X^i_{mod}) - log_{10}(X^i_{obs})]^2 \quad (1)$$

where $n_{obs}$ is the number of species detected at each position, $X^i_{mod}$ is the predicted abundance for the species $i$, and ($X^i_{obs}$) is the abundance of the species $i$ derived from GEMS observations. The results are used to construct the histograms shown in Fig. 3. We detect clear differences between the studied molecular cloud complexes. The majority of the positions in Taurus and Perseus are best fitted with *Age*=0.1 Myr, while the majority of the positions in Orion are fitted with *Age* > 1 Myr. Regarding [S/H] which is the most interesting parameter for us, most of the positions in Taurus and Perseus are fitted with [S/H] =

**Table 2.** Chemical models parameters.

| | |
|---|---|
| **Taurus** | |
| $n_H$ | $3.16\times10^3$ cm$^{-3}$ to $3.16\times10^6$ cm$^{-3}$ in steps of $\times$ 3.16 |
| $A_V$ | 1.5 mag to 11.5 mag in steps of 2 mag |
| $T_g$ | 10 K to 45 K in steps of 5 K |
| CRIR | (1.0, 5.0, 10, 50) x $10^{-17}$ s$^{-1}$ |
| $\chi_{UV}$ (Draine) | 5.0 |
| $[S/H]$ | (1.5, 0.15, 0.008) $\times10^{-5}$ |
| **Perseus** | |
| $n_H$ | $3.16\times10^3$ cm$^{-3}$ to $3.16\times10^6$ cm$^{-3}$ in steps of $\times$ 3.16 |
| $A_V$ | 1.5 mag to 13.5 mag in steps of 2 mag |
| $T_g$ | 10 K to 45 K in steps of 5 K |
| CRIR | (1.0, 5.0, 10, 50) x $10^{-17}$ s$^{-1}$ |
| $\chi_{UV}$ (Draine) | 25.0 |
| $[S/H]$ | (1.5, 0.15, 0.008) $\times10^{-5}$ |
| **Orion** | |
| $n_H$ | $3.16\times10^3$ cm$^{-3}$ to $3.16\times10^6$ cm$^{-3}$ in steps of $\times$ 3.16 |
| $A_V$ | 1.5 mag to 11.5 mag in steps of 2 mag ,15.5 mag, 20 mag |
| $T_g$ | 10 K to 45 K in steps of 5 K |
| CRIR | (1.0, 5.0, 10, 50) x $10^{-17}$ s$^{-1}$ |
| $\chi_{UV}$ (Draine) | 50 |
| $[S/H]$ | (1.5, 0.15, 0.008) $\times10^{-5}$ |

$1.5\times10^{-6}$. On the contrary, a large fraction of the positions in Orion are better fitted with [S/H]= $1.5\times10^{-5}$. This suggests that environment changes the sulphur elemental abundance, the harsh environment of Orion giving rise to the lowest sulphur depletion. It is also interesting to explore possible trends with visual extinction and density. There is no clear trend with visual extinction in our cuts. However, sulphur depletion seems to depend on density. The positions with lower densities are more likely associated with [S/H] = $1.5\times10^{-6}$, while the positions with higher densities are usually associated with [S/H] = $8\times10^{-8}$. In addition to the environment, the value of [S/H] depends on the local density, with sulphur depletion increasing with density. A word of caution should be put in this latter conclusion. Since only 53 out of the 244 positions belong to Orion, these histograms are dominated by the positions in Taurus and Perseus, and the extrapolation of this conclusion to Orion is risky. Additional observations are needed to have a statistically significant number of positions in Orion.

## 5 Summary

We present the molecular database and first results of the GEMS project. This program is focused on the observation of 29 starless cores located in proto-typical filaments of the nearby star-forming regions Taurus, Perseus, and Orion. These regions have different degrees of star formation activity, and therefore different physical conditions, providing a possibility to explore the effect of the environment on gas chemistry. The project includes observations towards 305 positions distributed in 29 cuts that have been selected to avoid recently formed stars and their associated bipolar outflows. The gas chemistry in these positions is therefore not affected by the shocks produced by these energetic phenomena and/or the UV radiation coming from internal sources. We have used chemical modeling with an updated sulphur network to determine the sulphur elemental abundance in the three studied regions. Our results suggest that environment is determining the sulphur elemental abundance, with the

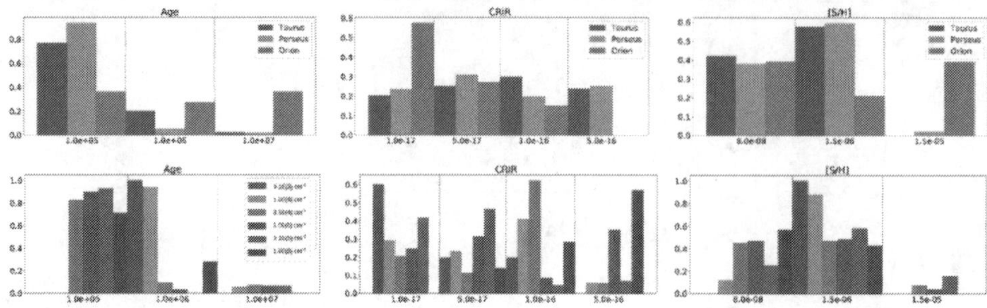

**Figure 3.** The panels show the statistics of the results of our fitting to the GEMS molecular database. The points are segregated in bins according with their localization (Taurus, Perseus, or Orion molecular cloud complexes) in the upper row, and in bins of density in the bottom row.

sulphur depletions decreasing in regions with higher star formation activity. In Taurus and Perseus, there is a trend with density that suggests that sulphur depletion increases towards the densest regions of the cloud. We would like to comment that the large number of positions observed in GEMS provides an unprecedented opportunity to investigate the chemistry of dark clouds.

## References

[1] Bulut, N., Roncero, O., et al., Astronomy & Astrophysics **646**, A5 (2021)
[2] Caselli, P., Walmsley, M., et al., The Astrophysical Journal Letters **523**, L165 (1999)
[3] Esplugues, G., Fuente, A., et al., Astronomy & Astrophysics, in press (2022)
[4] Fuente, A., Cernicharo, J., et al., Astronomy & Astrophysics **593**, A94 (2016)
[5] Fuente, A., Goicoechea, J. R., et al., The Astrophysical Journal Letters **851**, L49 (2017)
[6] Fuente, A., Navarro-Almaida, D., et al., Astronomy & Astrophysics **624**, A105 (2019)
[7] Navarro-Almaida, D., Le Gal, R., et al., Astronomy & Astrophysics **637**, A39 (2020)
[8] Navarro-Almaida, D., Fuente, A., et al., Astronomy & Astrophysics **653**, A15 (2021)
[9] Rivière-Marichalar, M., Fuente, A., et al., Astronomy & Astrophysics **628**, A16 (2019)
[10] Rodríguez-Baras, M., Fuente, A., et al., Astronomy & Astrophysics **648**, A120 (2021)
[11] Ruaud, M., Wakelam, V., et al., MNRAS **459**, 3756 (2016)
[12] Spezzano, S., Fuente, A., et al., Astronomy & Astrophysics **657**, A10 (2022)
[13] Zari, E., Lombardi, M., et al., Astronomy & Astrophysics **587**, A106 (2016)

# Isotopic ratios and fractionation in the local Universe

*Serena* Viti[1,2,*]

[1]Leiden Observatory, Leiden University, PO Box 9513, NL-2300 RA Leiden, The Netherlands

[2]Department of Physics and Astronomy, University College London, Gower Street, London WC1E 6BT, UK

**Abstract.** The knowledge of isotopic abundances is important in galaxy evolution studies because isotopes provide diagnostics for the chemical enrichment in galaxies over time. While measurements of isotopes in large sample of stars would be ideal to determine the fossil record of the enrichment history, in practice this is hampered by the need of very high resolution, high signal-to-noise spectroscopic data. A complementary, or alternative, method is to measure isotopic ratios from observations of gas-phase interstellar medium (ISM) isotopic abundances. In this proceedings I shall review the observations of the most abundant fractionated species in nearby galaxies and recent modeling efforts aimed at investigating the physical and chemical conditions that can lead to a large spread of isotopic ratios in external local galaxies.

## 1 Introduction

It is now well established that the cooler (< few thousands K) and denser (> 100 cm$^{-3}$) parts of our own as well as other galaxies are pervaded by molecules. Such dense and cool regions are of course important in themselves as they are the sites of star formation. But they are also important for our understanding of how galaxy evolve and in tracing the results of the nucleosynthesis in different types of galaxies where the star formation processes themselves may differ. Of particular use are the isotopologues of the main elements: carbon, nitrogen and oxygen.

In our own galaxy isotopes variations at different scales give clues about the link between the ISM and solar-like systems. On the other hand, isotopic measurements at larger scales are crucial for the understanding of the CNO cycles (nucleosynthesis), especially of how the yield differs between low mass and high mass stars and hence how the ISM enrichment is affected by such differences. Ideally one could measure isotopes in large sample of stars and determine the fossil record of the ISM enrichment history; in reality this is not possible due to limitations in resolution, and sensitivity. Instead we can measure isotopic ratios from observations of gas-phase ISM isotopic abundances [1]

In this procedings, we shall concentrate on carbon and nitrogen isotopic ratios in external galaxies. But before we move on to review our knowledge of isotopic ratios in other galaxies, it is worth noting that in the local ISM, variations in the $^{14}N/^{15}N$ ratio (from their solar value of $\sim$ 440, [2]) and in the $^{12}C/^{13}C$ ratio (from their solar ratio of $\sim$ 69, [3]) point to multiple origins as stellar nucleosynthesis alone can not lead to the observed range in the

*e-mail: viti@strw.leidenuniv.nl

Milky Way. A possible origin for the large ranges observed is chemical fractionation in the ISM. For example, the large spread in the measured $^{14}N/^{15}N$ ratio, ranging from ∼100 for meteorites, comets and protoplanetary disks to ∼1000 in pre-stellar and star-forming cores could be a combination of evolutionary effects (e.g. from protostellar cores to late stages of star formation) as well as different gas densities and temperatures ([4]; [5]; [6]; [7]).

Several theoretical "local" studies strongly confirm significant variations of the isotopic ratio as a function of time and gas density (e.g. [8], [9]). For example, [8] presented a chemical model where comprehensive $^{13}C$ and $^{14}N$ networks were included and found variations in the isotopic ratios both across time as well as molecules, and in particular they found that both nitriles and isonitriles are very depleted in $^{13}C$, a result with potential consequences for isotopic ratios derived using HCN, HNC and CN.

It may therefore not be surprising that extragalactic studies do not find agreement on a single value for either of these two ratios.

## 2 Nitrogen and Carbon Fractionation in external galaxies

Isotopic abundances are important in galaxy evolution studies because isotopes provide diagnostics for the chemical enrichment in galaxies over time.

In particular, let us consider the $^{14}N/^{15}N$ ratio: the latter is often used as an indicator of stellar nucleosynthesis because the two isotopes are synthesized in stars in different ways during the CNO cycle of massive stars (e.g. [10]), with $^{14}$ N being the primary product at low metallicity (e.g.[11]), while $^{15}N$ may be overproduced during the novae phase. Carbon, also, exists in the form of two stable isotopes, $^{12}C$ and $^{13}C$. The former is synthesised via Helium burning in both low-mass and massive stars in a short timescale while the latter is mainly expected to be synthesised through the CNO cycle in asymptotic giant branch (AGB) stars through slower processes and hence its 'distribution' to the ISM is less direct than in the case of $^{12}C$.

While of course differences in nuclesynthesis yields will lead to differences in isotopic ratios, other processes, unrelated to stellar nucleosynthesis, may be important too in affecting the isotopic composition of the ISM. For example, the cold gas in galaxies may favour low-temperature exothermic ion-neutral exchange reactions (e.g. [12], [13]) which will increase $^{13}C$ in species such as CO and $HCO^+$, although in some cases energy barriers do exist and in order to quantitatively determine the role of such reactions in augmenting the ISM of fractionate species, the presence of such barriers need to be systematically checked [8].

However, we note that the average gas in galaxies will not be at 10 K: much of the gas will be in photo-dissociation regions (PDRs). In PDRs, where the gas is exposed to UV radiation, isotope-selective photo-dissociation and other mechanisms may be important. Moreover, in some specific environments, e.g. in starburst regions or in the vicinity of AGN, the gas is dense ($\geq 10^4$ cm$^{-3}$) and warm ($> 50$ K).

In summary, the relative importance of these processes, and the existence of additional processes not yet considered, is still unclear. A first step towards determining the importance of these mechanisms requires an attempt at quantifying whether, for a fixed nucleosynthesis yield, chemical fractionation actually leads to any of the variations in isotopic ratios we observe in galaxies: Table 1 contains a (non exhaustive) summary of the observational values of $^{12}C/^{13}C$ and $^{14}N/^{15}N$ across different types of galaxies, combined from Tables 2 from [14] and Table 1 from [15].

**Table 1.** $^{12}C/^{13}C$ and $^{14}N/^{15}N$ measured in external galaxies.

| Galaxy | type | $^{12}C/^{13}C$ | Mol. | Ref. |
|---|---|---|---|---|
| NGC 253 | starburst | ~ 40; 30 – 67 | CN | [18]; [19] |
| NGC 253 | starburst | ~ 27 – 70 | CS | [20]; [21]; [18] |
| M82 | starburst | > 40 | CN | [22] |
| NGC 253 | starburst nucleus | ~ 21 | $C^{18}O^a$ | [23] |
| NGC 4945 | starburst nucleus | 6–44 | CN | [19] |
| VV 114 | LIRG | ~ 230 | CO | [24] |
| NGC 1614 | LIRG | ~ 130 | CO | [25] |
| Mrk 231 | ULIRG | ~ 100 | CO, CN | [18] |
| Arp 220 | ULIRG | ~ 100 | CO | [26] |
| Arp 193 | ULIRG | ~ 150 | CO | [27] |
| Cloverleaf | ULIRG $z = 2.5$ | 100 – 200 | CO | [28] |
| Eyelash | ULIRG $z = 2.3$ | ~ 100 | CO | [29] |
| LMC | 0.5 metal | ~ 49 | $H_2CO$ | [30] |
| NGC 1068 | AGN+starburst | ~ 50 ;24–62 | CN | [31]; [19] |
| NGC 4258 | AGN | ~ 46 | $HCO^+$ | [32] |
| NGC 3690 | AGN+starburst | ~ 40 | $HCO^+$ | [32] |
| NGC 6240 | AGN+starburst | ~ 41 | HCN | [32] |
| NGC 6240 | AGN+starburst | 300 – 500 | CO | [27]; [33] |
| IC 342 | spiral local | > 30 | CN | [22] |
| MA0.89 | spiral $z = 0.89$ | ~ 27 | HCN, $HCO^+$, HNC | [34] |
| MA0.68 | spiral $z = 0.68$ | ~ 40 | HCN, $HCO^+$, HNC | [35] |
| Galaxy | type | $^{14}N/^{15}N$ | Mol. | Ref. |
| NGC4945 | starburst | 200-500 | HCN | [36] |
| LMC | 0.5 metal | 111(± 17) | HCN | [37] |
| Arp220 | ULIRG | 440 (+140,-82) | HCN, HNC | [38] |
| NGC1068 | AGN+starburst | > 419 | HCN | [39] |
| IC694 | starburst | 200-400(?) | HCN | [32] |
| LMC | 0.5 metal | 91(± 21) | HCN | [30] |
| M82 | starburst | > 100 | HCN | [22] |
| Galactic Center | standard with high $\zeta$ | ≥164 | HNC | [40] |

## 3 The sensitivity of carbon and nitrogen fractionation to the physical conditions of a galaxy

As mentioned already, several theoretical studies have concentrated on the variations of carbon and nitrogen fractionation in low temperature (up to 50K) ISM ([8]; [16]; Colzi et al. 2022 this proceedings). All these studies in general do find that isotopic ratios display significant variations with density and cosmic ray ionization rates. In [15] and [14] we explored a higher temperature parameter space and performed a theoretical study of the ISM nitrogen and carbon fractionation to determine physical conditions that may lead to a spread in these ratios from the solar values. For those studies we used the publicly available ode UCLCHEM ([17]; https://uclchem.github.io/) which is a time dependent gas-grain chemical model. The already quite comprehensive chemical networks were augmented with the known exchange reactions from [8] and [9]. Large grids of models were ran spanning densities from $10^4$ to $10^6$ $cm^{-3}$, temperatures from 50 to 200 K, cosmic ray ionization rates from the galactic values of $\sim 10^{-17}$ $s^{-1}$ to a factor of $10^4$ higher, and radiation fields from 1 to 100 Draine (or $2.74\times10^{-3}$ $erg/s/cm^2$).

Two key results from these two studies are that (i) in general fractionation (of C and N) does not seem to be very sensitive to temperature in the explored range; and (ii) the CO

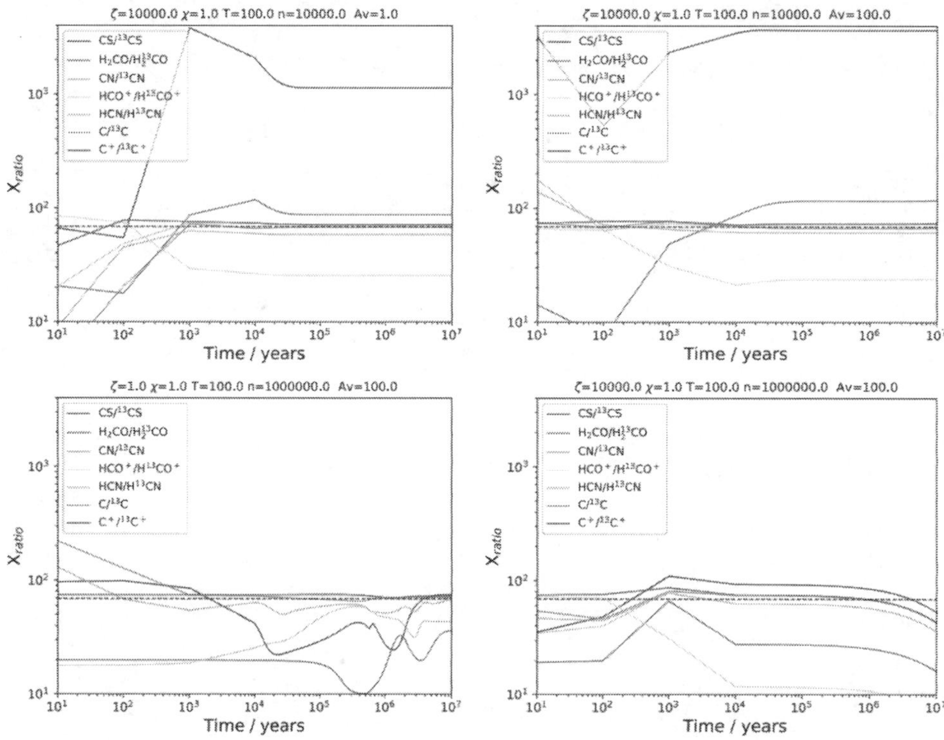


**Figure 1.** Figure 5 from [14] showing the effects of varying the cosmic ray ionization rate on the ISM chemistry of carbon isotopes.

fractionation is fairly constant implying that a high $^{12}C/^{13}C$ in CO (e.g. >100, see Table 1) must be a product of nucleosynthesis rather than ISM chemistry. Furthermore it was found that a high cosmic ray ionization rate and/or a high radiation field leads to a much more variable fractionation for most species (see Figure 1). This is particularly relevant for AGN-dominated galaxies or starburst galaxies. Finally there are clearly differences in fractionation among molecules: for example, the fractionation of CS is always lower than in other species (see Figure 2). More specific results, for carbon and/or nitrogen fractionation have been determined by these studies and can be found in the aforementioned papers. We can certainly conclude that, as previously found by other authors ([8]) there are always large variations in the isotopic ratios in the species investigate.

So does ISM fractionation in the ISM of galaxies contribute to the observed variations? It is clear that - with time, physical parameters, and across species - isotopic ratios vary in the ISM. The question then now is whether such changes are reflected in the observational ranges observed as reported for example in Table 1.

We already mentioned that the large $^{12}CO/^{13}CO$ observed for e.g. in starburst galaxies is a product of nucleosynthesis and not ISM fractionation. However, it is interesting to note that qualitatively it was found that different models can reproduce most of the observations. So for example, for starburst galaxies all the best fit models have a higher than galactic radiation field as well as cosmic ray ionization rate, while for AGN-dominated or composite galax-

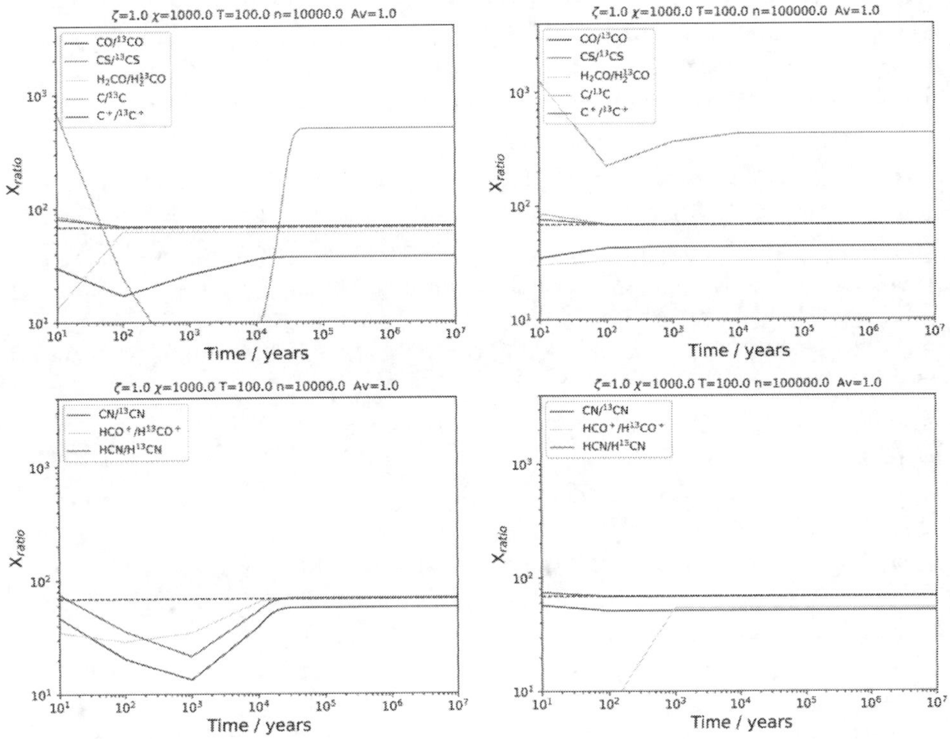


**Figure 2.** Figures 9 and 10 from [14]. Top: $^{12}C/^{13}C$ for selected species in extended gas in starburst galaxies. Bottom: as Top Figures but for dense gas in starburst galaxies.

ies the best fit models imply a large reservoir of PDR gas plus a considerably high density component.

It is worth to conclude with a note of caution when interpreting isotopic ratios in external galaxies: most current measurements are unable to determine such ratios for linear scales less than 50 pc or so. Only very high angular resolution observations will truly allow us to disentangle the effects of different nucleosynthesis yields from local chemical effects.

## References

[1] Wilson, T.L., & Rood, R., 1994, ARA&A, 32, 191
[2] Fouchet T., Irwin P. G. J., Parrish P., Calcutt S. B., Taylor F. W., Nixon C. A., Owen T., 2004, Icarus, 172, 50
[3] Wilson, T.L., 1999, RPPh, 62, 143
[4] Zeng S. et al., 2017, A&A, 603, A22
[5] De Simone M. et al., 2018, MNRAS, 476, 1982
[6] Redaelli, E.; Bizzocchi, L.; Caselli, P., 2020, A&A, 644, 29
[7] Benedettini, M.; Viti, S.; Codella, C.; Ceccarelli, C. et al, 2021, A&A, 645, 91
[8] Roueff, E., Loison, J.C., Hickson, K.M. 2015, A&A, 576, 99
[9] Loison J.-C., Wakelam V., Gratier P., Hickson K. M., 2019, MNRAS, 484, 2747

[10] Izzard R. G., Tout C. A., Karakas A. I., Pols O. R., 2004, MNRAS, 350, 407
[11] Matteucci F., 1986, MNRAS, 221, 911
[12] Langer, W. D., Graedel, T. E., Frerking, M. A., et al. 1984, ApJ, 277, 581
[13] Mladenovic, M., Roueff, E. 2014, A&A, 566, A144
[14] Viti, S.,; Fontani, F., Jiménez-Serra, I., 2020, MNRAS, 497, 4333
[15] Viti, S.,; Fontani, F., Jiménez-Serra, I., Holdship, J., 2019, MNRAS, 486, 4805
[16] Colzi, L., Sipila, O., Roueff, E., Caselli, P., and Fontani, F., 2020, A&A, 640, A51
[17] Holdship, J., Viti, S., Jimenez-Serra, I., Makrymallis, A., Priestley, F. 2017, AJ, 154, 38
[18] Henkel, C., Asiri, H., Ao, Y., Aalto, S. Danielson, A.L.R., Papadopoulos, P.P., García-Burillo, S. et al. 2014, A&A, 565, A3
[19] Tang, X.D., Henkel, C., Menten, K.M., Gong, Y., Martín, S., Mühle, S., Aalto, S., et al. 2019, A&A, 629, 6
[20] Martín, S., Martín-Pintado, J., Mauersberger, R., Henkel, C., García-Burillo, S. 2005, ApJ, 620, 210
[21] Martín, S., Mauersberger, R., Martín-Pintado, J., Henkel, C., García-Burillo, S. 2006, ApJS, 164, 450
[22] Henkel, C., Chin, Y.-N., Mauersberger, R., Whiteoak, J.B. 1998, A&A, 329, 443
[23] Martín, S., Muller, S., Henkel, C., Meier, D.S., Aladro, R., Sakamoto, K. van der Werf, P.P. 2019, A&A, 624, A125
[24] Sliwa, K., Wilson, C.D., Krips, M., Petitpas, G.R., Iono, D., Juvela, M., Matsushita, S., et al. 2013, ApJ, 777, 126
[25] Sliwa, K., Wilson, C.D., Iono, D., Peck, A., Matsushita, S. ApJL, 2014, 796, L15
[26] González-Alfonso, E., Fischer, J., Graciá-Carpio, et al. 2012, A&A, 541, A4
[27] Papadopoulos, P.P., Zhang, Z.-Y., Xilouris, E.M., Weiss, A., van der Werf, P., Israel, F.P., Greve, T.R., Isaak, K.G., Gao, Y. 2014, ApJ, 788, 153
[28] Spilker, J. S., Marrone, D. P., Aguirre, J. E., et al. 2014, ApJ, 785, 149
[29] Danielson, A.L.R., Swinbank, A.M., Smail, I., Bayet, E., van der Werf, P.P., Cox, P., Edge, A.C., et al. 2013, MNRAS, 436, 2793
[30] Wang, M., Chin, Y.-N., Henkel, C., Whiteoak, J.B., Cunningham, M. 2009, ApJ, 690, 580
[31] Aladro, R., Viti, S., Bayet, E., et al. 2013, A&A, 549, A39
[32] Jiang, X., Wang, J., Gu, Q. 2011, MNRAS, 418, 1753
[33] Pasquali, A., Gallagher, J. S., & de Grijs, R. 2004, A&A, 415, 103
[34] Muller, S., Guélin, M., Dumke, M., Lucas, R., Combes, F. 2006, A&A, 458, 417
[35] Wallström, S.H.J., Muller, S., Guélin, M. 2016, A&A, 595, A96
[36] Henkel, C., Mühle, S., Bendo, G., Józsa, G.I.G.; Gong, Y.; Viti, S.; Aalto, S.; Combes, F. et al. 2018, A&A, 615, 155
[37] Chin, Y.-N., Henkel, C., Langer, N., Mauersberger, R. 1999, ApJ, 512, L143
[38] Wang J., Zhang Z.-Y., Zhang J., Shi Y., Fang M., 2016, MNRAS, 455, 3986
[39] Wang J., Zhang Z.-Y., Qiu J., Shi Y., Zhang J., Fang M., 2014, ApJ, 796, 57
[40] Adande, G. R., & Ziurys, L. M. 2012, ApJ, 744, 194

# The IRAM-30m EMPIRE Nearby Galaxy Dense Gas Survey

*María Jesús* Jiménez-Donaire[1,2,*] and *Frank* Bigiel[3]

[1]Observatorio Astronómico Nacional (IGN), C/Alfonso XII, 3, 28014 Madrid, Spain
[2]Centro de Desarrollos Tecnológicos, Observatorio de Yebes (IGN), 19141 Yebes, Guadalajara, Spain
[3]Argelander-Institut für Astronomie, Universität Bonn, Auf dem Hügel 71, 53121 Bonn, Germany

**Abstract.** This work presents an overview of the main results from the EMIR Multi-Line Probe of the ISM Regulating Galaxy Evolution (EMPIRE) survey, an IRAM-30m Large Program ($\sim$ 500 h) which observed tracers of high density molecular gas across the disks of nine nearby, star-forming galaxies. EMPIRE is the first comprehensive and systematic study mapping high-density tracers including HCN, $HCO^+$ and HNC, as well as optically thin $J = 1-0$ transitions of $^{13}CO$ and $C^{18}O$. Such a combination of spectroscopic tracers offers the best way to study cold, immediately star-forming gas to address how dense gas fractions and star formation efficiencies vary across and among galaxies. The extensive and sensitive data collected from EMPIRE has allowed us to relate the fraction of star-forming gas and its ability to form stars to local interstellar medium (ISM) and dynamical conditions, such as stellar surface densities, ISM dynamical pressure or molecular gas surface densities. The main results from EMPIRE show that the star formation efficiency in the dense gas varies systematically in all galactic disks. Therefore, this provides support for a context-dependent role of gas density, where dense gas fractions follow interstellar pressure, but star formation only takes place in local over-densities. The EMPIRE survey has successfully turned into a stepping stone for on-going and future projects, aiming to link the large-scale EMPIRE extragalactic results to high-resolution measurements, accessible from our Milky Way.

## 1 Introduction

Only a few decades ago a rigorous study of the processes that lead to star formation could only be performed in detail in our Galaxy. From the extragalactic point of view, it was only possible to study entire star-forming regions or whole, averaged galaxies. At these large scales, most of our knowledge consists of scaling relations between the total gas mass forming stars, and the rate at which this is converted into stars. The pioneer work from [1] and [2, 3] on integrated galaxies found a tight relation between the disk-averaged gas surface density and the averaged star-formation rate in the form of a power-law.

Many observational studies started mapping high-resolution CO as a tracer for the bulk molecular gas, with the goal of exploring variability in this so-called *Kennicutt-Schmidt* relation. One of these pioneer studies was carried by [4], who combined H I maps with new $^{12}CO$ (2-1) from the IRAM-30m telescope in M51 and derived star formation rates (SFR) from radio-continuum at 20cm wavelength.

*e-mail: mdonaire@oan.es

Building up on these efforts, the Very Large Array large program THINGS (The H I Nearby Galaxy Survey, [5]) mapped 34 galaxies in the 21 cm line with high angular resolution and sensitivity. Later on, the IRAM large program HERACLES (Heterodyne Receiver Array CO Line Extragalactic Survey, [6]) mapped 48 galaxies in the CO (2-1) line. This was the first time that a wide, sensitive and high-resolution dataset was available to study the diversity of physical conditions in a large sample of nearby galaxies.

However, tracing easily accessible molecules such as CO emission, only allows us to probe the low-density regime in extragalactic systems. High-density tracers, such as hydrogen cyanide (HCN), have much higher dipole moments and larger critical densities, which allow us to probe the cold, dense sites of star formation. Triggered by the pioneer work by [7] in galaxy integrated measurements using HCN and $HCO^+$, the IRAM-30m Large program EMPIRE (PI Bigiel) was developed with a clear motivation: to map a suite of lines sensitive to different critical densities in diverse galactic environments . This allows to immediately:

- Look at empirical line ratios between different spectroscopic tracers, and analyze them across galaxies.
- Understand how the density distribution responds to galactic environment, as a function of molecular gas fractions, different stellar and SFR surface densities.
- Employ these lines for radiative transfer modeling, to better understand the underlying physical conditions of the gas.
- Link these measurements to high-z results, where the EMPIRE observed lines can also be detected. In addition, it allows us to provide more environment and context to the Galactic work, at lower resolution.

## 2 Survey and observations

The Large Program EMPIRE was designed to use the IRAM-30m telescope to systematically map the star-forming disks of a representative sample of nearby galaxies. With almost 500h of observing time, it provided full maps of a suite of high critical density lines, such as HCN (1-0), $HCO^+$ (1-0) and HNC (1-0), as well as carbon isotopologues such as $^{13}CO$ (1-0) and $C^{18}O$ (1-0) (see Table 1 for a complete overview of emission lines included in the bandpass). Complementary to EMPIRE, we obtained deep and homogeneous CO (1-0) maps for the entire sample (PI Jiménez-Donaire).

The EMPIRE sample consists of 9 well-resolved galaxies at $d \sim 10$ Mpc, and nearly face-on. All the galaxies were chosen to span a broad range of physical properties, morphological and dynamical features, representing strong spiral structures (e.g., NGC 628), heavily barred galaxies (e.g., NGC 3627), flocculent sources (e.g., NGC 5055), strong nuclear bursts (e.g., NGC 6946) and cluster galaxies (e.g., NGC 4321). Therefore this constitutes an excellent sample to study the dense gas emission as a function of environment across entire galaxy disks, with more than ~600 independent lines of sight across the sample, a major improvement over any previous observations.

Figure 1 displays the integrated intensity maps of HCN (1-0), $HCO^+$ (1-0), and HNC (1-0) in the galaxy M51, the pilot target for the Large Program. It is clear that the dense gas emission is detected across the disk of the galaxy, with the brightest emission in its center and along the spiral arms.

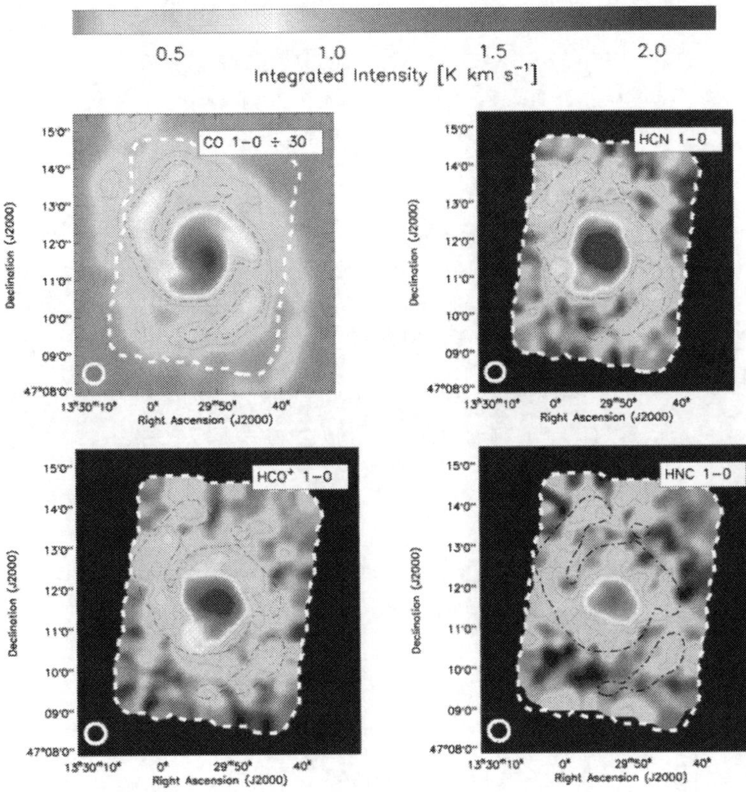


**Figure 1.** Integrated intensity maps from the EMPIRE survey of HCN (1-0) (top right), $HCO^+$ (1-0) (bottom left) and HNC (1-0) (bottom right) in the pilot galaxy M51, adapted from [8]. The $^{12}$CO (1-0) emission map from the PAWS survey [9] is shown for comparison in the upper left panel, convolved to a matching 30″ resolution.

# 3 Main Survey Results

## 3.1 Global correlations

Figure 2 shows the correlation between the IR luminosity, an indicator of the star-formation rate, as a function of the HCN luminosity tracing the dense gas content. The high signal-to-noise ratio measurements of HCN from EMPIRE are shown in comparison to literature measurements probing Galactic-scale cores out to unresolved entire galaxies. At a resolution of ~ 1 kpc, EMPIRE populates the intermediate regime corresponding to large parts of galaxy disks, between Giant Molecular Clouds and observations of entire galaxies.

This compilation shows that EMPIRE clearly follows the same relation found by [7] between, and it extends across almost ten orders of magnitude with a scatter of a factor of two. The large scatter is therefore an indication that the observed IR-to-HCN ratios are quite different from region to region, which suggests an environment-dependent efficiency of star formation in dense gas.

## 3.2 Galaxy-to-galaxy correlations

The left panel of Figure 3 shows the HCN-to-CO ratio for each EMPIRE galaxy, which is a good proxy for dense gas fractions, as a function of stellar surface density, one of several

**Table 1.** Main spectral lines covered by the EMPIRE setup.

| Species | Frequency GHz | $E_{up}$ K | $n_{crit}$ cm$^{-3}$ |
|---|---|---|---|
| SiO 2-1 | 86.45 | 6.25 | $1\times10^5$ |
| $C_2H$ 1-0 | 87.32 | 4.19 | $1\times10^5$ |
| HNCO 4-3 | 87.93 | 10.55 | $1\times10^4$ |
| HCN 1-0 | 88.63 | 4.25 | $2\times10^5$ |
| $HCO^+$ 1-0 | 89.19 | 4.28 | $3\times10^4$ |
| HNC 1-0 | 90.66 | 4.35 | $1\times10^5$ |
| $N_2H^+$ 1-0 | 93.20 | 4.47 | $4\times10^4$ |
| $C^{18}O$ 1-0 | 109.78 | 5.27 | $4\times10^2$ |
| HNCO 5-4 | 109.90 | 15.8 | $1\times10^7$ |
| $^{13}CO$ 1-0 | 110.20 | 5.29 | $4\times10^2$ |
| $^{12}CO$ 1-0 | 115.27 | 5.53 | $4\times10^2$ |

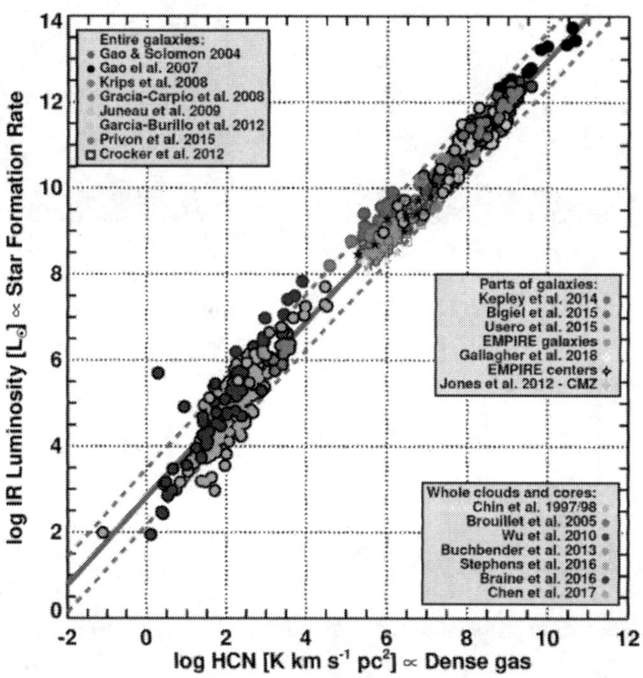


**Figure 2.** Figure adapted from [10]. Correlation between the measured luminosities of our tracers of the recent SFR and dense gas mass in EMPIRE, as well as a major literature compilation probing distinct scales and environments. The mean IR-to-HCN ratio is represented by the solid gray line, while the RMS scatter is indicated by the dashed gray lines.

physical quantities explored in [10]. EMPIRE indicates that there is more dense gas towards the centers of galaxies, an intuitive result since many physical quantities that facilitate the formation of dense gas (e.g., dust content, turbulence, mean gas density) also increase towards inner disk. The right panel of Figure 3 shows the IR-to-HCN ratio as a tracer for star formation efficiency in dense gas. The plot suggests a completely different behavior, where dense gas becomes more inefficient in those regions where there is more dense gas per unit mass. These findings appear to be the same for all galaxies in EMPIRE, leading to the same trends when using $HCO^+$ and HNC as dense gas tracers.

These interesting results directly match previous research in the center of our own Galaxy [11], where observations have shown abundant dense gas but no enhanced star formation efficiency. It is important to note, however, that these observational constraints are sensitive to the conversion factors needed to translate line emission into masses of dense molecular gas. In particular, it is well known that very different factors can be responsible for varying the emissivity of a given dense gas tracer (e.g., UV, X-rays, cosmic rays, and mechanical heating), which are linked to the chemistry of the region where this emission emerges. Taking all these factors into account is, therefore, an extremely complex task.

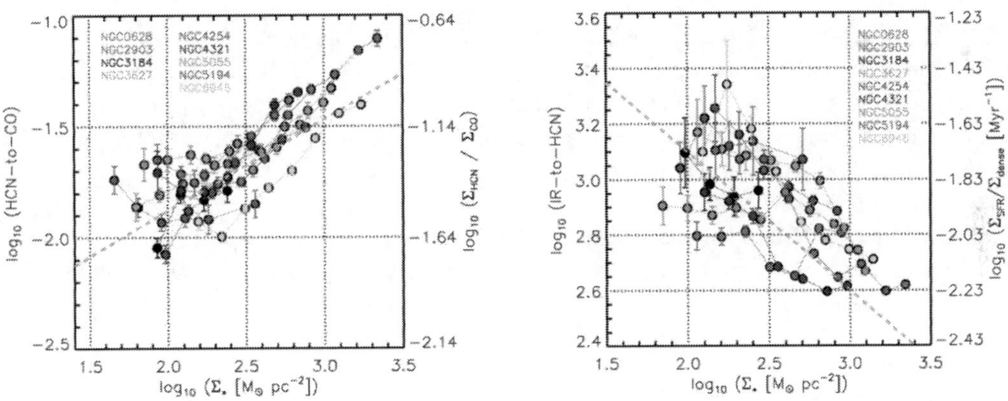


**Figure 3.** Figure adapted from [10]. The left panel shows the HCN-to-CO ratio as a proxy for the dense gas fraction, while the right panel shows the TIR-to-HCN ratio, tracing the efficiency of star formation in dense gas. Both quantities are plotted as a function of the stellar surface densities, for each galaxy. The gray dashed line indicates the fit to all EMPIRE data.

## 4 Conclusions and follow-up work

The EMPIRE survey results have constituted a major step forward into the characterization of the dense ISM in extragalactic sources. Systematic, resolved measurements of a set of dense gas tracers across disks of normal, nearby galaxies was a regime never explored before. The main results from this large program are:

- EMPIRE provided for the first time complete maps of high gas density tracers and CO isotopologues across 9 entire galaxy disks.
- Extends previous dense gas-star formation relations and finds systematic trends in IR-to-HCN (tracing star formation efficiency) and HCN-to-CO (tracing dense gas fraction) as a function of environment.
- The role of any particular gas density in star formation appears to be context dependent. However, environment alone is not sufficient to explain the variations seen in the star formation efficiency of dense gas.

In addition to the results presented in this work, EMPIRE has allowed us to constrain the optical depths of the commonly used dense gas tracers [12] as well as derive relative abundance variations of the main carbon isotopologues [13, 14]. Furthermore, the work led by [15] has allowed us to derive new constraints on the $^{12}$CO (2-1)/(1-0) line ratio across nearby disk galaxies.

Building up on EMPIRE grounds, much progress is currently being made in bridging the local and extragalactic scales, a key next step to build a complete picture of the interplay between ISM and star formation across a variety of environments. Our recent IRAM-NOEMA project is the perfect example to achieve this synergy. The work in [16] presents very high resolution and sensitivity interferometric measurements of dense gas tracers in NGC 3627, pushing the EMPIRE results to almost individual cloud scales. One of the main results of this research is that dense gas line ratios vary systematically with radius, and they show clear differences in distinct dynamic environments of the galaxy (e.g., center, bar, bar ends and spiral arms).

Linking the central galaxy environment to the ISM properties of galaxies is the focus of another EMPIRE follow-up project using the former IRAM-PdBI. As shown in [17], the center of NGC 6946 is extremely bright and rich in complex molecular emission. Several canonical line ratios targeted at high resolution can help us distinguish photon-dominated regions from X-ray dominated regions, and compare these environments to the many centers available from the EMPIRE survey.

This line of research has the advantage that it is also very important for studies of distant galaxies, therefore observation, analysis and theoretical modeling of many molecular lines and their correlations at high resolution are fundamental next steps.

## References

[1] M. Schmidt, The Astrophysical Journal **129**, 243 (1959)
[2] R. C. Kennicutt, The Astrophysical Journal **344**, 685 (1989)
[3] R. C. Kennicutt, The Astrophysical Journal **498**, 541 (1998)
[4] K. F. Schuster et al., Astronomy & Astrophysics **461**, 143 (2007)
[5] F. Walter et al., The Astronomical Journal **136**, 2563 (2008)
[6] A. K. Leroy et al., The Astronomical Journal **136**, 2782 (2009)
[7] Y. Gao & P. M. Solomon, The Astrophysical Journal **606**, 271 (2004)
[8] F. Bigiel et al., The Astrophysical Journal Letters **822**, L26 (2016)
[9] E. Schinnerer et al., The Astrophysical Journal **779**, 42 (2013)
[10] M. J. Jiménez-Donaire et al., The Astrophysical Journal **880**, 127 (2019)
[11] S. N. Longmore et al., Monthly Notices of the Royal Astronomical Society **429**, 987 (2013)
[12] M. J. Jiménez-Donaire et al., Monthly Notices of the Royal Astronomical Society **466**, 49 (2017)
[13] M. J. Jiménez-Donaire et al., The Astrophysical Journal Letters **836**, L29 (2017)
[14] D. Cormier et al., Monthly Notices of the Royal Astronomical Society **475**, 3909 (2018)
[15] J. S. de Brok et al., Monthly Notices of the Royal Astronomical Society **504**, 3221 (2021)
[16] I. Beslic et al., Monthly Notices of the Royal Astronomical Society **506**, 963 (2021)
[17] C. Eibensteiner et al., Astronomy & Astrophysics **659**, 173 (2022)

# Multi-line modelling in nearby galaxies: The link between dense gas and star formation in different environments

*Antonio* Usero[1,*], *Axel* García-Rodríguez[1,**], and *The PHANGS Collaboration*[***]

[1]Observatorio Astronómico Nacional (IGN), C/ Alfonso XII, 3, 28014 Madrid, Spain

**Abstract.**
The $J$=1–0 lines of HCN and $HCO^+$ have become the default tracers of dense molecular gas in external galaxies. To study the relation between HCN and $HCO^+$ emission and density, we have mapped their $J$=3–2 lines with the ALMA Compact Array (ACA) in the nearby star-forming galaxies NGC 3351, NGC 3627, and NGC 4321. Combined with archival line maps of HCN(1–0), $HCO^+$(1–0), CO(1–0), and CO(2–1), we use this data set to: (1) compare the excitation of dense gas tracers in normal and starburst/AGN galaxies; (2) explore how line ratios vary across galaxy disks; (3) infer the density distribution and other cloud properties on a pixel-by-pixel basis through a physically-motivated radiative transfer model.

## 1 Introduction – The HCN and $HCO^+$ $J$=3–2 observations in context

The $J$=1–0 lines of HCN and $HCO^+$ have become the default dense gas tracers in external galaxies [1–3], but in some Galactic clouds their luminosities have a sizeable contribution from gas at low column densities [4, 5]. Thanks to their higher critical densities, the $J$=3–2 lines should be less affected by low density gas. We have mapped with the ACA the HCN(3-2) and $HCO^+$(3-2) emission in the inner ~2-4 kpc disks of the nearby star-forming galaxies NGC 3351, NGC 3627, and NGC 4321. The HCN(3–2) images are shown in Fig. 1-bottom.

We combine these data with archival interferometer images of the HCN(1-0), $HCO^+$(1-0), CO(1-0), and CO(2-1) emission in each galaxy ([6, 7]). Fig. 2 compares our measurements of four HCN and $HCO^+$ line ratios (histograms) with the extragalactic literature (black circles), which for the $J$=3–2 lines is dominated by starburst and AGN samples. In the two leftmost panels, the significant overlap between the two data sets reminds us that HCN-to-$HCO^+$ ratios alone provide ambiguous diagnostics of AGN/starburst activity. In the two rightmost panels, our 3–2/1–0 ratios cluster in the lower range of the literature, as expected if the dense gas tracers were less excited in normal galaxies than in more luminous systems.

## 2 Empirical relations and radiative transfer modelling

In Fig. 3, we study the systematic variations across the targets of our two HCN/CO proxies for the dense gas fraction and the HCN 3–2/1–0 ratio. To first order, all panels suggest

*e-mail: a.usero@oan.es
**e-mail: axel.garcia@oan.es
***http://phangs.org

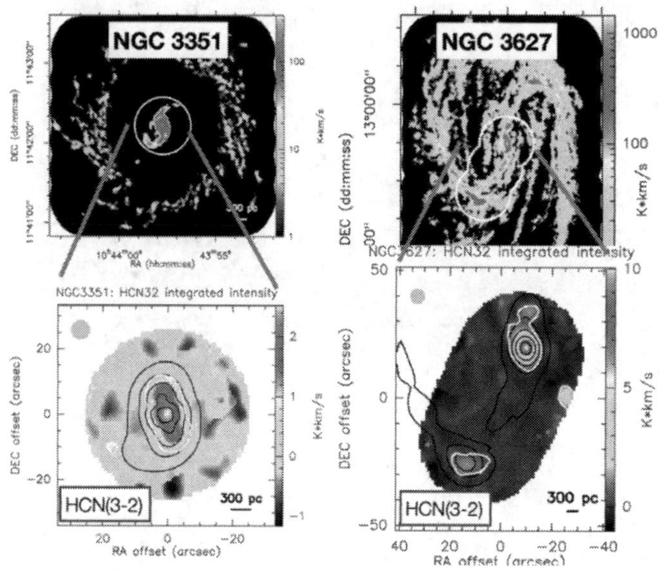


**Figure 1.** Top: CO(2-1) integrated intensity maps of our targets from PHANGS-ALMA [6] at ~1.5″ resolution. The footprint of our ACA observations is shown in white. Bottom: ACA HCN(3-2) integrated intensity maps at 6″ resolution (grey ellipses). The S/N≥5 threshold that we adopt for analysis is shown in white. Black contours represent the CO(2-1) integrated intensity at matching resolution.

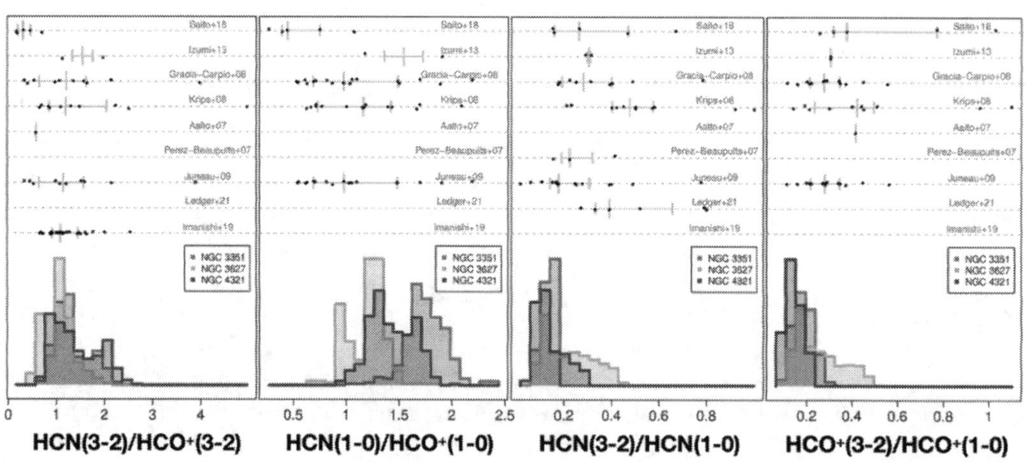


**Figure 2.** Histograms of four line ratios measured in our targets. From top to bottom, black circles correspond to AGN or starburst measurements by [8–16] (spatially resolved in the cases of [8, 9]).

that the dense gas fraction/excitation is enhanced when $\Sigma_*$ (an environment property; left) or $\langle I_{\rm CO}\rangle_{440\rm pc}^{1''}$ (an averaged, cloud-scale property; right) increase. The overall HCN(1-0)/CO(2-1) trends, in particular, are consistent with the fits by [2, 3, 7, 17]. In comparison, the HCN(3-2)/CO(2-1) relations are $\gtrsim 2$ times steeper, which could make them useful to diagnose changes in cloud properties, such as density. Finally, the negligible galaxy-to-galaxy offsets in the $\langle I_{\rm CO}\rangle_{440\rm pc}^{1''}$-vs-HCN(3-2)/HCN(1-0) panel (bottom-right) yield a neat relation between line ratios measured at ~400-pc scales and the sub-beam properties of the unresolved molecular clouds (also [17]).

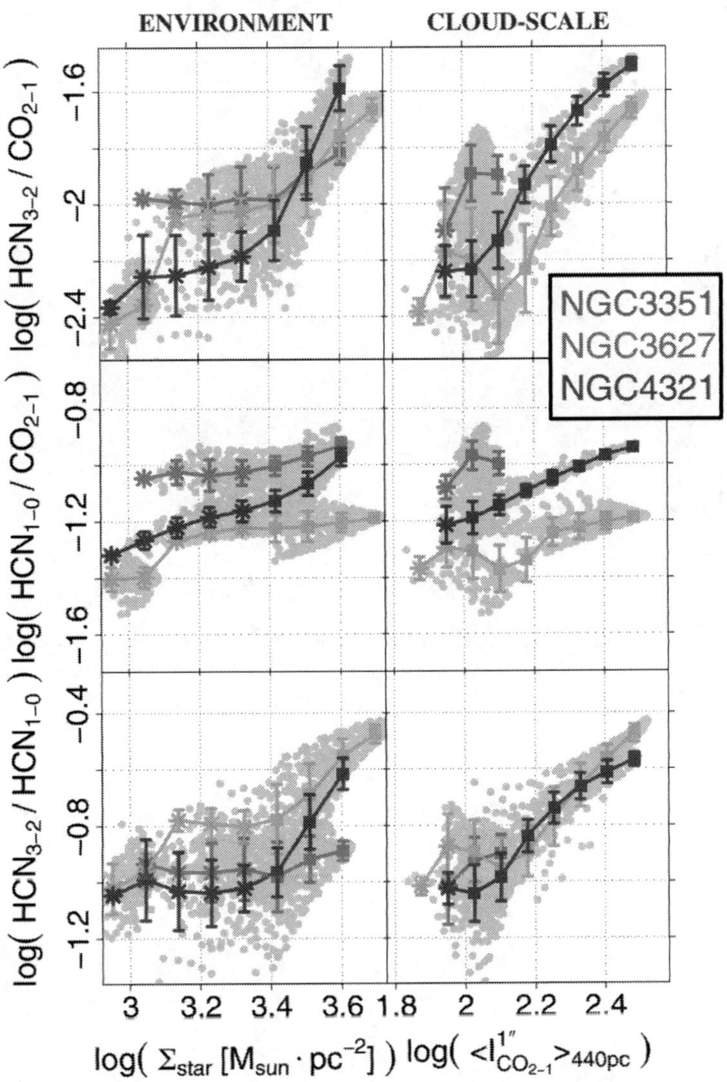


**Figure 3.** Line ratios at a common 440-pc resolution plotted on a pixel-by-pixel basis as a function of $\Sigma_*$ (stellar surface density; left) and $\langle I_{CO} \rangle^{1''}_{440pc}$ (PHANGS CO(2-1) intensity at $\sim 1''$ resolution, weighted by itself and averaged over the 440 pc beam; right). Coloured symbols, bars, and lines show binned average trends per galaxy. Star symbols indicate bins that might be affected by a completeness bias.

To get insights into the trends shown in Fig. 3 (also [2, 3]), we fit our suite of maps with a physically-motivated radiative transfer model based on [18]. We assume that, in each pixel: (1) the observed emission is a sum over a set of density layers that we treat as independent Large Velocity Gradient zones; (2) the layers are weighted by a lognormal+power-law density PDF inspired by ISM models and measurements of Galactic clouds [19]; (3) all layers share the same temperature and abundance per velocity gradient. The model yields maps of the PDF parameters (mean density, Mach number, index of the power-law high density tail), the gas temperature, and the chemical abundances. Fig. 4 shows some examples for NGC 4321.

## 3 Conclusions

Compared with those of their $J$=1–0 counterparts, the extragalactic measurements of HCN(3–2) and $HCO^+$(3–2) are still scarce and dominated by starburst and AGN samples. Our new ACA observations are an example of how current mm-wavelength facilities can map these lines in the fainter but more representative population of main-sequence galaxies. Multiline interferometer observations open a new opportunity to resolve the excitation con-

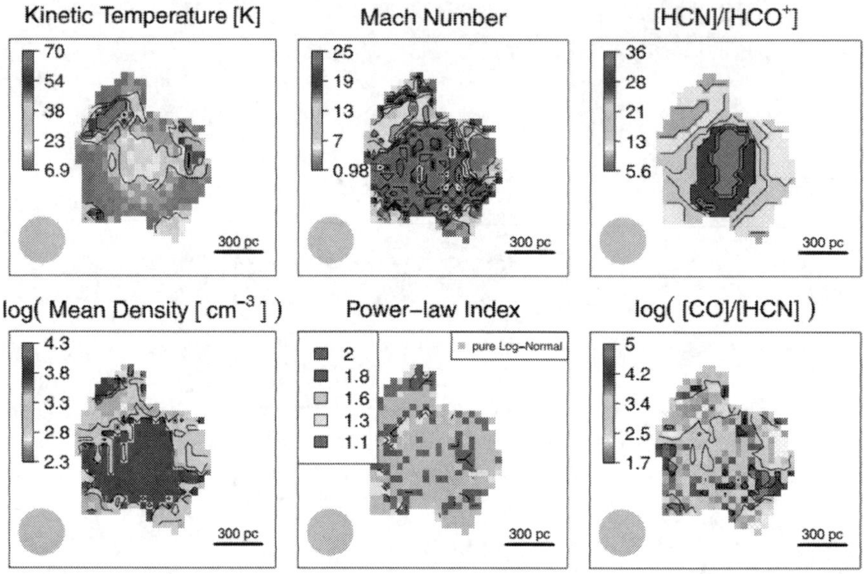


**Figure 4.** Maps of some gas properties inferred from the model in NGC 4321. We sample each beam (grey ellipse) with several pixels to assess model uncertainties (not discussed here). Contours correspond to 1.3 and 1.8 for the power-law index and to 20, 40, 60, and 80% of the maximum elsewhere.

ditions of the (dense) gas across galaxy disks. Once fed into physically-motivated radiative transfer models, these data sets can help us constrain key physical and chemical parameters of the molecular clouds that are relevant to many ISM and star formation theories.

## References

[1] Y. Gao & P.M. Solomon, ApJS **152**, 63 (2004)
[2] A. Usero et al., AJ **150**, 115 (2015)
[3] M.J. Jiménez-Donaire et al., ApJ **880** 127 (2019)
[4] J. Kauffmann et al., A&A **603**, A90 (2017)
[5] J. Pety et al., A&A **599**, A98 (2017)
[6] A. K. Leroy et al., ApJS **257**, 43 (2021)
[7] M. J. Gallagher et al., ApJ **858**, 90 (2018)
[8] T. Saito et al., ApJ **863**, 129 (2018)
[9] T. Izumi et al., PASJ **65**, 100 (2013)
[10] J. Graciá-Carpio et al., A&A **479**, 703 (2008)
[11] M. Krips et al., ApJ **677**, 262 (2008)
[12] S. Aalto et al., A&A **475**, 479 (2007)
[13] J. P. Pérez-Beaupuits et al., A&A **476**, 177 (2007)
[14] S. Juneau et al., ApJ **707**, 1207 (2009)
[15] B. Ledger et al., MNRAS **504**, 5863 (2021)
[16] M. Imanishi et al., ApJS **241** 19 (2019)
[17] M. J. Gallagher et al., ApJ **868**, L38 (2018)
[18] A. K. Leroy et al., ApJ **835**, 217 (2017)
[19] B. Burkhart, ApJ **863**, 118 (2018)

# Invertible Neural Networks in Astrophysics

*Ralf S.* Klessen[1,2,*]

[1]Universität Heidelberg, Zentrum für Astronomie, Institut für Theoretische Astrophysik, Albert-Ueberle-Str. 2, 69120 Heidelberg, Germany
[2]Universität Heidelberg, Interdisziplinäres Zentrum für Wissenschaftliches Rechnen, Im Neuenheimer Feld 205, 69120 Heidelberg, Germany

**Abstract.** Modern machine learning techniques have become indispensable in many fields of astronomy and astrophysics. Here we introduce a specific class of methods, invertible neural networks, and discuss two specific applications, the prediction of stellar parameters from photometric observations and the study of stellar feedback processes from on emission lines.

## 1 Introduction

Astronomy and astrophysics have always been highly data-intensive sciences. Large astronomical survey projects, conducted at modern Earth-bound or spaceborne observatories, or massively parallel astrophysical simulations, run at national and European supercomputing facilities, lie at the very forefront of the current "big data" deluge.

Whereas in the past, researchers were able to keep up with the amount of data pouring in by employing manual control and analysis methods, it became clear in recent years that this is no longer feasible, and that more automated and self-guided approaches are needed. Modern large-scale observational surveys or numerical simulation projects, such as those mentioned above, would not be possible without highly automated data-reduction pipelines that convert the raw data from the telescope or the supercomputer into a dimensionally-reduced and more science-ready form. Only this reduction enables an efficient analysis and astrophysical interpretation of the wealth of information available. Key concepts of artificial intelligence, driven by the ever increasing capabilities of modern machine learning techniques, are currently becoming a focal point of these developments. New neural network designs and supervised or unsupervised learning schemes allow for a comprehensive analysis of complex multi-scale astrophysical data with unprecedented accuracy and speed.

Here we introduce and discuss invertible neural networks (INNs) [1–3, 30], which have been successfully applied in the astronomical and astrophysical context to the analysis of star clusters [31], planetary systems [21], stellar feedback processes [25], or the analysis of galaxy mergers [14] in recent proof-of-concept studies.

## 2 Machine Learning in Astronomy and Astrophysics

Machine learning employs statistical models to predict the characteristics of a dataset using samples of previously collected data without relying on physical models of the system. The

*e-mail: klessen@uni-heidelberg.de

introduction of machine learning for solving regression, classification and clustering problems has revolutionized scientific research, and in particular has provided effective methods for analyzing big astronomical data [17, 24]. In order to construct a model from observed data, many methods rely on human-defined classifiers or 'feature extractors' [22]. However, complex problems require algorithms that automate feature extraction by learning from large amounts of data. Such self-learned feature extraction algorithms are an integral part of the deep learning family, which is based on the construction of artificial neural networks (NNs) [20]. While training NNs requires significant computational power, they achieve far higher levels of accuracy than classic machine learning for many non-linear problems.

There have been several recent studies that employ NN approaches to solve prediction tasks in astronomy and astrophysics. In the context of star cluster research, similar to the focus here, classical convolutional NNs have been employed to study stellar properties either from spectral [15, 41] or photometric [33, 44, 46] data, or they have been trained on data from the European astrometric Gaia satellite to predict properties of stellar clusters in the Milky Way [10, 28]. Classical NN methods have also been used for analyzing and classifying galaxy properties, either based on training data from large observational surveys [45] or from numerical simulations of cosmic structure formation [23, 47]. Other studies in this context have focused on determining the properties of the underlying dark-matter halos [12, 43] and on identifying merging galaxies or merger remnants from images [7, 8, 11, 19].

## 3 Invertible Neural Networks

A specific type of NN architecture based on the concept of normalizing flows [27] are invertible neural networks (INNs). They have been introduced to address complex and highly ambiguous inverse problems [1]. Unlike classical neural networks, which solve the inverse problem directly, INNs learn the forward process by using additional latent output variables to capture the information otherwise lost. Leveraging their invertible architecture, INNs then derive a solution for the inverse process without additional cost. Conditioned on the observations and the latent variable distribution, INNs can predict full posterior distributions, which is highly advantageous when studying multi-modal or degenerate problems, or when investigating complex correlations between parameters.

The advantage of invertible architectures is that the network automatically learns the inverse process when it is trained to approximate a known forward process. When considering degenerate problems or when taking uncertainties into account, an information loss is unavoidable in the forward process, such that different sets of physical parameters $\mathbf{x}$ are mapped onto identical observations $\mathbf{y}$. Consequently, the degenerate $\mathbf{y}$ cannot uniquely explain the corresponding $\mathbf{x}$. By introducing latent variables $\mathbf{z}$ that capture the information loss during the forward process, we can ensure a bijective mapping that could not be achieved with $\mathbf{x}$ and $\mathbf{y}$ alone. The original INN architecture links $\mathbf{x}$ and a unique pair of $[\mathbf{y}, \mathbf{z}]$, making a bijective forward mapping $f(\mathbf{x}) = [\mathbf{y}, \mathbf{z}]$ and a inverse mapping $\mathbf{x} = f^{-1}(\mathbf{y}, \mathbf{z}) = g(\mathbf{y}, \mathbf{z})$. The forward process has to be deterministic and there are certain requirements towards the intrinsic dimensionalities of $\mathbf{x}$ and $\mathbf{y}$. Zero padding is necessary if the dimension of $\mathbf{x}$ is smaller than the dimension of $[\mathbf{y}, \mathbf{z}]$.

The cINN architecture, as illustrated in Figure 1, avoids these problems [2, 3, 29, 30, 37]. It uses a different mapping system by considering the observations $\mathbf{y}$ in both the forward and inverse process as a condition $\mathbf{c}$: $f(\mathbf{x}; \mathbf{c} = \mathbf{y}) = \mathbf{z}$, $\mathbf{x} = g(\mathbf{z}; \mathbf{c} = \mathbf{y})$ [3]. This approach has the advantage that there are no assumptions or restrictions about the intrinsic dimensionalities of $\mathbf{x}$ and $\mathbf{y}$. It has the additional advantage that for very high-dimensional or complex datasets $\mathbf{y}$, we can include a feature extraction network in the conditioning block and fully integrate it in the training process [2, 3]. This allows to employ cINNs in image processing tasks [13, 32].

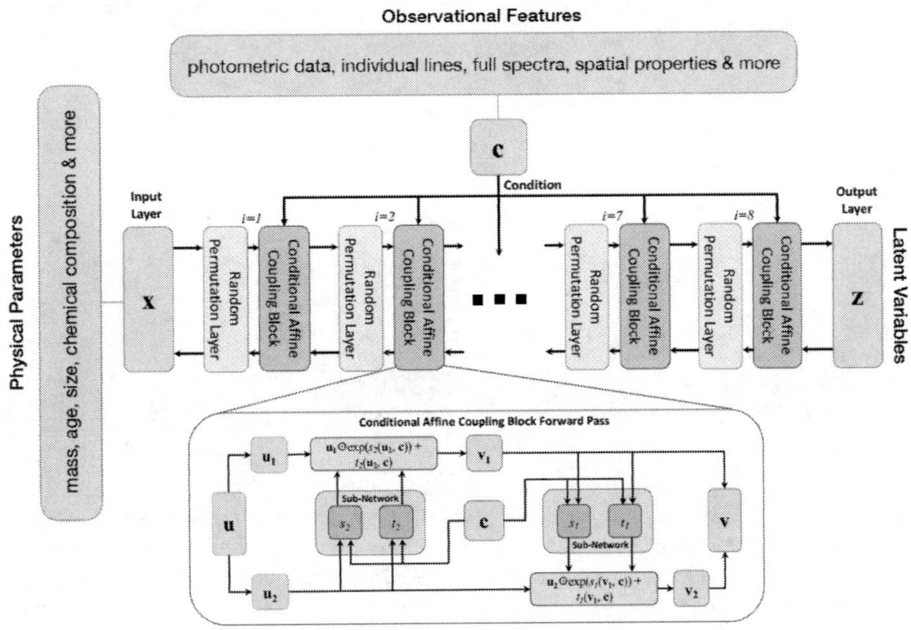


**Figure 1.** Schematic overview of the cINN architecture with physical input parameters **x** and observational features **y** as a condition **c**. The latent variables **z** capture the information loss during the (forward) training phase. The particular example depicted here consists of eight affine coupling blocks interchanged with permutation layers. It was developed as proof-of-concept for the analysis of diagnostic emission lines from young star clusters discussed by Kang and colleagues [25]. The zoom-in panel of the conditional affine coupling block shows how the information is passed through the block in the forward direction. Further details are provided by Ksoll and collaborators [31].

The posterior distribution of physical parameters, $p(\mathbf{x}|\mathbf{y})$, is estimated on the basis of the inverse mapping $f^{-1} = g$. During training, we prescribe the latent variables to have a standard normal probability distribution $p(\mathbf{z}) = N(\mathbf{z}, 0, \mathbf{I})$ with zero mean and unit standard deviation, where $\mathbf{I}$ is the identity matrix with a dimension of $\dim(\mathbf{z}) \times \dim(\mathbf{z})$. Following the inverse process $\mathbf{x} = g(\mathbf{z}; \mathbf{c})$, the posterior distribution is a transformation of the known distribution $p(\mathbf{z})$ to $\mathbf{x}$-space, conditioned on the observation.

## 4 Example 1: INN for Stellar Parameters

In the first example we employ the cINN approach to the task of predicting physical parameters of individual stars based on photometric observations of spatially resolved clusters [31]. In this pilot study we train and test the neural network on synthetic data from the PARSEC stellar evolutionary models [9] and perform a benchmark analysis on real observational data obtained by the Hubble Space Telescope for the young cluster Westerlund 2 [40] and the old globular cluster NGC 6397 [36]. These clusters are chosen to cover the extremes of the cluster range, i.e. very young and very old, in order to gain first insights into the systematics of our approach. We construct the synthetic training sets by adopting isochrone model tables of the correct metallicity for Westerlund 2 and NGC 6397, respectively. The prediction of stellar mass, luminosity, effective temperature and surface gravity works extraordinarily well with posterior distributions that are narrowly constrained around the true values. Determining the

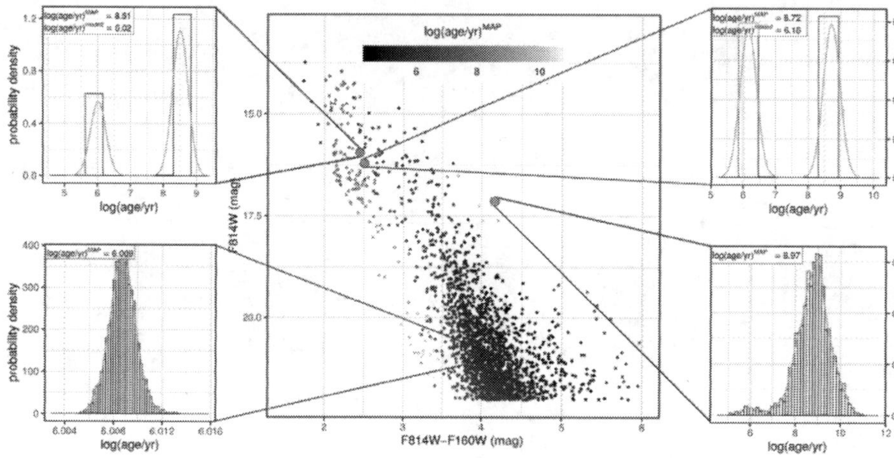


**Figure 2.** Illustration of the ability of the cINN to capture degeneracies in the physical model and properly cope with multi-modal posterior distribution functions, adopted from Ksoll and collaborators [31]. The middle panel shows a zoom-in of the optical color-magnitude diagram (CMD) of the young massive star cluster Westerlund 2 [40], with its stars color-coded according to the maximum a posteriori (MAP) estimates of log(age). The four smaller panels show the predicted age posterior distributions of highlighted stars. Note that stellar age is one of the most difficult stellar parameters to predict from photometric and spectroscopic observations. The bottom left panel is an example pre-main-sequence star for which our approach provides excellent results, returning a very narrow age distribution at the proposed cluster age. The remaining three cases are taken from stars likely on the turn-on of the main-sequence for which the MAP age estimate is significantly above the suggested age of Westerlund 2.

stellar age is a more difficult task, as illustrated in Figure 2. The predicted posteriors tend to be broader and often exhibit multi-modalities, revealing ample degeneracies in the age prediction. While we can confirm that the true value is part of the predicted distribution in more than 99% of the cases, there are several instances where it does not coincide with the most likely outcome of the posterior, falling into a second peak instead.

## 5 Example 2: INN for HII Region Diagnostics

Another application of the INN architecture is the study of physical properties of extragalactic star clusters and star-forming clouds from individual emission lines of HII regions. We present a cINN that predicts the posterior distribution of seven physical parameters (cloud mass, star formation efficiency, cloud density, cloud age as in the age of the first generation stars, age of the youngest cluster, the number of clusters, and the evolutionary phase of the cloud) from the luminosity of 12 optical emission lines, and test our network with synthetic models that are not used during training. The training database is constructed by using the WARPFIELD Emission Predictor [35], which allows us to collect both cloud properties and corresponding observable quantities (i.e. line luminosity). WARPFIELD-EMP describes the evolution of a cluster, expanding bubble, and the surrounding cloud using the 1D stellar feedback code WARPFIELD [38] and calculates detailed emission predictions based on the output from WARPFIELD with the help of CLOUDY [18] and the radiative transfer code POLARIS [39]. WARPFIELD takes into account several feedback mechanisms (i.e., stellar winds, radiation pressure, thermal gas pressure, supernovae, and gravity) self-consistently.

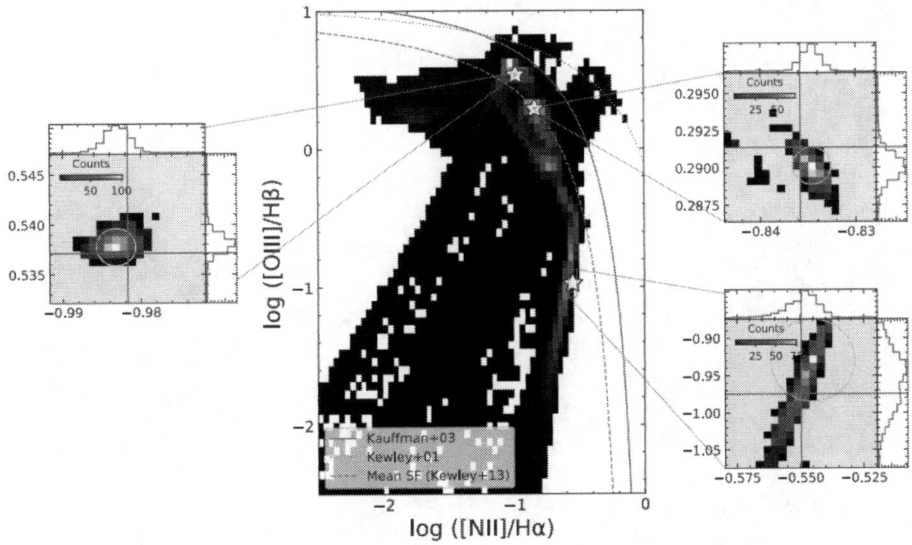


**Figure 3.** Two-dimensional histogram showing the ratios of diagnostic emission lines of singly ionized nitrogen, [NII], doubly ionized oxygen, [OIII], and atomic hydrogen, H$\alpha$ and H$\beta$, for all of the models in the test set, where brighter color indicates a higher number of models. Overlaid as yellow stars are the corresponding values of three representative cases. Zoom-in panels show the distribution of the line ratios that we recover if we sample the posterior distribution for each example model 1024 times and use the resulting values as input for new WARPFIELD-EMP calculations. The true line ratio values for each example model are represented by red lines in these zoom-in panels. Green circles in each zoom-in panel indicate the area in which 68% of the models are included from the centre of the distribution. The image is adopted from Kang et al. [25].

## 6 Summary

The proof-of-concept studies mentioned here have successfully demonstrated the large potential and versatility of invertible neural networks for astronomical and astrophysical applications. Despite addressing a wide range of different scales and physical systems, our approach has in common that the ground-truth sample used to train the neural network is based on synthetic data, generated either from large-scale numerical simulations, Markov-chain Monte Carlo (MCMC) methods, or a database of physical models to generate the physical feature space $\mathbf{x}$, combined with advanced post-processing and radiative transfer methods to produce the corresponding space of synthetic observables $\mathbf{y}$. Training on synthetic data is needed in many astrophysical applications, because it is often not possible to build a ground-truth sample based on observational data only, and even in those cases for which sufficiently well understood observations exist, the numbers are usually too low to adequately train a neural network. Furthermore, training on synthetic data gives us a high degree of control over the problem and allows us to better understand the flow of information through the network, so that we can validate the network performance with high precision and accuracy. We can address the important question of measurement errors and internal degeneracies in the astrophysical system, i.e. in the mapping from $\mathbf{x}$ to $\mathbf{y}$, and we can quantitatively assess how they influence the posterior distribution function. Once the INN is fully tested and characterized, its application to real astronomical observation allows us then also to assess the fidelity and accuracy of the underlying physical model that was used to train the network.

## References


[1] L. Ardizzone, et al., arXiv 1808.04730 (2018)
[2] L. Ardizzone, et al., arXiv 1907.02392 (2019)
[3] L. Ardizzone, et al., arXiv 2105.02104 (2021)
[4] M. Bellagente, et al., SciPost Physics **9**, 074 (2020)
[5] S. Bieringer, et al., SciPost Physics **10**, 126 (2021)
[6] T. Bister, M. Erdmann, U. Köthe, J. Schulte, arXiv 2110.09493 (2021)
[7] C. Bottrell, et al., MNRAS **490**, 5390 (2019)
[8] C. Bottrell, et al., MNRAS **511**, 100 (2022)
[9] A. Bressan, et al., MNRAS **427**, 127 (2012)
[10] T. Cantat-Gaudin, et al., A&A **640**, A1 (2020)
[11] A. Ćiprijanović, et al., MNRAS **506**, 677 (2021)
[12] M.E. de los Rios, et al., arXiv 2111.08725 (2021)
[13] A. Denker, M. Schmidt, J. Leuschner, P. Maass, Journal of Imaging **7** (2021)
[14] L. Eisert, et al., arXiv 2202.06967 (2022)
[15] S. Fabbro, et al., MNRAS **475**, 2978 (2018)
[16] C. Federrath, et al., Nature Astronomy **5**, 365 (2021)
[17] E.D. Feigelson, G.J. Babu, Modern Statistical Methods for Astronomy (2012)
[18] G.J. Ferland, et al., Revista Mexicana de Astronomia y Astrofisica **53**, 385 (2017)
[19] L. Ferreira, et al., ApJ **895**, 115 (2020)
[20] I.J. Goodfellow, Y. Bengio, A. Courville, Deep Learning (2016)
[21] J. Haldemann, et al., arXiv 2202.00027 (2022)
[22] T. Hastie, et al., The elements of statistical learning (2009)
[23] M. Huertas-Company, et al., MNRAS **499**, 814 (2020)
[24] Z. Ivezic, et al., Statistics, Data Mining, and Machine Learning in Astronomy (2014)
[25] D.E. Kang, et al., MNRAS **512**, 617 (2022)
[26] D.P. Kingma, P. Dhariwal, Advances in Neural Information Proc. Sys., 10215 (2018)
[27] I. Kobyzev, et al., IEEE T. on Pattern Analysis and Machine Intelligence **43**, 3964 (2021)
[28] M. Kounkel, K. Covey, K.G. Stassun, AJ **160**, 279 (2020)
[29] J. Kruse, G. Detommaso, U. Köthe, R. Scheichl, arXiv 1905.10687 (2019)
[30] J. Kruse, et al., arXiv 2101.10763 (2021)
[31] V.F. Ksoll, et al., MNRAS **499**, 5447 (2020)
[32] J.H. Nölke, et al. arXiv 2011.05110 (2020)
[33] R. Olney, et al., AJ **159**, 182 (2020)
[34] A. Paszke, et al., Advances in Neural Information Processing Systems **32**, 8024 (2019)
[35] E.W. Pellegrini, et al., MNRAS **496**, 339 (2020)
[36] G. Piotto, et al., AJ **149**, 91 (2015)
[37] S.T. Radev, U.K. Mertens, A. Voss, L. Ardizzone, U. Köthe, arXiv 2003.06281 (2020)
[38] D. Rahner, E.W. Pellegrini, S.C.O. Glover, R.S. Klessen, MNRAS **483**, 2547 (2019)
[39] S. Reissl, R. Brauer, R.S. Klessen, E.W. Pellegrini, ApJ **885**, 15 (2019)
[40] E. Sabbi, et al., ApJ **891**, 182 (2020)
[41] K. Sharma, et al., MNRAS **491**, 2280 (2020)
[42] D. Trofimova, et al., arXiv 2012.08195 (2020)
[43] R. von Marttens, et al., arXiv 2111.01185 (2021)
[44] W. Wei, et al., MNRAS **493**, 3178 (2020)
[45] C. Wu, et al., MNRAS **482**, 1211 (2019)
[46] L. Yang, et al., arXiv 2112.07304 (2021)
[47] L. Zanisi, et al., MNRAS **501**, 4359 (2021)

# A Cloud-Scale View of the Molecular Gas Disk in the Whirlpool Galaxy and Beyond

*Eva* Schinnerer[1,*], *Jérôme* Pety[2], *Ivana* Bešlić[3], *Cosima* Eibensteiner[3], *Sophia* Stuber[1], *Adam* Leroy[4], *Annie* Hughes[5], *Antonio* Usero[6], *Frank* Bigiel[3], and *PAWS & PHANGS* Collaborations

[1]Max Planck Institute for Astronomy, Königstuhl 17, 69117 Heidelberg, Germany
[2]IRAM, 300 rue de la Piscine, 38406 Saint-Martin d'Hères, France
[3]Argelander-Institut für Astronomie, Universität Bonn, Auf dem Hügel 71, 53121 Bonn, Germany
[4]Ohio State University, 140 West 18th Ave, Columbus, OH 43210, USA
[5]CNRS, IRAP, 9 Av. du Colonel Roche, BP 44346, 31028 Toulouse cedex 4, France
[6]Observatorio Astronómico Nacional (IGN), C/ Alfonso XII, 3, 28014 Madrid, Spain

**Abstract.** The nearby galaxy M51 (also known as the Whirlpool galaxy) hosts an iconic grand-design spiral pattern and both IRAM facilities conducted the first cloud-scale (∼50 pc resolution) survey of the molecular gas reservoir across a the disk of a massive star-forming galaxy (PAWS, PdBI+30m Arcsecond Whirlpool galaxy Survey) using the CO(1-0) line emission. PAWS showed that the various properties of the giant molecular cloud (GMC) population vary with galactic environment (center/bar, spiral arms, inter-arm). Recent observations of a ∼1000 pointing mosaic of the nearby late-type spiral galaxy IC342 using NOEMA resolved its GMC population at ∼70 pc resolution and find consistent trends. Investigation of the dense molecular gas phase at cloud-scales using tracers such as HCN(1-0) confirms the trends seen in kpc-scale surveys, namely that the dense gas star formation efficiency inn general apparently anti-correlates with the inferred dense gas fraction. Multi-line studies of the molecular gas in the galactic disks of nearby galaxies such as the ongoing large NOEMA+30m program to map the dense molecular gas phase in GMCs in the central part of M51 will allow for gaining new insights of the properties of this important molecular gas phase.

## 1 Introduction

Observations of molecular gas disks of nearby galaxies can provide key insights for our understanding of the physics driving the star formation process. In particular reaching resolutions of 100 pc or better allows to resolve the gas disks into giant molecular clouds (GMCs) and thus the study of their properties as a function of environment across the disks. While the low transitions of CO are excellent tracers of the bulk molecular, molecular emission lines of, e.g., HCN or HCO+, give access to the more dense molecular phase (e.g. [12]). IRAM facilities have been fundamental in progress in this area by enabling surveys such as HERACLES [11], PAWS [22], and EMPIRE [3] which have substantially increased our understanding of the molecular gas, its intrinsic properties and how these relate to the star formation process.

*e-mail: schinner@mpia.de

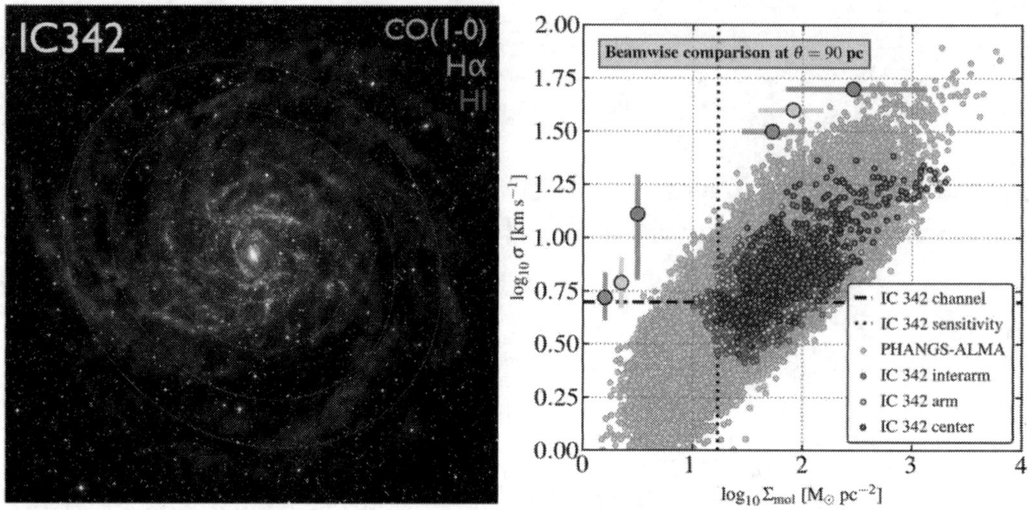


**Figure 1.** *Left*: Distribution of the atomic (*blue*), molecular (*green*), and ionized gas (*red*). Ellipses show galactocentric radii at 5, 10, 15, and 20 kpc. *Right*: Measurements of line width versus gas surface density at a fixed 90 pc resolution for each independent sight-line. IC 342 data are similar to the large sample of PHANGS–ALMA galaxies [14] with the center environment (*blue points*) scattering towards larger line widths and higher surface densities. Adopted from [19].

## 2 Resolving the GMC population in nearby galaxies

### 2.1 PAWS – Molecular clouds in the disk of M 51

The PdBI Arcsecond Whirlpool Survey (PAWS) by [22] provided the first complete ~40pc resolution imaging of the molecular gas disk in the nearby spiral galaxy M51 and allowed for the detection of about 1,500 individual GMCs [4]. PAWS has shown that (i) a significant fraction (~50%) of the molecular material is in a thick, diffuse disk component [18], (ii) the 3D distribution of the atomic and molecular gas phase is not identical [5], (iii) galactic structures (spiral arms, bar) strongly impact the organization of the molecular gas phase [4, 8], (iv) a GMC's ability to form stars depends on its exact location in the galactic disk [15], and (v) the typical lifetime of a GMC is about 20-30 Myr [16]. Taken together these results have challenged previous assumptions of molecular gas properties and the subsequent star formation, and imply that local conditions can have a significant impact on the star process (including the central AGN, [20]). Moreover, a low overall efficiency of star formation relative to the gravitational collapse time (~0.7%; [13]) has been inferred for M 51 – similar to theoretical expectations [10] where stellar feedback into the ISM plays a central role in regulating star formation in galaxies.

### 2.2 The molecular cloud population in the late-type galaxy IC 342

Using 941 pointings with NOEMA, we resolved the CO(1-0) emission from the very nearby ($d$ = 3.45 Mpc) spiral galaxy IC 342 (Fig. 1; [19]) into >600 individual giant molecular clouds and associations over $10.75 \times 10.75\,\mathrm{arcmin}^2$ ($\approx 11 \times 11\,\mathrm{kpc}^2$) with high resolution ($\theta = 4'' \approx 70\,\mathrm{pc}$) covering more than 1.5 effective radii. Overall the clouds show approximate virial balance, with typical virial parameters of $\alpha_{\mathrm{vir}} = 1{-}2$. The typical surface density and line width of molecular gas increase from the inter-arm region to the arm region and reach

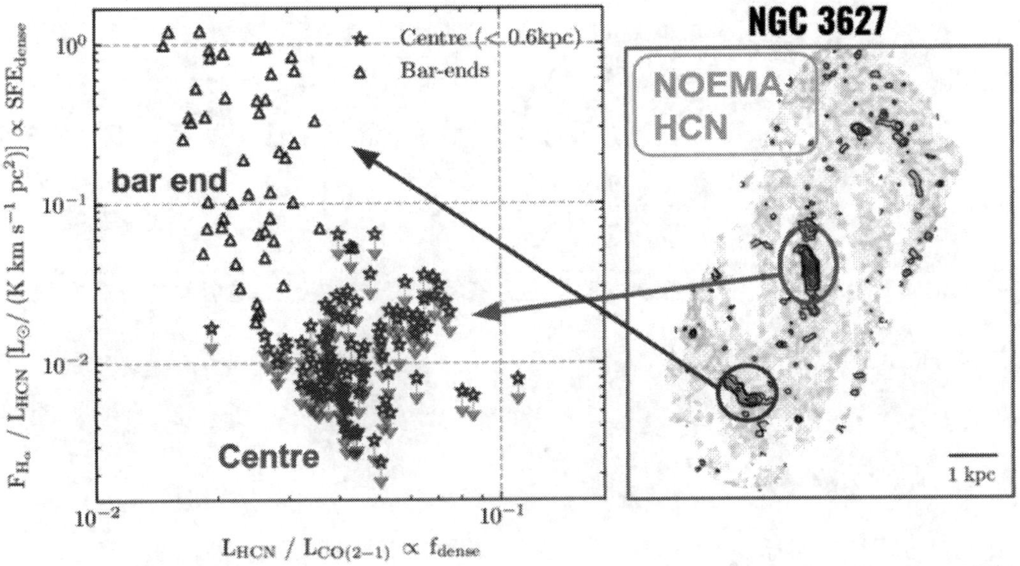


**Figure 2.** *R*ight: Distribution of the dense molecular gas as traced by HCN(1-0) across the inner disk of NGC 3627. *L*eft: Dense gas star formation efficiency ($SFE_{dense}$) versus dense molecular gas fraction ($f_{dense}$). Adopted from [1].

their highest values in the inner 3 kpc disk. Clouds in the central part of the galaxy show enhanced line width relative to their surface densities and evidence of additional sources of dynamical broadening (Fig. 1). This is in agreement with studies of clouds in more distant galaxies at similar physical resolution. This wide-area CO map of the closest face-on massive spiral galaxy demonstrates the current mapping power of NOEMA, the intriguing similarity between atomic, molecular, and ionized gas' large-scale distributions and the scientific potential of such large-scale maps of the molecular gas reservoir.

## 3 Toward the dense gas phase in GMC populations

### 3.1 Dense molecular gas at 100 pc across NGC 3627

It is still poorly constrained how the densest phase of the interstellar medium varies across galactic environments. This is partly because probing high (critical) density molecular tracers within extragalactic sources at high spatial resolution and covering a wide range of scales is challenging due to the extensive observing time required to achieve significant detections. In [1], we presented new IRAM NOrthern Extended Millimeter Array (NOEMA) observations of a range of dense molecular gas tracers (HCN, HNC, HCO+) and CO isotopologues ($^{13}CO$, $C^{18}O$) towards the nearby (11.3 Mpc), strongly barred galaxy NGC 3627. These data represent some of the highest resolution ($1.85'' \approx 100$ pc), large-scale mapping of the dense molecular phase across a nearby spiral galaxy. We find that the HCN/CO integrated intensity ratio, at face value tracing the dense gas fraction, does not correlate with the amount of recent star formation, yet it depends on the galactic environment (Fig. 2). In addition, stars appear to be forming less efficiently in the inner 600 pc of the galaxy than in the bar ends, despite the latter containing a lower fraction of denser molecular gas. The bar end regions show complex gas dynamics, seen as multiple velocity components of HCN emission, which were previously observed in CO(2-1) by [2]. This detection indicates that these regions are places

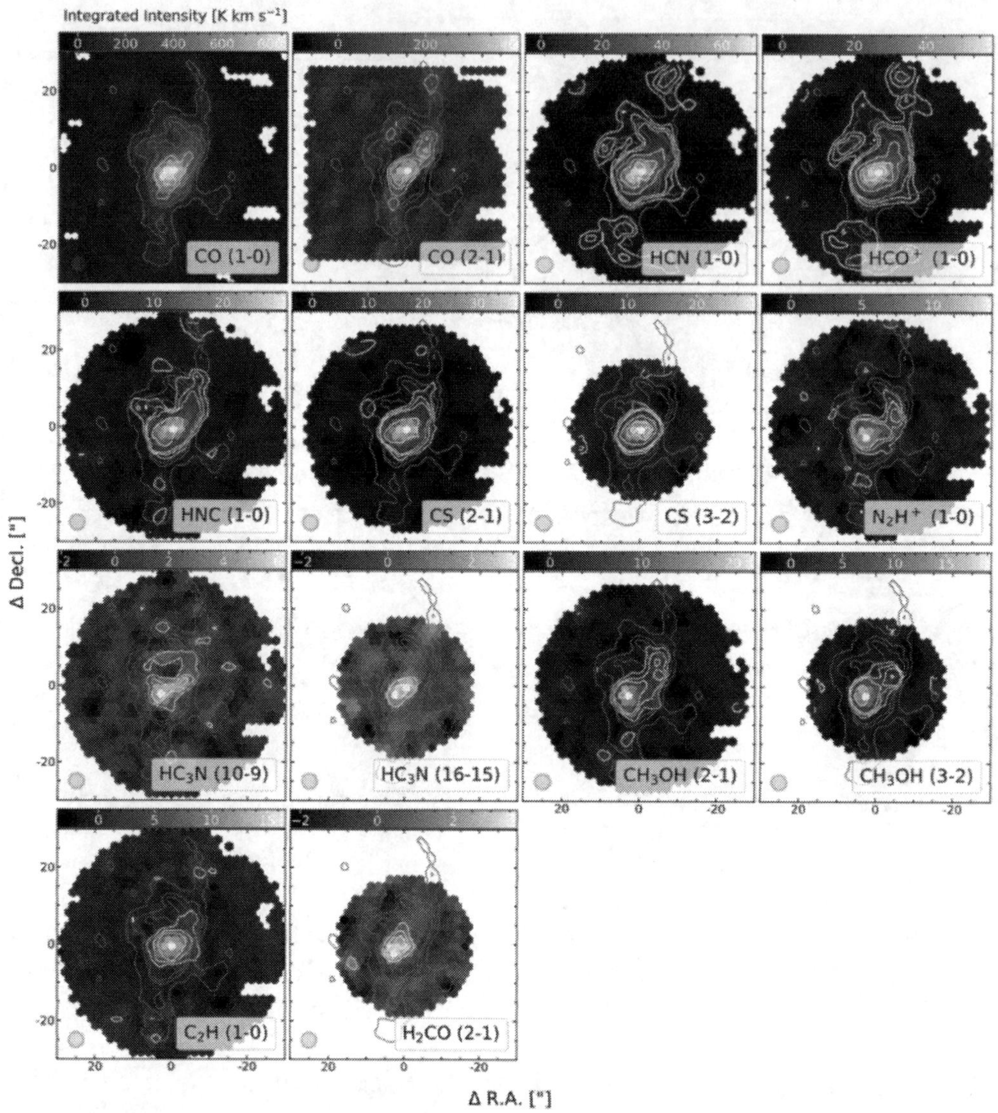


**Figure 3.** The central region of the "Fireworks" galaxy, NGC 6946, observed with the IRAM Plateau de Bure Interferometer (see [7]). Integrated intensity maps for 14 molecular emission lines at $4'' \approx 150$ pc resolution.

with colliding gas flows and possible cloud-cloud collisions, which can further enhance local star formation.

### 3.2 Molecular Fireworks: The high angular resolution 2-3 mm molecular line survey covering the central arc minute of the Fireworks Galaxy NGC 6946

NGC 6946 is a nearby, gas-rich, very actively star-forming double-barred spiral galaxy. We have observed a suite of 1–3 mm lines with the IRAM Plateau de Bure Interferometer (PdBI) [PI: Schinnerer] to assess the physical and chemical structure of the inner ~50″ of this galaxy. This gives us one of the most comprehensive, high resolution ($2$–$4'' \approx 75$–$150$ pc) molecular

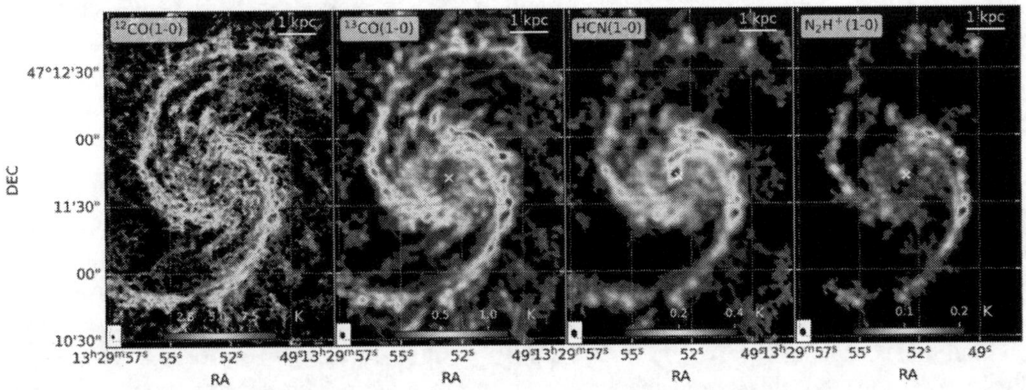


**Figure 4.** Peak temperature maps of dense gas tracers in the central ~5kpc of M51 at ~3″ resolution. From left to right: $^{12}$CO(1-0) high-resolution data from PAWS and preliminary (NOEMA-only) data from selected dense gas tracers ($^{13}$CO(1-0), HCN(1-0), and $N_2H^+$(1-0)).

line dataset for a nearby galaxy center in the northern hemisphere ([7]; see Fig. 3). In this work, we identify three distinct regions of prominent line emission in the inner small scale bar (based on the work of [21]): the nuclear region and the northern/southern inner bar ends. Our analysis suggests that the presence of this inner bar has a strong effect on the observed line emission. For example, higher molecular mean density and higher star formation rates correlate with the shock tracers in the southern bar end. The star formation efficiency of the dense gas ($SFE_{dense}$) shows a different behavior than expected from large-scale disk observations; $SFE_{dense}$ increases with increasing integrated intensity of CO(2-1). Adding 8 nearby galaxies (EMPIRE dense gas observations [9]; and high-resolution observations of M51 [20] and NGC3627 [1]) to our analysis, we speculate that (i) HCN/HNC is not an accurate probe for kinetic temperatures for sub-kpc and kpc sized extragalactic regions, and (ii) HCO+/HCN ratio might not be a unique indicator for AGN activity in galaxies at these scales.

### 3.3 A first cloud-by-cloud multi-wavelength view of M 51

IRAM's large program PAWS was the first to resolve the GMC population in a massive star-forming external spiral galaxy. With ~50pc resolution observations of the molecular gas reservoir in the inner ~8kpc in its CO(1-0) transition in the nearby galaxy M51, this allowed for a better understanding of the ongoing star formation processes (see 2.1). While $^{12}$CO is one of the most commonly used tracers of the dense molecular medium in external galaxies, other studies have shown how CO isotopologues (CLAWS, [6]) and additional tracers of the molecular medium such as HCN and $N_2H+$ (EMPIRE, [3]) are a key ingredient for understanding SF regulation. These studies, however, lack the spatial resolution and sensitivity for cloud-scale conclusions. The necessity for a high-resolution, high-sensitivity multi-wavelength view of M51 has led to a NOEMA+30m large program mapping emission from dense gas tracers in the 3mm window (e.g. $^{13}$CO, $C^{18}O$, HCN, HCO+, HNC, $N_2H^+$). We present the first results from the ongoing observations in comparison to the PAWS CO(1-0) map in Fig. 4. Using the resulting ~2" resolution dense-gas maps, we will expand the cloud-by-cloud analysis from PAWS to a set of important dense gas tracers which will bring critical insights into current models of star-formation theory by studying the dense gas properties of GMCs, their relation to the local star formation activity, and galactic environment of the morphologically rich central 5 kpc of M 51.

## 4 Conclusion

The IRAM facilities have been pioneering the study of the molecular gas reservoir of nearby galaxies over the past decade by providing information on molecular emission lines at the scales of GMCs and thus the properties of resolved cloud populations. Variations in GMC populations with galactic environment seen in M 51 by PAWS has resulted in a major ALMA effort [14] and motivated studies to measure the properties of the dense molecular gas phase at cloud scales. Recent observations are pointing to a more complex behavior of key emission lines – either due to fundamental changes in the star formation process or the behavior of the tracer molecules, i.e. HCN, HCO+. Combining NOEMA's large multiplexing capability with an efficient mapping mode (i.e. interferometric on-the-fly mapping) is a very promising way to better understand the conditions and excitation mechanisms in nearby galaxy disks.

## References

[1] Bešlic, I.,et al., MNRAS, 506, 963 (2021)
[2] Beuther H., et al., A&A, 597, A85
[3] Bigiel , F., et al., Ap.J., 822, L26 (2016)
[4] Colombo, D., et al., ApJ, 784, 3 (2014)
[5] Colombo, D., et al., ApJ, 784, 4 (2014)
[6] den Brok, J., et al., A&A, (2022)
[7] Eibensteiner, C. et al., A&A, Volume 659, id.A173, 37 pp. (2022)
[8] Hughes, A., et al., Ap.J., 779, 46 (2013)
[9] Jiménez-Donaire, M. J. et al., ApJ, 880, 127 (2019)
[10] Krumholz, M.R., Tan, J.C., Ap.J., 654, 304 (2007)
[11] Leroy, A., et al., AJ, 137, 4670 (2009)
[12] Leroy, A., et al., Ap.J., 835, 217 (2017)
[13] Leroy, A., et al., Ap.J., 849, 71 (2017)
[14] Leroy, A., et al., Ap.J.S., 257, 43 (2021)
[15] Meidt, S.E., et al., Ap.J., 779, 45 (2013)
[16] Meidt, S.E., et al., Ap.J., 788, 144 (2014)
[17] Meidt, S.E., et al., Ap.J., 806, 72 (2015)
[18] Pety, J., et al., Ap.J., 779, 43 (2013)
[19] Pety et al., A&A, subm. (2022)
[20] Querejeta, M., et al., A&A, 593, A118 (2016)
[21] Schinnerer, E. et al., Ap.J., 649, 181 (2006)
[22] Schinnerer, E., et al., Ap.J., 779, 42 (2013)

# CO-dark gas: What fuels the star formation in low metallicity dwarf galaxies?

*Suzanne C.* Madden[1,*]

[1] AIM, CEA Département d'Astrophysique, CNRS, Université Paris-Saclay, Université Paris Diderot, Sorbonne Paris Cité, 91191 Gif-sur-Yvette, France

**Abstract.** While star-forming dwarf galaxies have little molecular gas traced by CO, their extreme observed [C II]/CO (1-0) and [O III]/[C II] ratios set them apart from metal-rich star-forming galaxies. The decreased dust abundance, along with their star formation activity, encourages the photodissociation of CO and the presence of relatively prominent $C^+$ envelopes which can harbor a significant self-shielded $H_2$ reservoir, with CO being an inaccurate proxy for the total $H_2$ gas mass. Modeling the Dwarf Galaxy Survey allows us to quantify the mass of the CO-dark $H_2$ gas and derive a [C II]-to-$H_2$ gas mass conversion factor as well as a new CO-to-$H_2$ conversion factor as a function of metallicity.

## 1 Introduction

As ALMA and JWST push back the frontier to reach early star-forming galaxies, metal-poor galaxies should become more prevalent. Studying the interstellar medium (ISM) of the vast diversity of low-metallicity dwarf galaxies in the local universe can lend some insight into the process of the conversion of gas into stars in metal-poor environments. Their striking observational signatures set they apart in several ways from their metal-rich counterparts.

While CO has been a long-time favorite proxy to quantify the mass of molecular gas in galaxies, numerous surveys of dwarf galaxies have found that CO is faint or non-detected in low metallicity dwarf galaxies [e.g. 1–3], leaving us with uncertainties in quantifying the total molecular gas reservoir in dwarf galaxies. These effects are dramatically demonstrated [e.g. 4] in the Kennicutt-Schmidt relationship which is often used as a recipe to relate molecular gas to star formation. Directly translating the total molecular gas reservoir via CO may lead to the conclusion that they are super efficient in forming stars. Or, are we accurately measuring the total molecular gas reservoir via CO in dwarf galaxies?

Another distinct observational characteristic in star-forming dwarf galaxies surrounds the [C II]$\lambda 158\mu$m line which is the most luminous emission line in the FIR to mm wavelength range in metal-rich star-forming galaxies on global scales [e.g. 5, 6]. $C^+$ is easily excited by 11.3 eV photons and dominates the cooling of the cool ISM, primarily arising from photodissociation regions (PDRs). In low metallicity star-forming galaxies, however, it is the FIR [O III]$\lambda 88\mu$m line, requiring ionisation energies of 35 eV, which dominates on global scales, not the [C II] line [7, 8], an effect now also observed in some high-z galaxies [e.g. 9, 10]. This reveals the ease at which such hard photons, originating from young O stars, can traverse full galaxy scales, with a very porous ISM structure in low metallicity galaxies. While

*e-mail: suzanne.madden@cea.fr

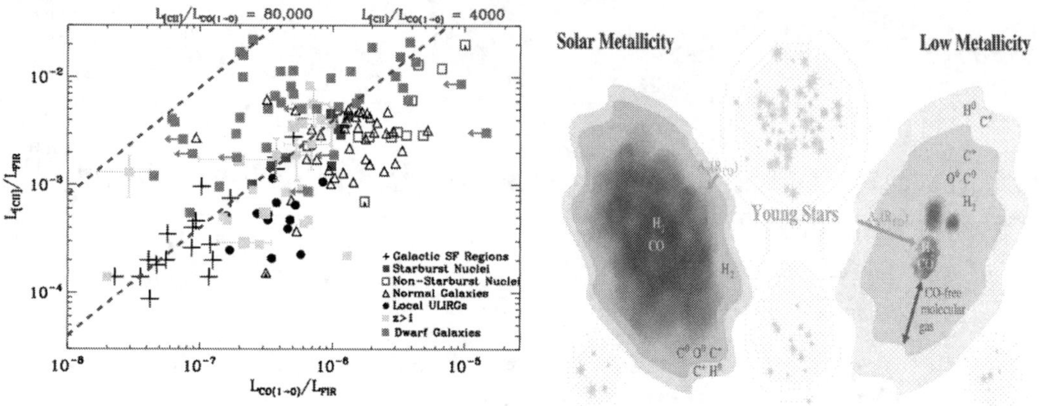


**Figure 1.** *left:* $L_{CO(1-0)}/L_{FIR}$ vs $L_{[C\,II]}/L_{FIR}$ observed in galaxies ranging widely in type, metallicity and in star formation properties. This is updated from [12, 14] to include dwarf galaxies [13]. The dashed lines show constant ratios of $L_{[C\,II]}/L_{CO(1-0)}$. Low metallicity dwarf galaxies (red squares) show extreme $L_{[C\,II]}/L_{CO(1-0)}$ values. *right:* Schematic comparing a metal-rich (solar) molecular cloud and a low-metallicity cloud impacted by UV photons of nearby star clusters. The decrease in dust shielding, in the case of the low metallicity cloud, leads to further photodissociation of molecular gas, leaving a layer of self-shielded $H_2$ outside of the small CO-emitting cores, associated with the $C^+$-emitting region. This CO-dark $H_2$ can be traced by [C II].

the observed $L_{[C\,II]}/L_{CO(1-0)}$ ratios in metal-rich disk and starburst galaxies have been shown from many surveys to range from ~1000 to 4000 [e.g. 11, 12], $L_{[C\,II]}/L_{CO(1-0)}$ is observed on global galaxy scales in dwarf galaxies to stand out at much higher levels [13, and references within] (Fig. 1). These are some observational signatures which are effects of the decreased dust abundance and active star formation in the metal-poor dwarf galaxies which result in low average ambient radiation fields over the large galaxy scales, high ionization parameters and low covering factor of PDR gas. Consequently, the CO clouds suffer photodissociation surrounded by large [C II] envelopes.

## 2 CO-dark gas

Such extreme observational distinctions observed in low metallicity galaxies may be indicative of the presence of a reservoir of molecular gas that is not traced by CO, but residing in the $C^+$-emitting region. As $A_V$ is reduced, photons photodissociate CO, while the $H_2$, which photodissociates via Lyman-Werner band photons, can become optically thick and self-shielded from photodissociation (Fig. 1). This CO-dark molecular gas reservoir [e.g. 15, 16] can be traced by [C II] [13, and references within] under low $A_V$ conditions, leaving CO a poor tracer of the total $H_2$ at low metallicities. We describe our modeling strategy to quantify the total mass of $H_2$ in the low metalliciy galaxies of the Dwarf Galaxy Survey [DGS; 17].

## 3 Modeling Strategy

The DGS observed a wide range of local universe (0.7 to 190 Mpc), low-metallicity star-forming galaxies of the in MIR and FIR fine structure lines with *Herschel* and *Spitzer* providing ideal constraints for multi-phase Cloudy modeling. While many of the compact local dwarf galaxies are unresolved at these wavelengths, we can uncover properties of different galactic phases by exploiting the many fine structure lines of widely varying critical densities

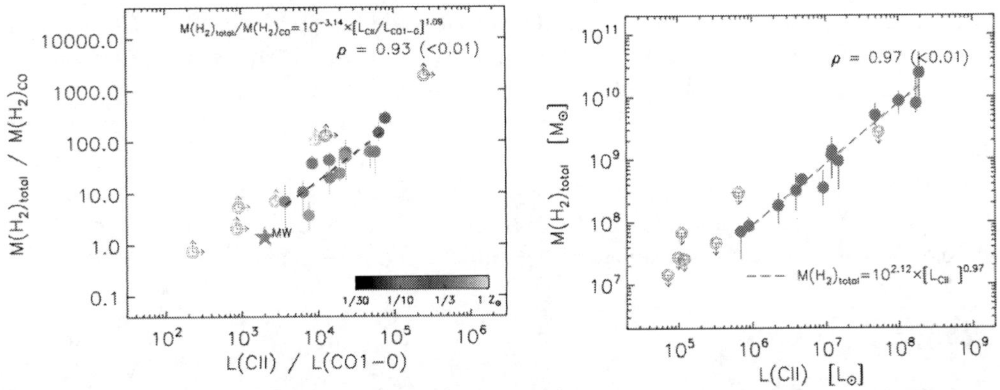


**Figure 2.** *left:* $M(H_2)_{total}$ / $M(H_2)_{CO}$ (total $M_{H_2}$ determined from the model over the $M_{H_2}$ determined from CO observations for the DGS, using the Galactic conversion factor) vs. observed $L_{[C\,II]}/L_{CO(1-0)}$ with color code for metallicity. *right:* $M(H_2)_{total}$ vs. observed [C II]. The open circles indicate limits in y-axis values due to upper limits in CO. The resulting fits for these relationships (dashed lines) are described in the panels with Spearman correlation coefficients ($\rho$) and p-values in parentheses [13].

and ionisation stages. The $L_{[C\,II]}/L_{CO(1-0)}$ relates to the depth of the molecular cloud and the $L_{[C\,II]}/L_{TIR}$ and $L_{[C\,II]}/L_{[O\,I]}$ ratios constrain the radiation field and density at the illuminated face of the PDR. The strategy is that once a well constrained Cloudy solution is reached, the self-consistent total molecular hydrogen mass from the model for each galaxy is extracted. The difference between the model solution for $H_2$ mass and the $H_2$ mass determined directly from the observed CO quantifies the mass of CO-dark gas. Application of the models to the wide range of environments of the DGS, using the observed [C II] and CO (1-0) observations (references in [13]), allows us to extract the total $H_2$ mass and determine the fraction of the total $H_2$ reservoir that is not traced by CO observations, investigating the role of galactic parameters (star formation rate, metallicity, $A_V$, etc) in the mass fraction of CO-dark gas. We find a fraction of CO-dark gas, $M(H_2)_{total}$ / $M(H_2)_{CO}$ based on $L_{[C\,II]}/L_{CO(1-0)}$ (Eq. 1) and a $L_{[C\,II]}$-to-$M(H_2)_{total}$ conversion factor (Eq. 2; Fig. 2). Acconting for the total mass of $H_2$ which includes the CO-dark gas mass, we find a new CO-to-$H_2$ conversion factor as a function of metallicity, which is steeper than previous conversion factors in the literature: $\alpha_{CO} = 10^{0.58} \times [Z/Z_\odot]^{-3.39}$ [13].

$$M(H_2)_{total}/M(H_2)_{CO} = 10^{-3.14} \times [L_{[C\,II]}/L_{CO(1-0)}]^{1.09} \quad (1)$$

$$M(H_2)_{total} = 10^{2.12} \times [L_{[C\,II]}]^{0.97} \quad (2)$$

## 4 Conclusion

We are studying local universe low metallicity environments to be able to relate their observational signatures to physical ISM properties and eventually lend insight into low metallicity early galaxies of the very high-z universe. We are able to carry out the detailed multi-phase modelling for local universe galaxies, where there is access to many observational spectroscopic tracers of different phases, which is not always the case for the high-z galaxies. From the global modeling of the Dwarf Galaxy Survey, we find that the combination of star formation and low dust abundance has a profound effect on the observational tracers, compared to their dustier counterparts. The ISM structure at low metallicity is very porous and clumpy with small filling factors of PDRs and high filling factors of ionized gas. Star forming dwarf galaxies stand out in their faint CO emission and uniquely high [C II]/CO (1-0) and [O III]/[C II] ratios. In some cases, up to ~ 90% of the total $H_2$ mass is missed when accounting for $H_2$ associated with the observed CO only. The bulk of the molecular gas, much of it CO-dark, can be traced by [C II]. Our modeling brings a new [C II]-to-$H_2$ mass conversion factor and a new CO-to-$H_2$ mass conversion factor. Taking into account this significant CO-dark $H_2$ reservoir, the star-forming dwarf galaxies shift to the Kennicutt-Schmidt relation found for dustier, star-forming disk galaxies. While our modeling has been applied to low metallicity galaxies, another step is to see if this can be applied in more general terms to galaxies.

## References


[1] D. Cormier, S.C. Madden, V. Lebouteiller, S. Hony, S. Aalto, F. Costagliola, A. Hughes, A. Rémy-Ruyer, N. Abel et al., Astronomy and Astrophysics **564**, A121 (2014)

[2] L.K. Hunt, S. García-Burillo, V. Casasola, P. Caselli, F. Combes, C. Henkel, A. Lundgren, R. Maiolino, K.M. Menten et al., Astronomy and Astrophysics **583**, A114 (2015)

[3] R. Amorín, C. Muñoz-Tuñón, J.A.L. Aguerri, P. Planesas, Astronomy and Astrophysics **588**, A23 (2016)

[4] D. Cormier, N.P. Abel, S. Hony, V. Lebouteiller, S.C. Madden, F.L. Polles, F. Galliano, I. De Looze, M. Galametz, A. Lambert-Huyghe, Astronomy and Astrophysics **626**, A23 (2019)

[5] J.D.T. Smith, K. Croxall, B. Draine, I. De Looze, K. Sandstrom, L. Armus, P. Beirão, A. Bolatto, M. Boquien, B. Brandl et al., Astrophysical Journal **834**, 5 (2017)

[6] T. Díaz-Santos, L. Armus, V. Charmandaris, N. Lu, S. Stierwalt, G. Stacey, S. Malhotra, P.P. van der Werf, J.H. Howell, G.C. Privon et al., Astrophysical Journal **846**, 32 (2017)

[7] D.A. Hunter, M. Kaufman, D.J. Hollenbach, R.H. Rubin, S. Malhotra, D.A. Dale, J.R. Brauher, N.A. Silbermann, G. Helou et al., Astrophysical Journal **553**, 121 (2001)

[8] D. Cormier, S.C. Madden, V. Lebouteiller, N. Abel, S. Hony, F. Galliano, A. Rémy-Ruyer, F. Bigiel, M. Baes et al., Astronomy and Astrophysics **578**, A53 (2015)

[9] T. Hashimoto, A.K. Inoue, K. Mawatari, Y. Tamura, H. Matsuo, H. Furusawa, Y. Harikane, T. Shibuya, K.K. Knudsen, K. Kohno et al., Publications of the Astronomical Society of Japan **71**, 71 (2019)

[10] Y. Harikane, M. Ouchi, A.K. Inoue, Y. Matsuoka, Y. Tamura, T. Bakx, S. Fujimoto, K. Moriwaki, Y. Ono, T. Nagao et al., Astrophysical Journal **896**, 93 (2020)

[11] G.J. Stacey, N. Geis, R. Genzel, J.B. Lugten, A. Poglitsch, A. Sternberg, C.H. Townes, Astrophysical Journal **373**, 423 (1991)

[12] S. Hailey-Dunsheath, T. Nikola, G.J. Stacey, T.E. Oberst, S.C. Parshley, D.J. Benford, J.G. Staguhn, C.E. Tucker, Astrophysical Journal Letters **714**, L162 (2010)

[13] S.C. Madden, D. Cormier, S. Hony, V. Lebouteiller, N. Abel, M. Galametz, I. De Looze, M. Chevance, F.L. Polles et al., Astronomy and Astrophysics **643**, A141 (2020)

[14] G.J. Stacey, S. Hailey-Dunsheath, C. Ferkinhoff, T. Nikola, S.C. Parshley, D.J. Benford, J.G. Staguhn, N. Fiolet, Astrophysical Journal **724**, 957 (2010)

[15] M.G. Wolfire, D. Hollenbach, C.F. McKee, Astrophysical Journal **716**, 1191 (2010)

[16] S.C.O. Glover, P.C. Clark, Monthly Notices of the Royal Astronomical Society **421**, 9 (2012)

[17] S.C. Madden, A. Rémy-Ruyer, M. Galametz, D. Cormier, V. Lebouteiller, F. Galliano, S. Hony, G.J. Bendo, M.W.L. Smith, M. Pohlen et al., Publications of the Astronomical Society of the Pacific **125**, 600 (2013)

EPJ Web of Conferences **265**, 00012 (2022) https://doi.org/10.1051/epjconf/202226500012
*Multi-line Diagnostics of the Interstellar Medium*

# A 30m Large Program: The CO Line Atlas of the Whirlpool Galaxy Survey (CLAWS)

*Jakob* den Brok[1,*], *Frank* Bigiel[1], and CLAWS collaboration

[1]Argelander-Institut für Astronomie, Universität Bonn, Auf dem Hügel 71, 53121 Bonn, Germany

**Abstract.** Robust knowledge of the distribution, amount, and physical/chemical state of the cold molecular ($H_2$) gas is key to understanding galaxy evolution. With the help of multi-CO line observations, it is possible to study the molecular gas distribution and disentangle numerous physical and chemical processes that shape and govern the molecular interstellar medium (ISM). For the first time, we obtain full-galaxy mapping data of faint CO isotopologues ($^{13}$CO, C$^{18}$O, C$^{17}$O) at 1mm and 3mm wavelengths across the disk of the nearby spiral galaxy M51. With the help of these CO isotopologues, it is possible to constrain the bulk physical and chemical conditions in the molecular gas. We study potential explanations for why CO isotopologue emission varies. Likely drivers include CO abundance variations due to selective nucleosynthesis and changes in the optical depth. Our analysis concludes that a combination of variation in opacity and relative abundances is the dominant driver for the observed CO isotopologue ratio trends on large (kpc) scales. In contrast, abundance variation due to selective photodissociation and chemical fractionation seem to only play a minor or negligible role on galaxy-wide scales.

## 1 Introduction

The *CO Line Atlas of the Whirlpool Galaxy Survey* (CLAWS), an IRAM 30-m large program (#055-17), targets the entire molecular disk of the nearby spiral galaxy M51. With the help of the Eight MIxer Receiver (EMIR; [1]), several molecular emission lines, in particular the CO isotopologues, are observed in the 1 mm (~220–230 GHz) and 3 mm (~90–110 GHz) with a total of 149 h (109.2 h on-source time) between 2017 and 2019. We published the survey paper in early 2022 [2]. The galaxy M51 is an excellent candidate for faint CO isotopologue studies: It is relatively nearby ($D \approx 8.6$ Mpc; [3]), oriented edge-on, and the molecular gas dominates the inner ~5–6 kpc region [4, 5]. Figure 1 shows the galaxy and indicates the observed IRAM 30m field-of-view. A wealth of ancillary data exists for the galaxy across all wavelength regimes. The key science question of the large program using the 30m observations are:

1. Investigate whether the CO isotopologue line ratios show radial and azimuthal variation across the galaxy's disk. Can we relate any changes to other tracers of galaxy conditions, such as the star formation rate or the distribution of stars and dust?
2. Using the wealth of molecular (CO) line emission, can we determine temperature, column and volume densities, and the CO-to-$H_2$ conversion factor in and across M51?

*e-mail: jdenbrok@astro.uni-bonn.de

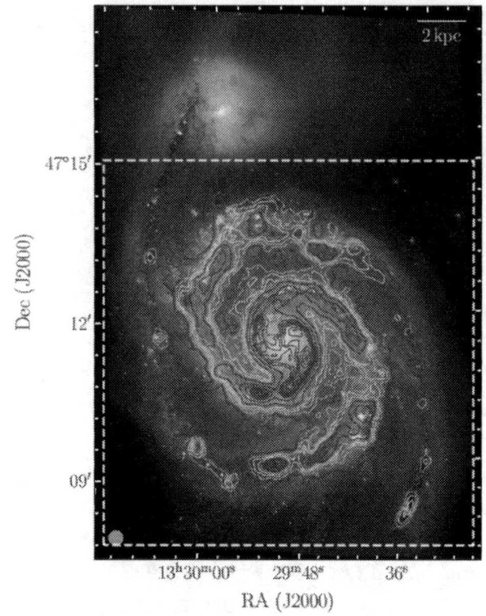


**Figure 1. SDSS Composite Image and $^{12}$CO(2-1) Contours.** The white box outlines the 4.4′ × 4.4′ map observed with the IRAM 30m telescope as part of CLAWS. The contours illustrate the $^{12}$CO(2-1) distribution and highlight various S/N levels from 7 – 70. Figure Credit: taken from [2].

## 2 Science with Multi-CO Line Ratios

While $H_2$ is the most abundant molecule in the interstellar medium (ISM), CO and its isotopologues have become a workhorse tracer of the overall molecular gas amount, distribution, and conditions (such as temperature, density, and opacity) [6]. Line ratios from the emission of different species and transitions capture information regarding the ISM environment and conditions. Ratios of the optical thick $^{12}$CO line of different $J \rightarrow J-1$ transitions are sensitive to changes in temperature and density of the emitting gas [7]. Furthermore, CO isotopologues provide constraints on optical depth and the abundance of the different species, since $^{13}$CO and C$^{18}$O remain optically thin across most of the galaxy's disk [8]. The different CO isotopologue species originate from various physical and chemical processes, such as selective nucleosynthesis, chemical fractionation, or selective photodissociation. Consequently, trends in abundance variations open a window into studying the enrichment history of the ISM [9].

### 2.1 The CO Line Ratio $R_{21}$

The $R_{21} \equiv {}^{12}\mathrm{CO}(2-1)/(1-0)$ line ratio has been studied on kpc-scales across M51 before with CO(1-0) observations from the Nobeyama Radio Observatory (NRO) 45m telescope [10] and CO(2-1) data from the IRAM 30m large program *HERA CO Line Extragalactic Survey* (HERACLES; [11]). Already [12] found a clear variation of $R_{21}$ between the arm and interarm region. Using the CLAWS $^{12}$CO(2-1) observations in combination with $^{12}$CO(1-0) data from PAWS [13], we found larger $R_{21}$ values in the interarm region (the volume weighted average and 16th to 84th percentile range is $\langle R_{21}^{\mathrm{int}} \rangle = 0.95^{+0.21}_{-0.10}$) as opposed to the spiral arm region ($\langle R_{21}^{\mathrm{arm}} \rangle = 0.86^{+0.10}_{-0.07}$). Based on a Kolmogorov-Smirnov test, the $R_{21}$ distributions for arm and interarm are different with a $p$-value of $4 \times 10^{-10}$. Fig. 2 illustrates the result of the binned line ratios by spiral phase. The red and blue bands indicate spiral phases that define the spiral arm. Our finding is in some tension with some of the prior work on this galaxy, e.g. [12], which found exactly the opposite arm-interarm trend. What separates this study is the comprehensive and detailed checking of the contribution of error beams and obtaining even IRAM 30m Director's Discretionary Time (DDT) observations of selected pointings in the arm and interarm regions of M51 [2]. We note that such a finding is actually not unexpected.

Previous studies studying other nearby galaxies also found higher $R_{21}$ values in the interarm region [14–16] and they are not unphysical: Potential explanations could be the presence of diffuse gas at higher excitation temperature and lower optical depths in the interarms which hence leads to brighter CO(2-1).

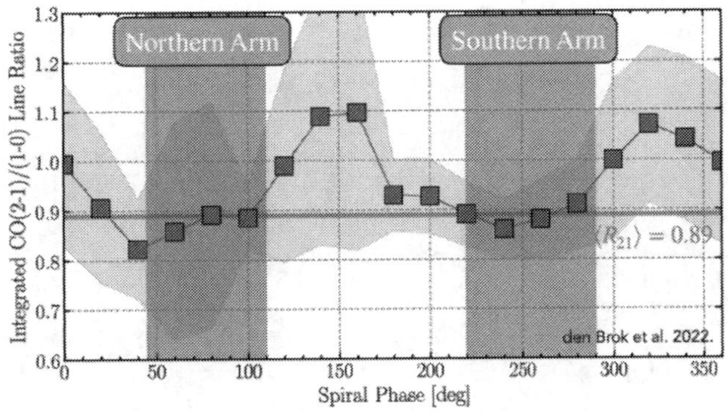


**Figure 2. Arm-Interarm Variation of $R_{21}$ in M51.** The $R_{21}$ values are binned by spiral phase following the prescription of [12]. The spiral phase values indicating the spiral arms are indicated in blue and red. The $1\sigma$ scatter per bin is indicated in light grey. The galaxy-wide CO(1-0) intensity weighted $R_{21}$ value is indicated by the blue horizontal line. Figure Credit: Taken from [2].

## 2.2 Physical and Chemical Drivers of CO Isotopologue Ratio Trends

Figure 3 shows line ratio trends as a function of the star formation rate (SFR) surface density, $\Sigma_{SFR}$, for four selected CO isotopologue ratios. $\Sigma_{SFR}$ scales with the average gas density and temperature in the molecular ISM [17]. Hence, $\Sigma_{SFR}$ is a potential proxy of varying environmental conditions. All selected ratios show a clear trend with SFR. Regarding global, galaxy-wide drivers, the observed trends agree with abundance variations from selective nucleosynthesis and changes in the opacity of the CO emitting gas. Since $^{13}$CO, C$^{18}$O, and C$^{17}$O are optically thin, their variation in the line ratio is mainly driven by changes in their relative abundance. Chemical fractionation which converts $^{12}$CO into $^{13}$CO is likely not a driver. This process increases the relative abundance of $^{13}$CO . Chemical fractionation is more efficient in the colder gas regions, hence at low $\Sigma_{SFR}$. So the trend we see in the $^{13}$CO/C$^{18}$O ratio would actually be in agreement with the process as the main driver. However, we would expect the opposite trend in $^{12}$CO/$^{13}$CO line ratio. In contrast, nucleosynthesis could explain the, relatively speaking, higher C$^{18}$O and lower $^{13}$CO abundances and hence the observed trends. In addition, the opacity could vary. This would affect mainly the optically thick $^{12}$CO emission line. Similarly to the explanation of the increased $R_{21}$ ratio, the fact that the ratios with $^{12}$CO(1-0) in the denominator drop toward higher $\Sigma_{SFR}$ could indicate the presence of diffuse, more optically thin CO gas in regions with lower SFR. A combination of the above mentioned effects likely drives the overall variation in CO isotopologue line ratios.

We also want to highlight the detection of C$^{17}$O(1-0) across a range of SFRs. The C$^{17}$O/C$^{18}$O(1-0) ratio can trac the primary and secondary processing of oxygen due to nucleosynthesis (similar to $^{13}$CO/$^{12}$CO, however, since $^{12}$CO is optically thick, that ratio does not necessarily trace changes in abundance). We find a C$^{17}$O/C$^{18}$O(1-0) ratio similar to the average in the Milky Way solar neighborhood [18]. From nucleosynthesis alone, we would, however, expect an opposite trend. Both isotopologues have very low abundance in general, so selective photodissociation likely plays a more relevant role for this ratio trend. Both species are not well shielded by the other, more abundant isotopologues (due to differences in the wavelengths of their ultraviolet absorption lines). The $C^{18}O$ isotopologue is more abundant than $C^{17}O$. Consequently, there can be regions within molecular clouds where $C^{18}O$ is still self-shielding, while $C^{17}O$ molecules are photodissociating. This could explain why we see an increase of relative $C^{18}O$ abundance where star formation is more active (and hence the photodissociating radiation field stronger).

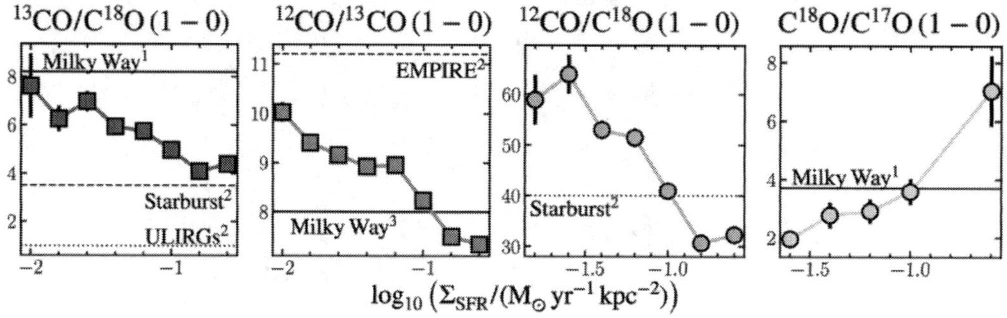


**Figure 3. CO Isotopologue Ratio Trends with SFR.** The trends are based on stacking by SFR surface density. For reference, average line ratio values for other types and samples of galaxies are indicated. The sources for the average values are: (1) from [18], (2) from [19], and (3) from [20].

## 3 Conclusion and Outlook

The rich dataset of multi-CO low-$J$ transitions offers the opportunity to study the chemical and physical conditions of the molecular gas across the spiral galaxy M51. We found apparent environmental-dependent line ratio variation, particularly between the arm and interarm regions. We trace that difference to the presence of more diffuse gas in the interarm region. Furthermore, the sense of the observed CO isotopologue line ratio trends agrees with a combination of mainly selective nucleosynthesis and opacity changes of the gas as main drivers. In an upcoming project, we will perform a non-local thermal equilibrium (non-LTE) modeling analysis that will help us obtain constraints on the CO gas temperatures, densities, and the CO-to-$H_2$ conversion factor.

## References

[1] Carter et al., A&A, 538, A89 (2012)
[2] den Brok et al., A&A, 662, A89 (2022)
[3] McQuinn et al., ApJ, 926, 21 (2016)
[4] Schuster et al., A&A, 461, 143 (2007)
[5] Leroy et al., AJ, 136, 2782 (2008)
[6] Bolatto et al., ARA&A, 51, 207 (2013)
[7] Peñaloza et al., MNRAS, 465, 2 (2017)
[8] Heyer & Dame, ARA&A, 53, 583 (2015)
[9] Davis, MNRAS, 445, 2378 (2014)
[10] Koda, et al., ApJS, 193, 19 (2011)
[11] Leroy et al., AJ, 137, 4670 (2009)
[12] Koda et al., ApJ, 761, 41 (2012)
[13] Pety et al., ApJ, 779, 43 (2013)
[14] Crosthwaite et al., AJ, 123, 1892 (2002)
[15] Lundgren et al., A&A, 413, 505 (2004)
[16] Crosthwaite & Turner, AJ, 134, 1827 (2007)
[17] Narayanan & Krumholz, MNRAS, 442, 1411 (2014)
[18] Wouterloot et al., A&A, 430, 549 (2005)
[19] Jimenez-Donaire et al., ApJ, 880, 127 (2019)
[20] Cormiere et al., MNRAS, 475, 3909 (2018)

# PDFchem: Fast simulations of the chemical ISM using probability distributions

*Thomas G.* Bisbas[1,2,*], *Ewine* van Dishoeck[3,4], *Chia-Yu* Hu[3], and *Andreas* Schruba[3]

[1]I. Physikalisches Institut, Universität zu Köln, Zülpicher Straße 77, 50923 Köln, Germany
[2]Department of Physics, Aristotle University of Thessaloniki, 54124 Thessaloniki, Greece
[3]Max-Planck-Institut für Extraterrestrische Physik, Giessenbachstrasse 1, 85748 Garching, Germany
[4]Leiden Observatory, Leiden University, P.O. Box 9513, 2300 RA Leiden, The Netherlands

**Abstract.** We present PDFchem, a new numerical method able to compute the photodissociation region (PDR) chemistry of large-scale inhomogeneous ISM regions using probability distributions of physical parameters as an input. We distinguish between two visual extinctions, namely the 'effective' ($A_{\rm V,eff}$) referring to the local extinction and the 'observed' ($A_{\rm V,obs}$) referring to the extinction taken from observations. Using 1,200 pre-calculated PDR simulations, we apply PDFchem to two hypothetical $A_{\rm V,obs}$–PDFs, representing a diffuse region and a Giant Molecular Cloud. PDFchem is fast and can replace the computationally expensive hydrodynamical models in understanding the chemistry of the ISM in different environments, including at low metallicities and high cosmic-ray ionization rates.

## 1 Introduction

To understand the evolution of the cosmic star-formation process, we need to know the corresponding molecular mass content of the interstellar medium (ISM) [15]. One of the key issues is to determine the distribution of species of large-scale inhomogeneous ISM objects, including galaxies. Classical studies using photodissociation region chemistry (PDR; [10, 16]), approach these systems with one-dimensional slabs of some uniform density. However, the ISM is very complex with varying densities and temperatures and, thus, far from a simple one-dimensional slab. To include the observed ISM complexity in simulations, the community has developed advanced astrophysical and cosmological codes which couple the magneto-hydrodynamical ISM evolution with PDR chemistry. Although this is a significant advancement, the downside is that it is computationally very expensive and takes a prohibitively long time to use such codes to study the ISM in a variety of environmental combinations (e.g. cosmic-ray ionization rates, FUV intensities, metallicities).

On the other hand, a complex ISM structure can often be represented by a characteristic column-density distribution, the so-called $N$-PDF or $A_{\rm V}$–PDF [1, 12, 14] if we refer to the visual extinction. This provides an opportunity to use these distributions to mimic the ISM complex structure and calculate their corresponding chemistry with much simpler PDR models. Towards this, a new approach has been presented recently by [5]. Here, we extend this approach by introducing two new important features: i) by reducing the grid of

*e-mail: bisbas@ph1.uni-koeln.de | thomasbisbas.com

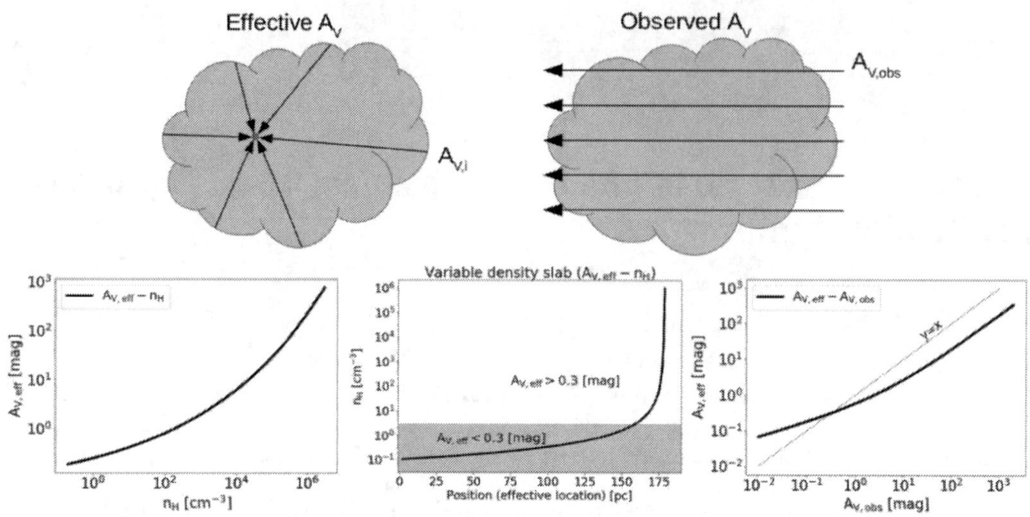


**Figure 1.** Top row: schematic showing the effective ($A_{V,eff}$, left cloud) and the observed ($A_{V,obs}$, right cloud) visual extinctions. Bottom row: the $A_{V,eff} - n_H$ (left panel), the variable density slab (middle panel) and $A_{V,eff} - A_{V,obs}$ relations (right panel). In the latter, the $y = x$ function is additionally plotted to guide the eye. This function is to represent the classical approach of $A_{V,obs} \equiv A_{V,eff}$.

one-dimensional uniform density slabs with *only one* variable density slab that represents a known connection between the local $A_V$ and the local number density, and ii) by including radiative transfer calculations to compute the line emission of the most important coolants for any inputted $A_V$−PDF under different combinations of ISM environmental parameters.

## 2 Numerical method and results

We distinguish between two visual extinctions: the 'effective' $A_V$ (denoted as $A_{V,eff}$), which represents the local visual extinction and it is the average of all $A_V$ emanated from a computational cell of density $n_H$ (see [9, 13]), and the 'observed' $A_V$ (denoted as $A_{V,obs}$), which represents the visual extinction along the line-of-sight of an observer. The top row of Fig. 1 shows a schematic of how $A_{V,eff}$ and $A_{V,obs}$ are defined. The bottom left panel shows the correlation of $A_{V,eff} - n_H$, which connects the local density with a most probable local visual extinction used for solving for the PDR chemistry (see [5, 11] for more details). From this relation, the variable density slab is constructed, which is shown in the bottom middle panel. The bottom right panel shows the $A_{V,eff} - A_{V,obs}$ relation. In the classical PDR approach, there is no distinction between $A_{V,eff}$ and $A_{V,obs}$. Instead they are assumed to be equivalent. This case is shown in dotted lines ($y = x$ function) in the latter panel and it deviates significantly from the presented $A_{V,eff} - A_{V,obs}$ relation [7]. It, therefore, alerts that *more special care must be taken when using observed column densities to perform PDR calculations.*

The publicly available code 3D-PDR[1] [2, 5], which treats the chemistry of PDRs in one- and three-dimensional clouds of arbitrary density distribution, has been used to construct a grid of 1,200 PDR simulations. For this grid, the variable density slab has been embedded in different combinations of ISM environmental parameters. In particular, we considered a cosmic-ray ionization rate in the range of $\zeta_{CR} = 10^{-17} - 10^{-14}\,s^{-1}$, an FUV intensity of $\chi/\chi_0 = 0.1 - 10^3$ normalized to the spectral shape of [8], and a metallicity of $Z = 1\,Z_\odot$.

[1] https://uclchem.github.io/3dpdr/

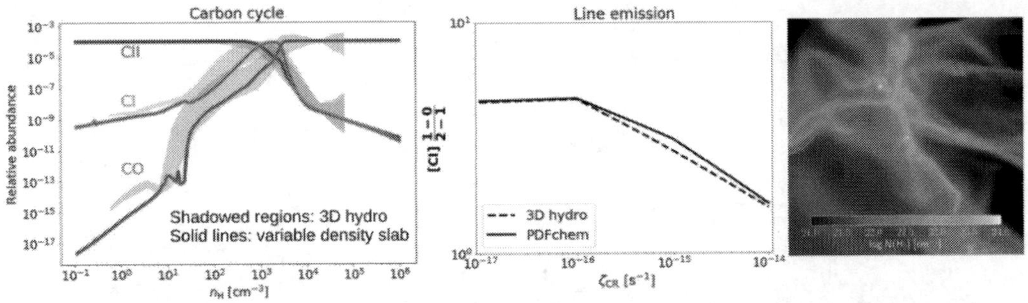


**Figure 2.** Comparison between the PDR calculations from the 3D hydrodynamical post-processed distribution. Left panel shows the carbon cycle (CII, CI and CO). The shadowed regions correspond to the 3D hydro calculations whereas solid lines show the corresponding results using the variable density slab. The middle panel shows the comparison between the 3D hydro (dashed) and PDFCHEM (solid) for the atomic carbon line ratio. The agreement is remarkable in all cases. The right panel shows the cloud used for the 3D hydro models (see [6] for further details on the 3D models).

PDFCHEM performs calculations in the following way: a) the user inputs the $A_{V,obs}$−PDF, b) each of the $A_{V,obs}$ is connected with a most probable $n_H$ following the $A_{V,obs} - A_{V,eff}$ and $A_{V,eff} - n_H$ relations of Fig. 1, c) using the density distribution resulting from the $A_{V,eff} - n_H$ relation, each $n_H$ is in turn connected to an effective position, $r$, up to which the column density of species and the radiative transfer calculations are evaluated for the requested combinations of ISM environmental parameters, d) steps (b)–(c) are repeated for every $A_{V,obs}$ followed by a weighted-average summation expressed in Eqn. (6) of [5].

Figure 2 benchmarks PDFCHEM against a 3D hydrodynamical post-processed simulation (see [6]), the density distribution of which is shown on the right panel. The left panel demonstrates the validity of using the variable density slab to replace the 3D-hydro PDR models. The middle panel shows the resultant ratio of atomic carbon lines for the 3D hydro model (dashed) and PDFCHEM (solid) as a function of $\zeta_{CR}$. For the PDFCHEM, the $A_{V,obs}$−PDF relation of the 3D-hydro cloud was used (not shown). The agreement is remarkable in all cases.

Figure 3 shows preliminary results for two hypothetical distributions. The first is to represent a diffuse ISM region with $\overline{A_{V,obs}} = \overline{1.0}$ mag and a width of $\sigma = 0.5$ (top row). The second is a Giant Molecular Cloud with $\overline{A_{V,obs}} = 9.0$ mag and $\sigma = 0.8$ (bottom row; see Eqns.(2–5) of [5]). First and second columns show how the HI/2H$_2$ average abundance and the CO(1-0)/[CI](1-0) antenna temperature ratios, change for any $\zeta_{CR} - \chi/\chi_0$ combinations at $Z = 1\,Z_\odot$, respectively. The black solid line marks the HI/2H$_2$=1 ratio. The diffuse medium remains molecular only for $\chi/\chi_0 \lesssim 1$ and $\zeta_{CR} \lesssim 10^{-16}\,s^{-1}$. For these conditions, it is in general brighter in CO(1-0) than in [CI](1-0). On the other hand, the GMC remains molecular even for the highest values of $\zeta_{CR}$ and FUV intensities considered. It is brighter in CO(1-0) for $\zeta_{CR} \lesssim 10^{-16}\,s^{-1}$, however it becomes brighter in [CI](1-0) as $\zeta_{CR}$ increases. This effect of cosmic-ray induced destruction of CO has been studied by [3, 4, 6] and it makes the [CI](1-0) line an excellent alternative H$_2$ gas tracer particularly in systems permeated by high $\zeta_{CR}$.

## 3 Conclusions

We present PDFCHEM, a new and fast method for determining the atomic and molecular mass contents of large-scale inhomogeneous clouds characterized by $A_{V,obs}$−PDFs. We distinguish between the effective (local) and the observed $A_V$ and demonstrate that for more accurate PDR comparison with observations, a non-equivalent relationship between these two must be

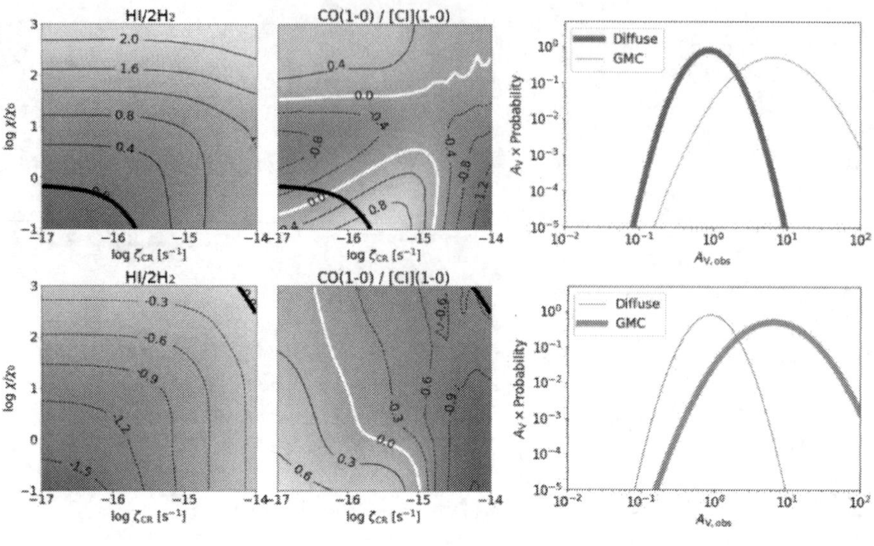


**Figure 3.** Top row corresponds to a hypothetical diffuse region and bottom row to a Giant Molecular Cloud. Their $A_{\rm V,obs}$–PDFs are shown in the right column. The left column shows the PDF-averaged HI/2H$_2$ abundance ratio. The black solid line marks the condition HI/2H$_2$=1. Underneath of it, the distribution remains molecular. The middle column shows the antenna ratio of CO(1-0)/[CI](1-0). The white solid line shows the condition in which this ratio is equal to one. All contours are in log-space.

considered. Our simple 1D PDR models can reproduce the abundances and line emission results of complicated 3D hydro simulations, thus offering a much quicker alternative approach to study PDRs of complex systems. Our statistical method presented in [5] and expanded in this work can be used to estimate the PDR properties of large-scale ISM regions.

TGB acknowledges support from DFG grant No. 424563772. CYH acknowledges support from the DFG via German-Israel Project Cooperation grant STE1869/2-1 GE625/17-1.

## References

[1] Abreu-Vicente J., Kainulainen J., Stutz A., et al. 2015, A&A, 581, A74
[2] Bisbas T. G., Bell T. A., Viti S., Yates J., Barlow M. J., 2012, MNRAS, 427, 2100
[3] Bisbas T. G., Papadopoulos P. P., Viti S., 2015, ApJ, 803, 37
[4] Bisbas T. G., van Dishoeck E. F., Papadopoulos P. P., et al., 2017, ApJ, 839, 90
[5] Bisbas T. G., Schruba A., van Dishoeck E. F., 2019, MNRAS, 485, 3097
[6] Bisbas T. G., Tan J. C., Tanaka K. E. I., 2021, MNRAS, 502, 2701
[7] Clark P. C., Glover S. C. O., 2014, MNRAS, 444, 2396
[8] Draine B. T., 1978, ApJS, 36, 595
[9] Glover S. C. O., Federrath C., Mac Low M.-M., Klessen R. S., 2010, MNRAS, 404, 2
[10] Hollenbach D. J., Tielens A. G. G. M., 1999, RvMP, 71, 173
[11] Hu C.-Y., Sternberg A., van Dishoeck E. F., 2021, ApJ, 920, 44
[12] Kainulainen J., Beuther H., Henning T., Plume R., 2009, A&A, 508, L35
[13] Offner S. S. R., Bisbas T. G., Viti S., Bell T. A., 2013, ApJ, 770, 49
[14] Schneider N., Bontemps S., Motte F., et al., 2016, A&A, 587, A74
[15] Tacconi L. J., Genzel R., Sternberg A., 2020, ARA&A, 58, 157
[16] Wolfire M. G., Vallini L., Chevance M., 2022, arXiv, arXiv:2202.05867

# ALCHEMI: Results from the ALMA Comprehensive High-resolution Extragalactic Molecular Inventory of NGC 253

*Jeffrey G.* Mangum[1,*], *Serena* Viti[2,3,**], and *The ALCHEMI* Collaboration[a]

[1]National Radio Astronomy Observatory, 520 Edgemont Road, Charlottesville, VA 22903-2475, USA
[2]Leiden Observatory, Leiden University, PO Box 9513, 2300 RA Leiden, The Netherlands
[3]Department of Physics and Astronomy, University College London, Gower Street, London WC1E 6BT, UK

**Abstract.** The star formation process in galaxies drives their evolution. The physical conditions which drive the star formation process in galaxies are studied using measurements of atomic and molecular species found in the dense has from which stars form in galaxies. Molecular emission measurements at millimeter/submillimeter wavelengths have revealed the molecular complexity of the star formation regions in galaxies. In a recent study of the nearby starburst galaxy NGC 253 using the Atacama Large Millimeter/submillimeter Array (ALMA), the ALMA Comprehensive High-resolution Extragalactic Molecular Inventory (ALCHEMI) large program imaged the continuum and spectral line emission from the central molecular zone (CMZ) of this starburst galaxy. In this article we summarize the current results derived from the ALCHEMI large program. Many of these studies have focused on clarifying the state of dense gas heating processes in the NGC 253 CMZ.

## 1 Introduction

The study of the star formation process in galaxies requires a wide array of diverse measurements at all wavelengths. Central to the early phases of the star formation process are measurements of the atomic and molecular spectral line and continuum emission toward those regions in galaxies which are, or will in the near future, form stars [10]. Starburst galaxies sit at the extreme end of the spectrum of star formation activities in galaxies due to their extreme star-forming environments as compared to the Milky Way. Observing starburst galaxies allows us to study how stars form in regions with high densities and temperatures. The proper characterization of the physical conditions in extragalactic star-forming regions requires high spatial resolution and sensitivity which will allow for the study of structures on giant molecular cloud (GMC) scales at millimeter/submillimeter wavelengths. Molecular spectral line emission in galaxies is driven by processes associated with star formation (mechanical heating in the form of shocks and turbulence from supernova explosions, radiative heating from massive stars, ionization by cosmic rays from supernova remnants, etc.) that have competing effects on the interstellar medium (ISM). Determining the influence of each of these physical processes on extragalactic star-forming regions is crucial to our understanding of the chemical and physical processes that guide star formation in starburst environments.

*e-mail: jeff.mangum@nrao.edu
**e-mail: viti@strw.leidenuniv.nl

https://doi.org/10.1051/epjconf/202226500014

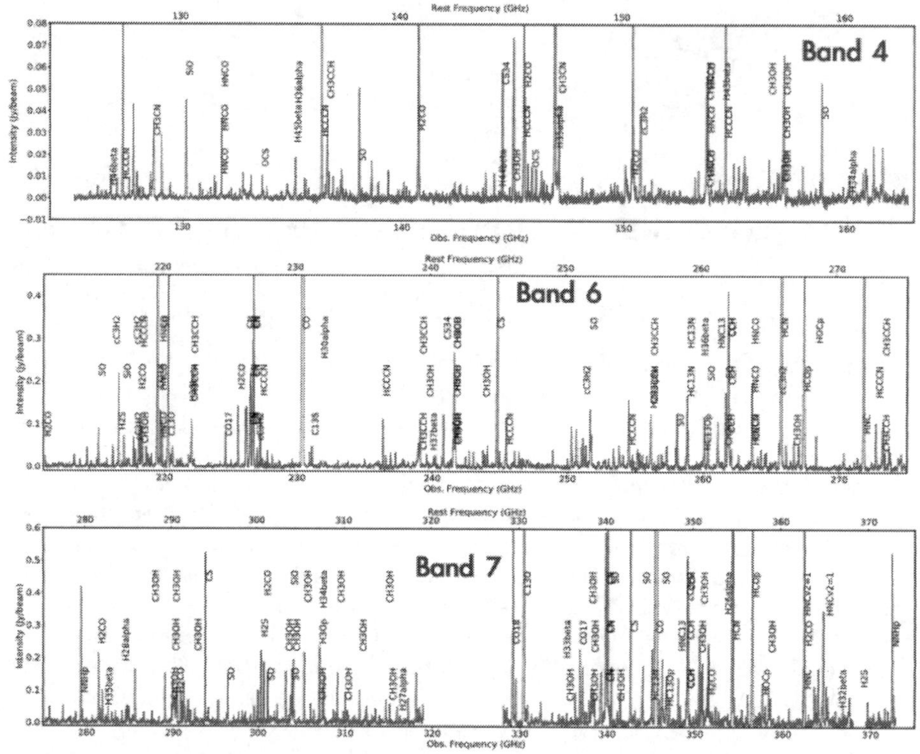


**Figure 1.** Focused ALMA Band 4, 6, and 7 spectra from the ALCHEMI survey. Molecular species associated with each spectral line are indicated.

The ALMA large program ALCHEMI [11] imaged the continuum and spectral line emission from the CMZ ($\sim 50 \times 20$ arcsec $\sim 850 \times 340$ pc) of NGC 253 covering a nearly contiguous 289 GHz frequency range from 84.2 to 373.2 GHz at 1.6 arcsec ($\sim$ 28 pc) spatial resolution. The imaging sensitivity for these measurements is 30 to 50 mK with sensitivity to spatial scales up to 15 arcsec ($\sim$ 250 pc). These images have allowed the ALCHEMI collaboration to characterize the physical and chemical properties of the dense star-forming gas in the NGC 253 CMZ. In this article we summarize a few of the ALCHEMI studies that have been completed to-date. These studies [1–4, 6–8, 11] have begun to clarify the chemical state and dominant heating processes within the heart of the NGC 253 starburst galaxy.

## 2 The Chemical State of the NGC 253 CMZ

As the sample spectra from the ALCHEMI survey (Figure 1) show, the chemical complexity and richness of the NGC 253 CMZ is outstanding. Spectral line LTE modelling identified 78 molecular species, including isotopologues and vibrational states. Newly detected species in the extragalactic ISM include complex organic species and isotopologues, namely $H_2^{13}CO$, ethanol ($C_2H_5OH$), $^{13}CCH$, $C^{13}CH$, HOCN, the three $^{13}C$ isotopologues of $CH_3CCH$, propynal ($HC_3HO$), and tentatively $Si^{17}O$. The ALCHEMI survey can also be used as a template for measurements of high-redshift galaxies that can be used to estimate the number of individual molecular species that are potentially detectable in a starburst environment. Our estimate, based on a stacked spectrum from 22 high-z sources at z = 2.0 – 5.7 by [12] yields 3 – 4 expected species detections, in agreement with their reported identification.

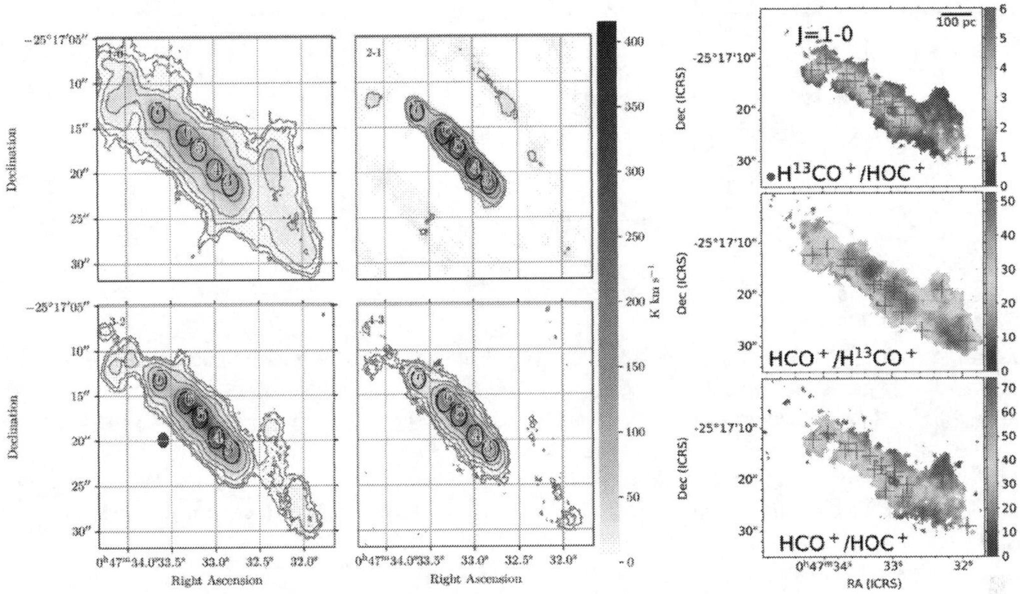


**Figure 2.** Left: Primary beam-correntegrated emission maps of $C_2H$ in NGC 253 (from [7]). The beam size for all panels is 1.6" and is shown in the bottom left panel. Contours at the 3, 5, 10, 20 and 50$\sigma$ level are drawn in black. The noise levels in order of increasing N are 1.4, 8.5, 2.0 and 1.6 K km/s. Each ellipse shows the position and size of the Gaussian fits to the four individual clumps described in [7]. The numbers indicate the position of the GMCs. Right: Ratios of velocity-integrated intensity images of the $HCO^+$, $HOC^+$, and $H^{13}CO^+$ J=1−0 emission toward NGC 253 (from [4]). GMC positions indicated by + symbols.

## 3 Heating Processes Driving Star Formation in NGC 253

Several studies of the physical and chemical conditions in the NGC 253 CMZ have investigated the processes which heat the dense gas in the primary source of the starburst in this galaxy. $C_2H$ [Figure 2; 7], $H_3O^+$/SO [Figure 3; 6], and HCN/HNC [2] emission characteristics have been used to constrain chemical (UCLCHEM; [5]) and physical (SpectralRadex[1]) models of the dense gas. A similar analysis of the $HCO^+$/$HOC^+$ emission [4] provided additional information on the CMZ physical conditions. All of these studies found that $T_K > 50$ K and $n(H_2) > 10^5$ cm$^{-3}$, with the high kinetic temperatures driven by cosmic ray ionization rates (CRIRs) $10^{-15}$ to $10^{-12}$ s$^{-1}$, more than 100 times larger than the average CRIR found in the Milky Way. Furthermore, the low measured HCN/HNC abundance ratios are inconsistent with the existence of substantial mechanical heating in the NGC 253 CMZ.

## References

[1] Barrientos, A., Holdship, J. Solar, M., *et al.*, Experimental Astronomy, **52**, 157 (2021)
[2] Behrens, E., Mangum, J. G., Holdship, J., *et al.*, ApJ, **submitted** (2022)
[3] Haasler, D., Rivilla, V. M., Martín, S., *et al.*, A&A, **659**, A158 (2022)
[4] Harada, N., Martín, S., Mangum, J. G., *et al.*, ApJ, **923**, A24 (2021)
[5] Holdship, J., Viti, S., Jimeénez-Serra, I., *et al.*, AJ, **154**, 38 (2017)

[1] https://spectralradex.readthedocs.io/

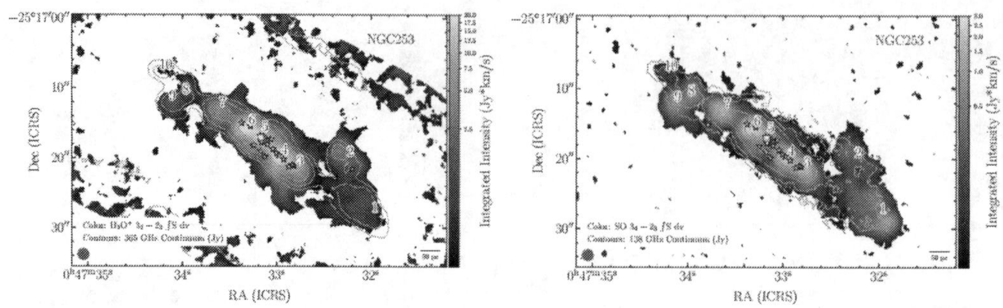


**Figure 3.** $H_3O^+$ $3_2 - 2_2$ (left) and SO $3_4 - 2_3$ (right) integrated intensity (moment 0) images toward NGC 253. The green ellipse and black scale bar in the lower-left and lower-right corners show the final imaged beam size (1.6 arcsec) and physical scale, respectively. Black numbers indicate the locations of the GMCs. Star shaped markers locate the positions of the 2 cm radio continuum emission peaks [13], with a square indicating the position of the strongest radio continuum peak. The lower integrated intensity limit for each transition is set to $3\sigma$ in the integrated intensity. Overlain in contours is the associated continuum emission distribution for each transition. From [6].

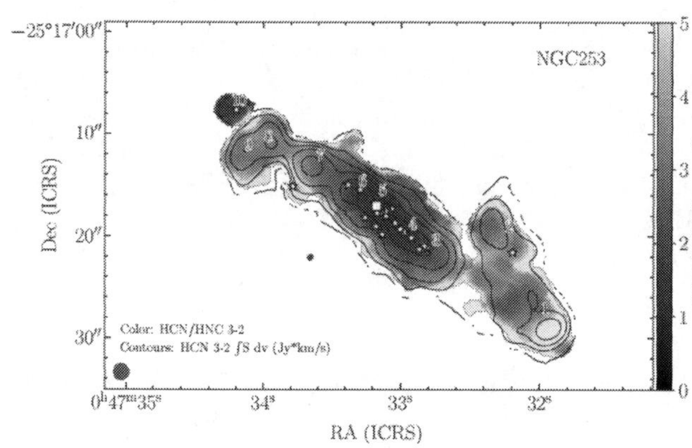


**Figure 4.** HCN/HNC J=3 − 2 integrated intensity ratio (from [2]). Symbols and contours same as Figure 3.

[6] Holdship, J., Mangum, J. G., Viti, S., *et al.*, ApJ, **in press** (2022)
[7] Holdship, J., Viti, S., Martín, S., *et al.*, A&A, **654**, A55 (2021)
[8] Humire, P. K., Henkel, C., Hernández-Gómez, A., *et al.*, A&A, **in press** (2022)
[9] Leroy, A. K., Bolatto, A. D., Ostriker, E. C., *et al.*, ApJ, **801**, 25 (2015)
[10] Leroy, A. K., Bolatto, A. D., Ostriker, E. C., *et al.*, ApJ, **869**, 126 (2018)
[11] Martín, S. Mangum, J. G., Harada, N., *et al.*, A&A, **656**, A46 (2021)
[12] Spilker, J. S., Marrone, D. P., Aguirre, J. E., *et al.*, ApJ, **785**, 149 (2014)
[13] Ulvestad, J. S., & Antonucci, R. R. J., ApJ, 488, 621 (1997)

[a]Sergio Martín (ESO); Nanase Harada (NAOJ); Rebeca Aladro (MPIfR); Erica Behrens (Department of Astronomy, University of Virginia); Laura Colzi (INAF); Kimberly L. Emig (NRAO); Santiago García-Burillo (OAN-IGN); Christian Henkel (MPIfR); Rubén Herrero-Illana (ESO); Jonathan Holdship (Leiden Observatory/UCL); Ko-Yun Huang (Leiden Observatory); P. K. Humire (MPIfR); Leslie K. Hunt (INAF); Takuma Izumi (NAOJ); Kotaro Kohno (University of Tokyo); David S. Meier (NMT); Sebastien Muller (Chalmers University of Technology); Kouichiro Nakanishi (NAOJ); Víctor M. Rivilla (CSIC-INTA); Kazushi Sakamoto (ASIAA); Kunihiko Tanaka (Keio Univer-sity); Paul P. van der Werf (Leiden Observatory); Yuki Yoshimura (University of Tokyo)·

# Gas condensation in Brightest Group Galaxies unveiled with MUSE

*Valeria* Olivares[1,*]

[1]Department of Physics and Astronomy, University of Kentucky, 505 Rose Street, Lexington, KY 40506, USA

**Abstract.** How central galaxies in groups acquired their cold gas is still not fully understood. To unfold this unknown, we map the kinematics and distribution of the optical emission-line gas using MUSE observations of 18 optically selected local brightest group galaxies (BGGs). The observations reveal a distribution of gas morphologies, including complex networks of filaments extending up to ~10 kpc to compact (<3 kpc) and extended (>5 kpc) disk-dominated structures. By exploring the thermodynamical properties of the X-ray atmospheres, we find most of the filaments and compact have are likely cooled out from the intragroup medium (IGrM).

## 1 Introduction

Traditionally elliptical galaxies have been described as "red and dead" galaxies. Observational evidence, however, has challenged this view revealing that a significant fraction of the population of ellipticals contains large quantities of cold gas, which can fuel both their Active Galactic Nuclei (AGN) and star formation activity. Massive elliptical galaxies located at the cluster cores show extended multiphase filamentary structures [e.g. 1], that follow the cavities dug by the radio jet in the hot intracluster medium (ICM) [2, 3]. Extensive multiwavelength observations of Brightest Cluster Galaxies (BCGs) are consistent with predictions of chaotic cold accretion (CCA), precipitation from the hot ICM, and stimulated feedback models [e.g., 4–6].

While there is considerable evidence from observations and simulations that cold gas in BCGs forms through ICM cooling, less attention has been paid to their lower-mass counterparts in galaxy groups. Overall, BGGs have remained relatively unstudied as a population, compared to both BCGs and the general population of elliptical galaxies, despite representing the intermediate-mass range and being the building blocks of massive clusters. Several groups show remarkable similarities to clusters. They host a hot intragroup medium (IGrM) [e.g., 7], that cools through X-ray emission in a similar form to the ICM in clusters. However, groups are a more diverse class, and their low-velocity dispersions and small galaxy separation make mergers more common than in clusters. This allows a greater fraction of the cold gas from infalling gas-rich spirals to reach the core, affecting the overall gas content. Moreover, cold molecular gas is more common in BGGs galaxies than in other ellipticals. The single-dish survey of 53 BGGs found that about 40% of the galaxies contain cold molecular gas, although the cold gas mass is usually low, on the order of a few $10^8$ M⊙ [8].

*valeria.olivares@uky.edu

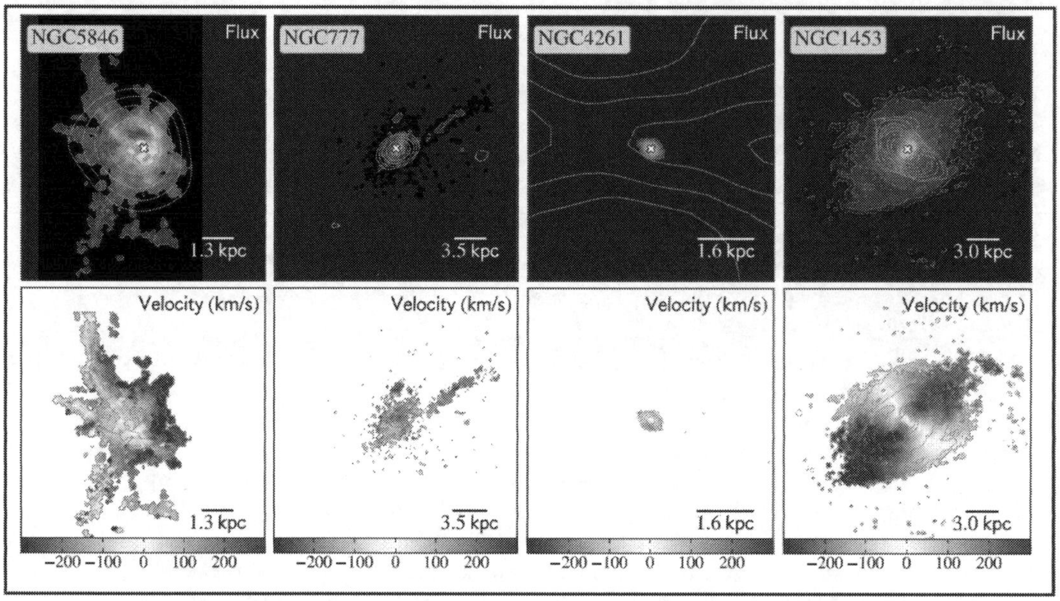


**Figure 1.** Example of ionized gas distribution (top row) and velocity (bottom row) for systems with extended and compact filaments and disks. The radio emission is displayed with green contours, while the center of the galaxy is marked with a cross.

## 2 CLoGS - Complete Local-Volume Groups Sample

To study AGN feedback in galaxy groups and identify groups with molecular gas reservoirs, a survey has been constructed: the Complete Local-Volume Groups Sample (CLoGS, [7]). CLoGS is the first statistically complete optically selected, representative sample of local galaxy groups. The sample consists of 53 groups in the local universe, covering a Richness (R)[1] between 2 and 8. All have been studied its molecular gas content using single-dish observations [8, 9], and have X-ray and radio observations [10, 11].

## 3 A large variety of ionized gas distribution

To explore the nature of the optical line-emitting gas in nearby BGGs observed with MUSE 18 sources drawn from the CLoGS high-richness subsample [12]. These observations reveal optical emission lines in 17/18 systems, with filamentary structures in most (10/18) BGGs (see Fig. 1 for a few examples). The projected sizes of the filaments in groups are shorter than those found in central cluster galaxies, consistent with the different sizes of the condensation regions of the hot halos [13]. Another significant difference with cluster galaxies is the presence of several (7/18) disks and ring-dominated structures found in our sample, of which two disks are very compact, ~1–3 kpc. In contrast, the rest of the disks can reach projected sizes of up to ~21 kpc. The extended disks also reveal clumpy rings and extended structures. The velocity structure of the ionized gas in the sources with filaments is often chaotic, but some shallow velocity gradients are noticed along some filaments that typically host extended filamentary nebulae.

[1]Number of large galaxies within the galaxy group.

## 4 Origin of the gas

Several scenarios have been discussed for the origin of cold gas in elliptical galaxies over the past years. Those scenarios are the acquisition of cold gas from the cooling of gas ejected by the stellar population (stellar-mass loss), through mergers or interactions with gas-rich galaxies, or cooling from the hot atmosphere (cooling-flows). A mixture of these different processes is also and likely possible. [14] suggested that in elliptical galaxies, the gas produced by the stellar-mass loss should form a kiloparsec-scale rotating disk aligned with the stellar component. In contrast, the gas brought into the system through mergers is likely to be misaligned or to create multiple tails, rings, or disks. Numerical simulations of cool-core clusters and groups also found kiloparsec rotating disks of raw material at the center of the galaxy. The formation of the kiloparsec cold rotating disk may thus also occur via the condensation of the hot gas via thermal instabilities [e.g., 4, 6] or merger events.

### 4.1 IGrM cooling

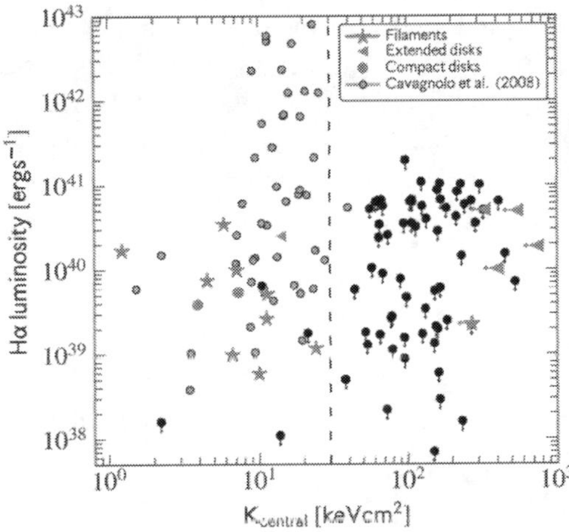


**Figure 2.** H$\alpha$ luminosity versus central entropy values at 1 kpc. Filamentary sources are shown with blue stars, compact disks with orange circles, and extended disks with orange left-pointing triangles. As a matter of comparison, we have included the sample from [15]. The dashed gray line is the entropy value of 30 keV cm$^2$.

A few systems own extended X-ray halos (>65 kpc) with luminosities and temperatures typical of group-scale haloes, while others have X-ray distributions with galaxy-scale sizes of 10–65 kpc [7]). These systems reveal mostly ionized gaseous filaments or compact disks. On the other hand, extended gaseous disks usually lack detection of an IGrM. The disturbed kinematics and filamentary distribution, combined with low central entropy hot gas and short central cooling times, indicates that the cold gas is an outcome of thermally unstable cooling from the X-ray emitting gas, driven possibly by the AGN feedback, as shown in central cluster galaxies [e.g., 16, 17]. For the entropy, we use K = kT/n$_e$, where $kT$ is the temperature of the hot gas and $n_e$ the electron number density. From the systems with detected IGrM, we found that almost all filamentary sources (expect NGC584) and compact disks, plus one extended rotating disk (NGC1453), have low central entropy, <30 keV cm$^2$ (Fig. 2), and cooling times, Consistent with being a by-product of IGrM condensation. On the contrary, extended disks have larger central entropy values that are inconsistent with hot gas cooling.

Another interesting cooling criteria is the C-ratio, defined as the ratio of the gyration time scale of turbulent eddies, $t_{eddy}$, and the cooling time, $t_{cool}$, which is expected to be around unity for the onset of gas condensation via turbulence [4]. With the filamentary sources and one compact disk (NGC4261) have a C-ratio (0.5–1.6 at 10 kpc) consistent with the predicted values by numerical simulations (C=0.6–1.8, [4]).

### 4.2 Mergers and interaction

**Figure 3.** Optical image of NGC 1587 overlied with he distribution of the optical emission-line gas shown with red contours.

Two sources show signs of interaction showing how diverse galaxy groups are as a population. A clear example of a tidal interaction is seen in NGC 1587 with its elliptical neighbor (Fig. 3), as the ionized gas appears to be spirally inflowing towards the central galaxy [12]. Accordingly, we suspect that at least a fraction, if not all, of the colder gas detected in this source may have been acquired through this interaction.

Extended gas disks lack radio and are found in systems where the current X-ray observations do not detect the IGrM. The latter may indicate cold gas formed through gas-rich mergers or galaxy interactions, which could also potentially form central disks and rings in the final stages of mergers, as suggested by studies on elliptical galaxies [14]. The larger fraction of rotating disks in central group galaxies than in clusters may hint toward a non-negligible contribution of mergers or gas stripped from satellites that are more prone to happen in low-mass halos.

## References

[1] V. Olivares, **631**, A22 (2019), 1902.09164
[2] P. Salomé, **454**, 437 (2006), astro-ph/0603350
[3] H.R. Russell, **490**, 3025 (2019), 1902.09227
[4] M. Gaspari, **854**, 167 (2018)
[5] B.R. McNamara, **830**, 79 (2016), 1604.04629
[6] R.S. Beckmann, **631**, A60 (2019), 1909.01329
[7] E. O'Sullivan, **472**, 1482 (2017)
[8] E. O'Sullivan, **573**, A111 (2015), 1408.7106
[9] E. O'Sullivan, Astronomy & Astrophysics **618**, A126 (2018)
[10] K. Kolokythas, **481**, 1550 (2018), 1807.11095
[11] K. Kolokythas, **489**, 2488 (2019), 1907.10768
[12] V. Olivares, arXiv e-prints arXiv:2201.07838 (2022), 2201.07838
[13] M. McDonald, **731**, 33 (2011), 1102.1972
[14] T.A. Davis, **417**, 882 (2011), 1107.0002
[15] K.W. Cavagnolo, **683**, L107 (2008)
[16] Y. Li, **789**, 153 (2014)
[17] Y. Qiu, **917**, L7 (2021), 2108.04247

# Multi-scale dynamics in star-forming regions: The interplay between gravity and turbulence

*Alessio* Traficante[1,*], *Gary A.* Fuller[2], *Ana* Duarte-Cabral[3], *Davide* Elia[1], *Mark H.* Heyer[4], *Sergio* Molinari[1], *Nicolas* Peretto[3], and *Eugenio* Schisano[1]

[1]IAPS - INAF, via Fosso del Cavaliere, 100, 00133 Roma, Italy
[2]Jodrell Bank Centre for Astrophysics, Department of Physics and Astronomy, The University of Manchester, Oxford Road, Manchester M13 9PL, UK
[3]School of Physics and Astronomy, Cardiff University, Queens Buildings, The Parade, Cardiff CF24 3AA, UK
[4]Department of Astronomy, University of Massachusetts, Amherst, MA 01003, USA

**Abstract.** In the multi-scale view of the star formation process the material flows from large molecular clouds down to clumps and cores. In this paradigm it is still unclear if it is gravity or turbulence that drives the observed supersonic non-thermal motions during the collapse, in particular in high-mass regions, and at which scales gravity becomes eventually dominant over the turbulence of the interstellar medium. To investigate this problem we have combined the dynamics of a sample of 70 $\mu$m-quiet clumps, selected to cover a wide range of masses and surface densities, with the dynamics of the parent filaments in which they are embedded. We observe a continuous interplay between turbulence and gravity, where the former creates structures at all scales and the latter takes the lead when a critical value of the surface density is reached, $\Sigma_{th} = 0.1$ g cm$^{-2}$. In the densest filaments this transition can occur at the parsec, or even larger scales, leading to a global collapse of the whole region and most likely to the formation of the massive objects.

## 1 Introduction

The formation of stars is a multi-scale process which starts in giant molecular clouds, structures extended tens of parsecs. Within these clouds the process continues through dense, elongated structures called filaments and, within these filaments, parsec-scale dense clumps that are the progenitors of sub-parsec cores and fragments, finally leading to the formation of (massive) cluster of stars. In this scenario it is still unclear how the interplay between gravity and turbulence regulates this continuous flow of energy and gas across several orders of magnitude of spatial scales and gas densities. The theoretical scenarios are still divided in models where the turbulent flow is primarily responsible for the formation of the parsec and sub-parsec structures [1], and opposite approaches in which gravity takes over the turbulent cascade already at the scales of several parsecs [2].

Observationally, the study of such flows consider two critical aspects: 1) the observations of star forming regions with various physical properties, such as mass and densities, to be

*e-mail: alessio.traficante@inaf.it

able to probe the interplay of turbulence and gravity under different gravitational potentials; 2) the observations of tracers sensitive to the progressively denser layers of each star-forming region, which allow us to observe the gas dynamics from the (relatively) diffuse clouds observed with low-J transitions of CO to the dense, massive clumps observed with high-density tracers such as $N_2H^+$ (1-0). Several works have been dedicated to the study of the gas dynamics in statistically significant sample of objects traced with a specific tracer, such as the GMCs observed with CO (1-0) [3], filaments with e.g. CO (2-1) [4] or CO (3-2) [5] and massive clumps with $N_2H^+$ (1-0) [6]. In this work we present a different approach: we show the results obtained from the combination of different gas tracers used to look within the same star-forming regions with the aim of investigating the turbulent cascade and the role of the gravitationally driven motions at the different density layers in each region.

## 2 Dataset

In order to study the turbulent cascade and its interplay with the gravitational potential, it is mandatory to minimize any possible processes that can alter the gas kinematics. The starting requirement to perform our study has been the choice of a sample of 70 $\mu$m-quiet clumps, i.e. objects at the earliest phases of their evolution and therefore likely to be unaffected by internal feedback from newborn stars [7]. The lack of the 70 $\mu$m emission indeed guarantees very low value of the luminosity over mass (L/M) parameter in these objects, which is a strong indicator that the clump is almost pristine, with very little or no internal star formation yet started [8]. Our final sample consists of 22 clumps which have been further chosen to be divided into three different surface density bins: $\Sigma_{low}$, i.e. clumps with surface density $\Sigma < 0.05$ g cm$^{-2}$ (5 clumps); $\Sigma_{int}$, i.e. clumps with $0.05 \leq \Sigma \leq 0.1$ g cm$^{-2}$ (6 clumps); $\Sigma_{high}$, i.e. clumps with $\Sigma > 0.1$ g cm$^{-2}$ (11 clumps). These values are all above the suggested threshold for star-formation ($\simeq 0.02$ g cm$^{-2}$, [9]) and the subdivision follows the suggested thresholds for massive star-formation ($\Sigma \geq 0.05$ g cm$^{-2}$, [10]; $\Sigma \geq 0.1$ g cm$^{-2}$, [11]). This sample therefore allows us to explore the interplay between gravity and turbulence in different potential wells. The kinematics of these clumps have been evaluated with $N_2H^+$ (1-0) observations collected by the IRAM 30m single dish telescope. In addition, for each clump we identify the parent filament in the Hi-GAL filaments catalogue [12], from which we extracted the mass and equivalent radius of each object. The kinematics of the filaments are evaluated using the $^{13}$CO (1-0) emission. We carefully mask the positions and solid angles of all clumps where the CO emission is likely optically thick in order to measure the average velocity dispersion of the filament itself as opposed to the kinematics of the inner clumps evaluated using the $N_2H^+$ (1-0) emission. Furthermore, we estimate and subtract the contribution of the large-scale velocity gradients from each filament to derive the velocity dispersion from turbulence and/or gravitational motions.

## 3 Results

The first result to notice is that the high density clumps are embedded in the densest filaments of our sample [7]. In addition, the internal velocity dispersion of the $\Sigma_{int}$ and $\Sigma_{high}$ clumps correlates very well with the velocity dispersion of the parent filament after the gradient subtraction [7, Pearson correlation coefficient $\rho = 0.79$]. These results suggest that the environment plays a significant role in the formation of the massive objects and the dynamics of the clumps are inherited from the dynamics of their parent cloud. The different scales implies a continuous flow of material from the cloud down to the star-formation sites.

In order to investigate the interplay between gravity and turbulence at the different scales, we built the $\sigma$ vs. radius (R) relation for both filaments and clumps separately [7]. These

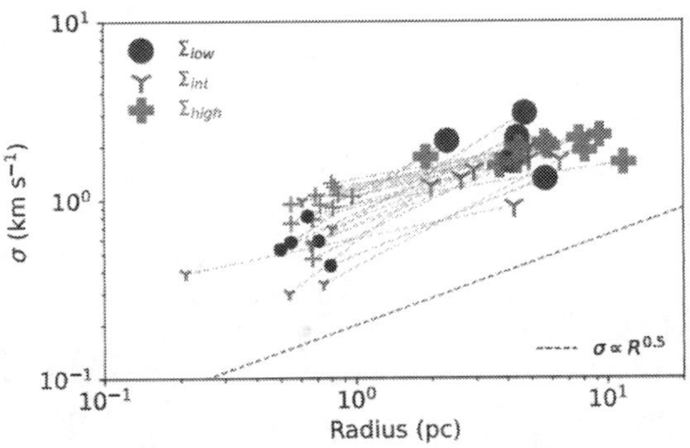


**Figure 1.** Larson relation for the clumps and the parent filaments belonging to the three groups. The dotted lines connect each clump (left-hand points) with the parent filament (right-hand points). Objects associated with clumps in the $\Sigma_{low}$, $\Sigma_{int}$ and $\Sigma_{high}$ groups are represented as blue circles, green tri_downs and red crosses, respectively. The black-dotted line is the $\sigma \propto R^{0.5}$ relation in the case of a supersonic turbulent cascade of energy from large down to small scales [7].

plot do not show the expected relation for supersonic turbulence, i.e. $\sigma \propto R^{\delta}$ with $\delta = 0.5$ [13]. The break of this relation was already noticed at the clump scales in previous surveys of massive clumps [6, 14], and it has been interpreted as evidence of gravity dominating over turbulence at these scales. With this work we point out that the role of environment can actually hide the turbulent cascade in some cases, in particular in regions where the local environment is naturally less turbulent than others. To demonstrate this conjecture we show the $\sigma$ vs. R plot but this time connecting the clump position in this plot with the position of its parent filament (see Figure 1) . As discussed more in details in [7], the average values of $\delta$ of the lines connecting $\Sigma_{low}$, $\Sigma_{int}$ and $\Sigma_{high}$ clumps to their parent filaments are, respectively, $\delta_{low} = 0.68$, $\delta_{int} = 0.60$ and $\delta_{high} = 0.38$. These numbers indicate that when the turbulent cascade is investigated within the same object, and therefore within the same local conditions of the environment, the effect of a turbulent cascade may be identified. The average increasing of the exponent in denser and denser regions is interpreted as evidence that in clumps with $\Sigma > 0.1$ g cm$^{-2}$ there must be an extra contribution to the kinetic energy, likely driven by the gravitational collapse, which dominates over the energy that could arise from a pure turbulent cascade [7]. The critical threshold above which gravity overtakes the turbulent cascade may br driven by the local density of the medium, rather than by a specific spatial scale.

## 4 Discussion and conclusions

This work concludes that the dynamics of the filaments and of their embedded clumps are strongly correlated. The flow of energy and gas is continuous from the filament down to the clumps, and the interplay between gravity and turbulence depends on the density of the objects under investigation. We propose a scenario where the transfer of energy via the turbulent cascade is the dominant mechanism on the largest, and less dense, scales, with a gradual variation in the dynamics: the gravity gains more and more importance as we move into denser and denser regions (e.g. [15]). Above the critical threshold of $\Sigma_{th} = 0.1$ g cm$^{-2}$ gravity dominates the motions and the kinematics over the turbulent cascade, driving the global collapse of the region above this density threshold. This critical value of the surface density can be reached at any spatial scales, depending on the initial conditions of the turbulent cloud. Gravity can therefore start to drive the motions at the core, the clump, or even the filament scales in star-forming regions embedded in significantly dense clouds. The cartoon in Figure 2 describes the scenario that we propose, which is detailed in [7].

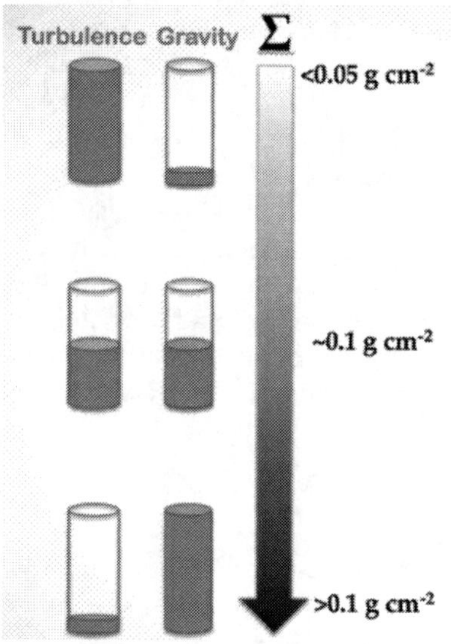


**Figure 2.** Cartoon representation of the interplay between gravity and turbulence in star-forming regions. A cloud begins to be compressed due to the effect of the strong interstellar medium turbulence, with a minimal contribution from gravity. As the material starts to accumulate in different eddies due to turbulence, these regions increase their surface density and with it the role of gravity in driving the motions. As the density reaches the critical value of $\Sigma_{th} = 0.1$ g cm$^{-2}$, the motions induced by gravity dominate over the turbulence, which, from this point, becomes sub-dominant [7].

## References

[1] Padoan, P. et al., ApJ, 900, 82 (2021)
[2] Vazquez-Semadeni, E. et al., MNRAS, 490, 3061 (2019)
[3] Miville-Deschenes, M. A. et al, ApJ, 834, 57 (2017)
[4] Duarte-Cabral, A. et al., MNRAS, 500, 3027 (2021)
[5] Rigby, A. et al., A&A, 632, A58 (2019)
[6] Traficante, A. et al., MNRAS, 477, 2220 (2018)
[7] Traficante, A. et al., MNRAS, 491, 4310 (2020)
[8] Molinari, S. et al, ApJL, 826, L8 (2016)
[9] Lada, C. J. et al., ApJ, 724, 687 (2010)
[10] Urquhart, J. et al., MNRAS, 443, 1555 (2014)
[11] Tan, J. et al. 2014, PPVI, 149 (2014)
[12] Schisano, E. et al., MNRAS, 492, 5420 (2020)
[13] Larson, R. B., MNRAS, 194, 809 (1981)
[14] Ballesteros-Paredes, J. et al., MNRAS, 4111, 65 (2011)
[15] Pingel, N. M. et al. 2018, ApJ, 856, 136 (2018)

# Mapping the high ionization rate of the GC starburst Sgr B2 through low $HCO^+$ / $N_2H^+$ $J$=1-0 intensity ratios

*Miriam* G. Santa-Maria[1,*] and *Javier* R. Goicoechea[1]

[1]Instituto de Física Fundamental (IFF), CSIC. Calle Serrano 121-123, 28006 Madrid, Spain

**Abstract.** We still do not understand which mechanisms dominate the heating and ionization of the extended molecular gas in galactic nuclei. The starburst Sgr B2, in the Galactic Center (GC), is an excellent template to spatially resolve the high-mass star-forming cores from the extended cloud environment, and to study the properties of the warm neutral gas in conditions likely prevalent in star-forming galaxies. We mapped $\sim$1000 pc$^2$ of Sgr B2 complex, using the IRAM 30m telescope, in the $N_2H^+$, $HCO^+$ $J$=1-0 and SiO $J$=2-1 line emission. The extended nature of the $N_2H^+$ $J$=1-0 emission is remarkable. Compared to molecular clouds in the disk of the galaxy, the $N_2H^+$ $J$=1-0 emission is not confined to cold and dense cores and filaments. This can be explained by the high ionization rate ($\zeta \gtrsim 10^{-15}$ s$^{-1}$), leading to overabundant $H_3^+$, $He^+$, and $N_2H^+$. The enhanced ionization rate is likely responsible of the much lower line intensity ratio $R_I = HCO^+/N_2H^+$ $J$=1-0 observed in Sgr B2 ($R_I \approx 2 \pm 2$), Arp 220 ($R_I \approx 2$), and NGC 253 ($R_I \approx 5$), compared to disk clouds such as Orion B ($R_I \approx 24$) and starburst galaxies such as M82 ($R_I \approx 21$).

## 1 Introduction

The conditions prevailing in the interstellar medium (ISM) of galactic nuclei can be extreme compared to those in the more quiescent disk of galaxies. Such conditions are regulated by the feedback from OB-type stars: strong UV radiation fields, stellar winds, and supernovae explosions [1]; cloud-cloud collisions [2], enhanced cosmic ray (CR) fluxes [3], and X-ray emission (from accretion onto a super-massive black hole, SMBH) [4]. To understand star formation in these environments, and galaxy evolution in general, it is critical to characterize the gas heating and ionization mechanisms in the ISM of galactic nuclei.

Sagittarius B2 is the most massive and far-infrared (FIR) luminous ($\sim 10^7$ $L_\odot$) molecular cloud in the Galactic Center (GC). It shows ongoing high-mass star formation and line-of-sight column densities above $N_{H_2} \gtrsim 10^{24}$ cm$^{-2}$ [3]. Sgr B2 is located at a projected distance of about 100 pc from the dynamical center of the Milky Way (Sgr A$^*$) [5]. It contains three main high-mass star-forming cores (named N, M, and S), with more than 49 compact H II regions [6] embedded in a moderately high density ($n(H_2) \gtrsim 10^5$ cm$^{-3}$) molecular cloud which is surrounded by a more extended and lower density envelope, hereafter "Sgr B2 envelope" [7].

It is a long-standing debate whether radiation, CRs, or shocks dominate the heating of the widespread warm molecular gas in the Sgr B2 complex. Plausible candidates are: cloud-cloud collisions [2, 8], UV-irradiated low-velocity shocks [9], X-rays from brief outburst periods of Sgr A$^*$, or low-energy cosmic ray (LECRs) showers [4].

*e-mail: miriam.g.sm@csic.es

Variations in the abundances of particular molecular ions, such as the $HCO^+$ over $N_2H^+$ abundance ratio, help to constrain the ionization processes that result from X-ray and LECRs irradiation [10]. In this paper we briefly summarize the main results of our IRAM 30m mapping study of Sgr B2 complex, focusing on the characteristics of the $N_2H^+$ and $HCO^+$ $J$=1-0 emission.

## 2 Mapping observations

We mapped Sgr B2 in the 3mm band (~90 GHz) using the IRAM 30m telescope. The size of our maps is $12' \times 12'$ (see Fig. 1). The angular resolution at the target frequencies is $\sim 27''$ ($\sim$ 1 pc at the distance to Sgr B2). We complement these observations with *NuSTAR* hard X-rays (3-78 keV) continuum observations. However, these X-rays observations are not conclusive of whether Sgr B2 is (still) reprocessing X-rays from past flaring activity of Sgr A* or if (the observed) X-rays arise from LECRs [4, 12]. In this work we refer to the ionization rate ($\zeta$), without distinction between X-rays and CRs ionization rate.

## 3 Widespread $N_2H^+$ emission far from star-forming cores

The observed $N_2H^+$ $J$=1-0 emission is extended at pc scales. This is an unexpected result as in nearby molecular clouds of the Galactic disk the $N_2H^+$ emission is confined to dense filaments and cold prestellar cores inside them [13]. The common interpretation is that $N_2H^+$ traces very dense gas at high extinction depths, that is, cold gas ($T_k < 20$ K) shielded from

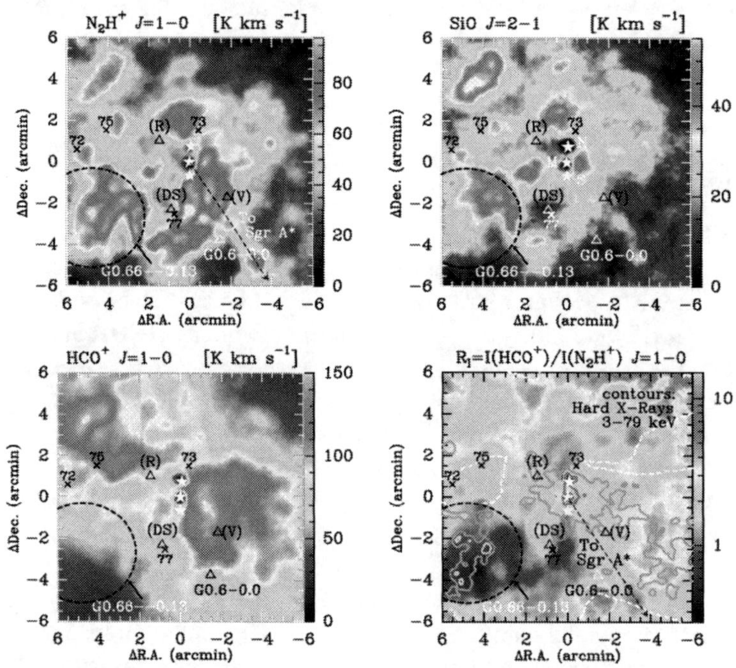


**Figure 1.** IRAM 30m total line intensity maps (in K km s$^{-1}$): $N_2H^+$, $HCO^+$ $J$=1-0 and SiO $J$=2-1 integrated in the LSR velocity range [-20, 120] km s$^{-1}$. *Bottom-right*: $R_I$=I($HCO^+$)/I($N_2H^+$) $J$=1-0 intensity ratio. The three white stars mark the position of the main star-forming cores (N, M, S). Triangles mark other H II regions. Black dashed circle marks an X-ray irradiated region [4]. Contours on the $R_I$ map show the hard X-ray emission published in 2015 [4]. Dashed white lines delineate the regions with no X-ray observations. The X symbols and the numbers are *NuSTAR* X-ray point sources [11].

stellar FUV radiation where CO freezes-out on dust grains [14]. In Sgr B2, however, the $N_2H^+$ emission is remarkably widespread and extended. In addition, it spatially correlates with the SiO $J$=2-1 emission (see Fig. 1; $\rho_{sp}=0.7$ [3]), considered to be tracer of warm shocked gas.

The main formation route of $N_2H^+$ is reaction of $H_3^+$ with $N_2$ (the major gas-phase nitrogen reservoir inside molecular clouds). Its main destruction route is reacting with CO,

$$N_2 + H_3^+ \rightarrow N_2H^+ + H_2, \quad (1)$$
$$N_2H^+ + CO \rightarrow HCO^+ + N_2. \quad (2)$$

$H_3^+$ forms after $H_2$ ionization and drives the production of molecular ions in dense interstellar clouds [15]. The ionization rate in Sgr B2 is high, $\zeta \simeq (1\text{-}10)\times10^{-15}\ s^{-1}$ [4], about 10-100 times higher than that in typical molecular clouds of the Galactic disk. This enhanced rate increases the $H_3^+$ and $He^+$ production rate, but also the destruction rate of CO through gas-phase reactions with $H_3^+$ and $He^+$:

$$H_3^+ + CO \rightarrow HCO^+ + H_2, \quad (3)$$
$$He^+ + CO \rightarrow C^+ + O + He \ \text{ or } \ C + O^+ + He. \quad (4)$$

In addition, $He^+$ reacts with $N_2$ to form $N_2^+$, which then reacts with $H_2$ to form more $N_2H^+$. For high values of $\zeta$, the ionization fraction, $x_e = n_e/n_H$, is also high. This implies that $N_2H^+$ and $HCO^+$ start to be destroyed by dissociative recombinations with electrons. In consequence, the column density ratio $R_N$=$N(HCO^+)/N(N_2H^+)$ decreases as $\zeta$ increases (see also [10]). This reasoning applies as long as the destruction of $H_3^+$ is dominated by reactions with CO (so-called "low ionization phase", LIP) and not by recombinations with electrons [16, 17]. In this phase, protonation reactions with $H_3^+$ dominate the formation of abundant molecular ions such as $N_2H^+$. Our photochemical models of the Sgr B2 envelope (with $G_0 = 10^3$ and $\zeta \gtrsim 10^{-15}\ s^{-1}$; Fig. 2 *left*) show that the $N_2H^+$ emission must arise from moderately dense gas ($n_H \gtrsim 10^5\ cm^{-3}$) and from moderate cloud depths ($A_V$) so that the CO abundance is higher than $x_e$.

At low densities ($n_H < 10^4\ cm^{-3}$) and high ionization rates ($\zeta \gtrsim 10^{-15}\ s^{-1}$) the gas chemistry switches to a phase in which destruction of $H_3^+$ via electron recombinations is comparable to destruction via reactions with CO. This phase is called "high ionization phase" (HIP). The abundances of molecular ions such as $HCO^+$ and $N_2H^+$ are much lower in this phase. It is unlikely that the observed extended $N_2H^+$ emission in Sgr B2 is in the HIP.

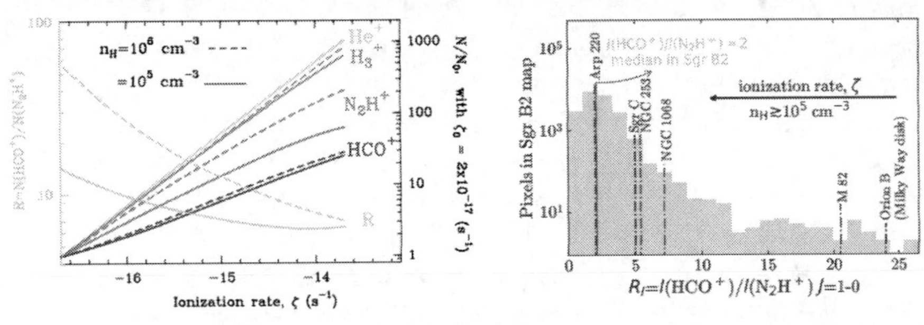


**Figure 2.** *Left:* Model predictions for different ionization rates, $\zeta$, (constant $G_0$=$10^3$, $n_H$=$10^5$ and $10^6$ cm$^{-3}$, and $A_{V,tot}$=15 mag). Green curves show the column density ratio $R_N$. Colored curves (blue, pink, gray, and cyan) show the column density enhancement, $N/N_0$, with respect to models with $\zeta_0$=$2\times10^{-17}$ s$^{-1}$ (average ionization rate in Galactic disk molecular clouds) [3]. *Right:* Histogram of all observed $R_I = I(HCO^+)/I(N_2H^+)$ $J$=1-0 line intensity ratios in our map of Sgr B2 (green). The black arrow indicates the direction of increasing ionization rate. The red line marks the median value of the intensity ratio in Sgr B2 complex. Dashed blue lines mark the line intensity ratios $R_I$ observed in Orion B at large spatial scales [13], in M 82, NGC 1068, NGC 253, Arp 220 [18], and in Sgr C [19]. Recall that, in general, the value $R_N$ (*left*) is different than $R_I$ (*right*).

### 3.1 $HCO^+/N_2H^+$ intensity ratio in Sgr B2 envelope and in other environments

At sufficiently high densities, $n_H \gtrsim 10^5$ cm$^{-3}$ and in the LIP phase, the column density ratio $R_N$ ($HCO^+/N_2H^+$) inversely scales with the ionization rate $\zeta$, and it scales with gas density. In Sgr B2, the $R_I$($HCO^+/N_2H^+$) map (Fig. 1, *bottom-right*) shows low ratios (<1) in specific X-rays irradiated regions (e.g.,G0.66–0.13) but higher ratios in others (e.g.,western regions). The average value of the line intensity ratio $R_I$ is low, $R_I = 2 \pm 2$. This is much lower than that the ratio observed in Orion B averaged over large spatial scales, $R_I \sim 24$ [13]. However, the observed intensity ratio matches that of Arp 220 (a merger ULIRG galaxy). This match suggests similar environmental conditions (Arp 220 possibly hosts thousands of embedded Sgr B2-like starburst regions). NGC 253 (a spiral galaxy with widespread shocks) shows a ratio of $R_I \sim 5$, whereas the starburst galaxy M 82 shows a much higher ratio, $R_I \sim 21$, closer to the values observed in molecular clouds of the galactic disk (i.e., affected by lower ionization rates).

In this study we find that low $R_I$($HCO^+/N_2H^+$) line intensity ratios trace relatively dense molecular gas and high ionization rates (produced by enhanced doses of cosmic rays and X-rays). In Sgr B2, the intensity ratio $R_I$ is low (<1) over large spatial scales. This result suggests that part of the rich and characteristic chemistry of Sgr B2 is produced by the high ionization rate in the GC. This also implies that the emissivity and behavior of certain molecules differ from that in typical molecular clouds of the disk (or in normal galaxies).

In order to determine the column densities of $HCO^+$ and $N_2H^+$, and to constrain $R_N$ accurately, one would need to map their higher frequency rotationally excited lines and study their excitation and radiative transfer. Much higher angular resolution maps will allow us to spatially separate the different environments in these complex regions and to dissect their dominant heating and ionization mechanisms.

## References

[1] Goicoechea, J. R., Santa-Maria, M. G., Teyssier, D., et al. 2018, A&A, 616, L1
[2] Tsuboi, M., Miyazaki, A., & Uehara, K. 2015, PASJ, 67, 90
[3] Santa-Maria, M. G., Goicoechea, J. R., Etxaluze, M., et al. 2021, A&A, 649, A32
[4] Zhang, S., Hailey, C. J., Mori, K., et al. 2015, ApJ, 815, 132
[5] Molinari, S., Bally, J., Noriega-Crespo, A., et al. 2011, ApJ, 735, L33
[6] Ginsburg, A., Bally, J., Barnes, A., et al. 2018, ApJ, 853, 171
[7] Huttemeister, S., Wilson, T. L., Henkel, C., et al. 1993, A&A, 276, 445
[8] Hasegawa, T., Sato, F., Whiteoak, J. B., et al. 1994, ApJ, 429, L77
[9] Goicoechea, J. R., Rodríguez-Fernández, N. J., & Cernicharo, J. 2004, ApJ, 600, 214
[10] Ceccarelli, C., Dominik, C., López-Sepulcre, A., et al. 2014, ApJL, 790, L1
[11] Hong, J., Mori, K., Hailey, C. J., et al. 2016, ApJ, 825, 132
[12] Kuznetsova, E., Krivonos, R., Lutovinov, A., et al. 2022, MNRAS, 509, 1605
[13] Pety, J., Guzmán, V. V., Orkisz, J. H., et al. 2017, A&A, 599, A98
[14] Caselli, P., Benson, P. J., Myers, P. C., et al. 2002, ApJ, 572, 238
[15] Wolfire, M. G., Vallini, L., & Chevance, M. 2022
[16] Pineau des Forets, G., Roueff, E., & Flower, D. R. 1992, MNRAS, 258, 45P
[17] Le Bourlot, J., Pineau des Forets, G., Roueff, E., et al. 1993, ApJ, 416, L87
[18] Aladro, R., Martín, S., Riquelme, D., et al. 2015, A&A, 579, A101
[19] Jones, P. A., Burton, M. G., Cunningham, M. R., et al. 2012, MNRAS, 419, 2961

# Dynamically regulated star formation in the strongly interacting Taffy galaxies

*Axel* García-Rodríguez[1,*], *Antonio* Usero[1,**], *Santiago* García-Burillo[1], *Frank* Bigiel[2], *Elias* Brinks[3], *Asunción* Fuente[1], *Adam K.* Leroy[4], and *Miguel* Querejeta[1]

[1]Observatorio Astronómico Nacional (IGN), C/ Alfonso XII, 3, Madrid 28014, Spain
[2]Argelander-Institut für Astronomie, Universität Bonn, Auf dem Hügel 71, 53121 Bonn, Germany
[3]Centre for Astrophysics Research, University of Hertfordshire, Hatfield AL10 9AB, UK
[4]Department of Astronomy, The Ohio State University, 140 West 18th Ave, Columbus, OH 43210, USA

**Abstract.** The Taffy system stands out as one of the strongest gas-rich galaxy mergers with little star formation (SF) activity. We study the causes of this SF inefficiency by observing with the IRAM 30m telescope the line emission of several species in 6 positions. We report clear detections of tracers of bulk ($^{12}$CO and $^{13}$CO) and dense (HCN and HCO$^+$) molecular gas in all regions, as well as evidence for shocks (SiO $J$=2–1) over the intergalactic bridge. Our observations not only confirm the SF inefficiency of the bulk gas, but they also show that the SF efficiency of the dense gas phase is abnormally low too (~1 dex below ULIRGs). The dense gas fraction ($\propto$ HCN/CO) only shows small variations across the entire system, with a typical value of 4%. Although this fraction is somewhat low compared to other interacting/merging systems, it is similar to the values measured in the disk of nearby, normal star-forming galaxies. Finally, we use the outstanding properties of the Taffy system to place constraints on some of the turbulence-regulated SF models from the literature.

## 1 Introduction

The Taffy system was formed ~23 Myr ago after the high-speed, head-on collision of the galaxy pair UGC 12914/5 [1]. This strong interaction has created a 20 kpc-long massive and turbulent intergalactic bridge of atomic and molecular gas [1–3]. The mass of molecular gas in the bridge ($\sim 2\times10^9 M_\odot$) is similar to that of the entire Milky Way. However, unlike in other gas-rich interacting/merging systems [4], the star formation (SF) in Taffy is not enhanced but rather inhibited [5–7]. The Taffy system is thus an ideal testbed for understanding SF in violent environments. In particular, it is interesting to understand which mechanisms are hindering the SF in Taffy despite having perfect conditions to trigger strong starburst events.

## 2 Observations

We carried out line observations at 3 mm using the IRAM-30m radiotelescope. Specifically, we selected 6 representative pointings that cover the main environments (see Figure 1 – left),

*e-mail: axel.garcia@oan.es
**e-mail: a.usero@oan.es

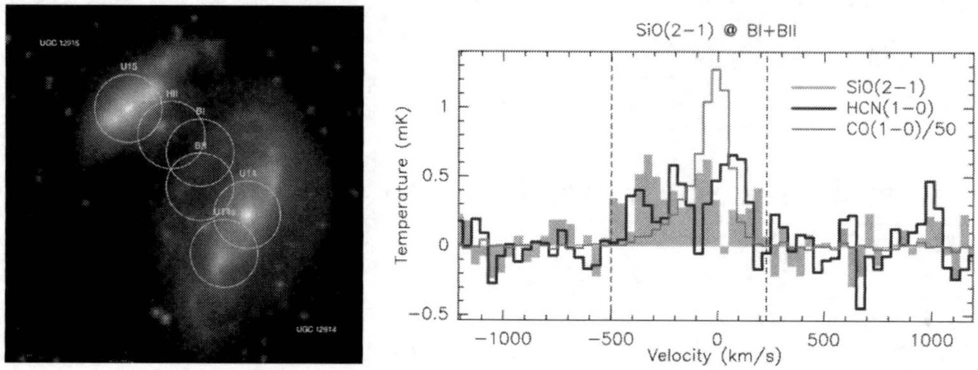


**Figure 1.** *Left:* IRAM-30m pointings (green circles) overlaid on an archival RGB image of Taffy from IRAC (R: 5.7 $\mu$m; G: 4.5 $\mu$m; B: 3.6 $\mu$m). The size of the circles is 28″ (~ 8.3 kpc), which corresponds to the beam size of the IRAM-30m at ~90 GHz. *Right:* SiO (2–1) combined spectrum from regions BI and BII. Re-scaled versions of CO(1–0) and HCN(1–0) spectra are superimposed in red and blue.

including both galaxy centers (U14 and U15 in Figure 1). The extragalactic HII region (HII), which is the only spot within the bridge with intense SF, was also observed to compare it with the other two bridge position (BI and BII; Figure 1). We successfully detected several tracers of bulk molecular gas (e.g., $^{12}CO(1–0)$, $^{13}CO(1–0)$), dense gas (e.g., HCN(1–0), $HCO^+$(1–0)), and even shocks (SiO(2–1); see Figure 1–right). The data reduction process of all these molecular lines has followed a standard procedure similar to that adopted by [8].

## 3 Causes of the SF inefficiency

According to previous works, the star formation efficiency (SFE $\propto$ SFR/CO) of the Taffy system is considerably lower than that of other highly interacting systems like ULIRGS [5–7]. We confirm this behaviour using CO(1–0) as mass tracer, and GALEX far-UV + MIPS 24 $\mu$m as a hybrid star formation rate (SFR) indicator. We see that the SFE of all our bridge observations fall roughly a factor 4-8 below the disk pointings (cyan circles and squares in Figure 3). These disk environments, in turn, are in good agreement with the efficiencies found in nearby Star-Forming Galaxies (SFG) [8], and ~1 dex below the SFE of ULIRGs [9]. In general, the SFE can be expressed as a product of two different factors: the dense gas fraction ($f_{dense}$ $\propto$ HCN/CO) and the star formation efficiency of the dense gas ($SFE_{dense}$ $\propto$ SFR/HCN). A decrease in any of them could explain why the SFE of Taffy is depressed. Figure 2 left and right represent, respectively, our estimations for $f_{dense}$ and $SFE_{dense}$ (derived as in [8]) against the stellar surface density ($\Sigma_{star}$). For comparison purposes, we have also added as black points the sample of SFG from [8], and, as orange bands, we show the fitted trends for the EMPIRE galaxies [10].

Dense gas fractions (coloured points in Fig. 2–left) appear to be quite homogeneous across Taffy, with a typical value of ~4%. This fraction is compatible with the low-end value of the SFG sample, making all Taffy environments consistent with the overall trend followed by SFG. In comparison, $f_{dense}$ is systematically higher in ULIRGs, reaching typical values of 20% [9]. To certain degree, these results suggest that at least part of the SF inefficiency found in Taffy can be interpreted as a lack of dense gas within the system.

In contrast, the right-hand panel of Fig. 2 shows very different results. The $SFE_{dense}$ of the Taffy system presents nearly the opposite behaviour than the trends followed by both SFG samples. In fact, although the disk environments in Taffy remain comparable to other SFGs,

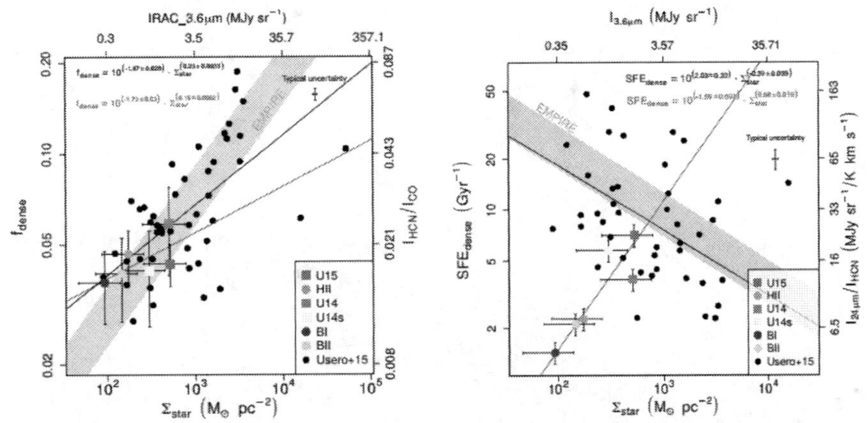


**Figure 2.** Dense gas fraction (left) and star formation efficiency of the dense gas (right) as a function of the stellar surface density. Coloured points represent our Taffy observations, whereas the black ones stand for different SFG [8] or (U)LIRGs [9]. Orange bands show the fitted trend and scatter of the EMPIRE galaxies [10] in the same parameter space. Within the Taffy sample, squares illustrate the galactic pointing, while circles show the extragalactic ones. Finally, the black and red lines correspond, respectively, to the fitted deming regression model to the SFG sample and to the Taffy sample.

all bridge pointings show extremely low $SFE_{dense}$ for their corresponding $\Sigma_{star}$. Indeed, these regions have ~1 dex lower $SFE_{dense}$ than expected from the EMPIRE sample (orange band). Altogether, we conclude there exists a global lack of dense gas in Taffy compared to other highly interacting systems. However, the relative contrast in the SF activity of both galaxy disks and that of the intergalactic bridge can only be explained by an unusually low $SFE_{dense}$.

We propose that the enhanced level of turbulence in the bridge [3, 11, 12] lowers the $SFE_{dense}$ in this region. In this regard, we report the first detection of SiO(2–1) in this system. The SiO line, one of the most reliable shocks tracers, likely signposts strong cloud-cloud collisions. In Figure 1–right, we show the average SiO(2–1) spectrum in BI and BII, which are the pointings with the lowest SFR. We postulate that this SiO detection is an indirect evidence of an enhanced turbulence likely driven by the undergoing galaxy interaction.

## 4 Multi-free-fall SF models cannot explain observations

We can compare our observations with the predictions from the family of turbulence-regulated SF models [13]. Our Taffy sample complements other single-dish surveys of SFGs [8] or (U)LIRGs [9]. In general, these SF models differ in their assumptions about the density for SF onset, the density PDF, and the multi free-fall factor (multi-ff). The latter basically determines the rate of collapse of each density layer. In the end, the main free parameters are: the mean of the density PDF ($\bar{n}$), the virial parameter ($\alpha_{vir}$), and the Mach number ($\mathcal{M}$). The coloured regions of Fig. 3 show the SFE-$f_{dense}$ pairs predicted by the models over a generous range of conditions. Most models show very narrow ranges in SFE compared to the scatter in the observations. In fact, only the KM05 model [13], which is the only one with a fixed multi-ff=1, is able to encompass the broad empirical range in SFE.

## 5 Summary and conclusions

We have studied the dense molecular gas in the highly interacting Taffy system, finding that its low SFE is mainly attributed to both a lack of dense gas and a low SF efficiency of this

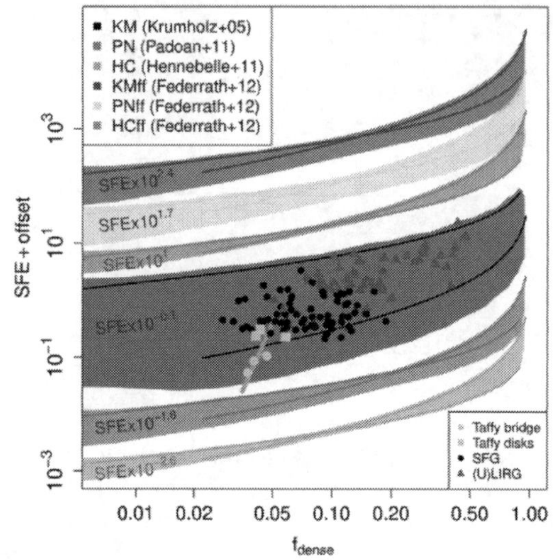


**Figure 3.** Comparison of different turbulence-regulated SF models with IRAM-30m observations (Taffy, SFG [8], (U)LIRGs [9]). Each coloured region represents the model predictions assuming $\bar{n} \in \left[10^0, 10^5\right]$ cm$^{-3}$, $\alpha_{vir} \in [0.8, 15]$, and $\mathcal{M} \in [10, 150]$. Each region is arbitrarily offset for the sake of visualisation. The colored curves correspond to minimum and maximum Mach numbers (10 and 150) for a fiducial virial parameter $\alpha_{vir} = 1.3$. The orange line is a linear fit to the Taffy data.

dense gas. Nevertheless, the relative difference in the SF activity between the galaxy disks and the intergalactic bridge can only be explained by a significantly lower $\mathrm{SFE_{dense}}$ in the latter. Our detection of SiO supports the hypothesis that an enhanced turbulence is hindering the SF in the bridge. Finally, we combine our Taffy observations with literature data to place constraints on the family of turbulence-regulated SF models. Models with non-unity multi-ff [13] have problems to account for the observed scatter in the SFE – $f_{dense}$ parameter space.

## References

[1] Gao, Y., Zhu, M. & Seaquist, E. R., AJ **126**, 2171G (2003)
[2] Braine, J., Davoust, E., Zhu, M. et al., A&A **516**, A40 (2003)
[3] Vollmer, B., Braine, J. & Soida, M., A&A **547**, A39 (2012)
[4] Genzel, R.; Tacconi, L. J.; Gracia-Carpio, J. et al., MNRAS **407**, 2091G (2010)
[5] Peterson, B. W., Appleton, P. N., Bitsakis, T., et al., ApJ **855**, 141P (2018)
[6] Joshi, B. A., Appleton, P. N., Blanc, G. A., et al., ApJ **878**, 161J (2019)
[7] Appleton, P. N., Lanz, L., Bitsakis, T. et al., ApJ **812**, 118A (2015)
[8] Usero, A., Leroy, A. K., Walter, F., et al., AJ **150**, 115 (2015)
[9] García-Burillo, S., Usero, A., Alonso-Herrero, A. et al., A&A **539**, A8 (2012)
[10] Jiménez-Donaire, M. J., Bigiel, F., Leroy, A. K. et al., ApJ **880**, 127J (2019)
[11] Vollmer, B., Braine, J., Mazzilli-Ciraulo, B. et al., A&A **647**, A138 (2021)
[12] Appleton, P. N. ; Emonts, B. ; Lisenfeld, U. et al., ApJ **931**, 121A (2022)
[13] Federrath, C., and Klessen, R. S., ApJ **606**, 156 (2012)

# The LEGO Large Program: Constraining the Physics of Line Emission in Galaxy Observations

*Jens* Kauffmann[1,*] and *Ashley* Barnes[2]

[1]Haystack Observatory, Massachusetts Institute of Technology, 99 Millstone Road, Westford, MA 01886, USA
[2]Argelander-Institut für Astronomie, Universität Bonn, Auf dem Hügel 71, 53121 Bonn, Germany

**Abstract.** The IRAM Large Program LEGO studies molecular clouds in the Milky Way to constrain the physics controlling molecular line emission in galaxies. This is done by imaging two dozen clouds with setups that provide continuous spectral coverage of about 85–115 GHz. To give one example, research in this area permits to "calibrate" extragalactic observations of molecules like HCN and $N_2H^+$ to explore how the star formation activity in galaxies depends on their dense gas contents (i.e., Gao & Solomon relation, [1]). Interestingly, LEGO and other studies now reveal a substantial cloud-to-cloud variation in line ratios. Once understood properly, this diversity can be used to constrain the properties of extragalactic molecular clouds at great detail. Here we outline the LEGO sample and describe the project status.

## 1 Introduction

Investigations into the formation of stars over cosmic time continue to constitute one of the central research areas of astronomy. One important area in this field of study is how the star formation (SF) rate per unit gas mass depends on the properties of the molecular clouds. Gao & Solomon [1], e.g., argued in an influential paper that the SF rate of a galaxy is directly proportional to the mass of dense gas that could support star formation. That said, current modeling indicates that the outcome of the SF process sensitively depends on the physical structure of the star-forming clouds (e.g., [2]), and it is known that molecular cloud structure varies substantially inside and between galaxies (e.g., [3]). Detailed research into the connection of molecular cloud properties and SF activity thus continues to be important.

Unfortunately it is typically impossible to resolve the substructure of extragalactic clouds. To give one example, ALMA is limited to resolutions $\vartheta_{\rm beam} \gtrsim 1''$ at useful sensitivities (e.g., [4]), corresponding to spatial scales $\ell_{\rm beam} = 5~{\rm pc} \cdot (\vartheta_{\rm beam}/1'') \cdot (d/1~{\rm Mpc})$ in nearby galaxies that just marginally resolve individual molecular clouds of $\sim$ 10 pc. Research into extragalactic SF therefore routinely relies on indirect means to characterize cloud substructure. Molecular emission lines at $\sim$ 100 GHz frequency are particularly useful for such assessments. To give examples, emission from the $^{12}$CO molecule is generally used to sense the total molecular gas reservoir, while it has been reasoned that the luminosity in the HCN (1–0) line is a probe of the enclosed mass of dense gas (i.e., at $\gg 10^4~{\rm cm^{-3}}$; e.g., [1]). Molecules like $CH_3OH$, which prefer to stick in non-radiating states on dust grains unless disturbed by energetic events, often indicate the presence of shocks (e.g., [5]). Some of the emission lines

*e-mail: jens.kauffmann@mit.edu

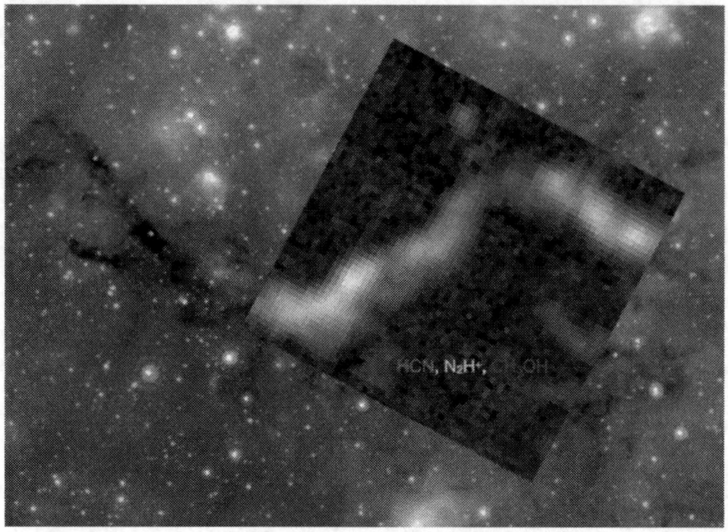


**Figure 1.** Example map of G11.11−0.12, a.k.a. the "Snake". LEGO data are shown as an inset on top of a Spitzer image obtained at 3.6–24 $\mu$m wavelength. A remarkable aspect is the relatively faint emission of HCN in this cloud of high gas density. Another noteworthy aspect is the substantial emission from $CH_3OH$ in a cloud largely devoid of star formation feedback.

are sensitive to the impact of star formation feedback. For example, it has been observed that the HCN-to-HNC line ratio is sensitive to the temperature of the gas heated by stars (e.g., [6]), while UV-radiation from hot young stars impacts the HCN-to-$HCO^+$ line ratio (e.g., [7]).

Interestingly, while molecular line ratios are frequently employed to study extragalactic star formation, very little observational work has been done in the Milky Way to "calibrate" emission lines and their ratios as probes of molecular cloud structure. This work is important: recent observational work by LEGO and other efforts shows for example that the HCN (1–0) transition is *not* a tracer of high-density gas at $\gg 10^4$ cm$^{-3}$ as reasoned before (e.g., [1]), but is a probe of gas at rather modest densities $\sim 10^3$ cm$^{-3}$ (e.g., [8]).

The LEGO ("Line Emission in Galaxy Observations") project was therefore specifically designed to advance the interpretation of extragalactic molecular line emission via the observational "calibration" of emission line properties via detailed studies of well-characterized molecular clouds in the Milky Way. This is enabled by imaging of large fields of $30' \times 30'$ and more size with the IRAM 30m-telescope in spectral setups that cover a large number of molecular emission lines. To give examples, LEGO can determine how a specific molecular emission line responds to variations in gas temperature, density, or star formation activity. This is possible because all of these physical properties can be gleaned from rich reference data, such as dust emission maps from Herschel or infrared surveys for young stars.

This article focuses on providing an overview of LEGO's targets, the observational setup, and the processing strategy (Sec. 2). It also outlines selected initial results, describes the project's status and schedule, as well the expected long-term impact of the project on the field (Sec. 3). A summary is presented in Sec. 4.

## 2 Project Overview

### 2.1 Sample Selection

LEGO is designed to obtain data for a comprehensive sample of targets. The resulting list is outlined in Table 1. Some targets are covered with large maps of $30' \times 30'$ size or larger. Maps of this scale are produced from data obtained at very high telescope scan speeds, resulting in

**Table 1.** Summary of molecular clouds targeted by the LEGO survey.

| Targets | Map Sizes |
|---|---|
| *nearby clouds: study of detailed physics* | |
| Orion ONC, Orion L1641, IC5146, Rosette | $\geq 30' \times 30'$ |
| *luminous complexes: impact of star formation* | |
| W49A, W43, G45.1+0.1 | $30' \times 30'$ |
| *galactic disk: variation between inner and outer Milky Way* | |
| G6.67−0.20, G8.15+0.27, G11.11−0.12, G25.34−0.20, G28.34+0.06, G28.67+0.08, G35.18−0.77, G45.48+0.06, G51.38−0.02, G125.6+2.1, G144.8+0.4, G149.0+0.4, G160.2+0.8, G168.6+1.0, G170.6−0.3 | $10' \times 10'$ |
| *low metallicity: impact of low metallicity* | |
| Sh 2–208, Sh 2–266, Sh 2–284 | $30' \times 30'$ |

an effective angular resolution of $\vartheta_{\rm beam} \sim 60''$. Other targets are covered with smaller maps of $10' \times 10'$ size that are processed to $\vartheta_{\rm beam} = 35''$ resolution.

The project covers selected nearby clouds out to $d \sim 1.5$ kpc with maps of $30' \times 30'$ size or larger. These data have sub-parsec physical resolutions of $\ell_{\rm beam} = 0.29~{\rm pc} \cdot (d/{\rm kpc}) \cdot (\vartheta_{\rm beam}/60'')$. Such data permits to moderately resolve the physical substructure of molecular clouds, so that the dependence of molecular line emission on the density and temperature structure of a molecular cloud (e.g., warm and diffuse at the surface, and cold and dense in the center) can be investigated. The sample further includes some of the most luminous star-forming regions in the Milky Way (taken from [9]). Data on these regions can be combined with observations of less luminous clouds in order to probe the impact of star formation and stellar luminosity on molecular line emission. The high-luminosity complexes are of large angular size, and so they are covered with large maps of $30' \times 30'$ size. A first LEGO study of a high-luminosity complex is presented in [10]. Similarly, several clouds believed to have metallicities as low as 1/10 of the solar value are observed in the outer Milky Way (see [11–13]), in order to probe the impact of metallicity on molecular line emission. Finally, 15 clouds at galactic longitudes of $\ell = 6°–170°$ are sampled with maps of $10' \times 10'$ to probe the dependence of line emission on the galactic environment. The selection of this latter sample heavily builds on [14].

The key aspect of the LEGO sample is the substantial angular size of the maps. These maps correspond to physical scales $\ell_{\rm map} = 6~{\rm pc} \cdot (d/2~{\rm kpc}) \cdot (\vartheta_{\rm map}/10')$. Given the chosen combinations of target distance $d$ and map size $\vartheta_{\rm map}$, the derived maps probe *large representative sections of molecular clouds, or clouds in their entirety*. LEGO achieves this by making use of substantial time allocations. Including overheads, imaging of one region of $10' \times 10'$ size in two setups at $\vartheta_{\rm beam} = 35''$ resolution for example takes 10.4 h, while such observations of a field of $30' \times 30'$ size with $\vartheta_{\rm beam} = 60''$ take 16.9 h.

## 2.2 Spectral Setup

LEGO primarily targets the 85–115 GHz spectral window. This can be achieved with two tunings of the EMIR receiver system on the IRAM 30m-telescope. LEGO employs a spectral resolution $\Delta\nu = 200$ kHz, corresponding to a velocity resolution of $0.6~{\rm km/s} \cdot (\Delta\nu/200~{\rm kHz}) \cdot (\nu/100~{\rm GHz})^{-1}$. The survey targets a noise temperature of 0.1 K in the $T_{\rm mb}$-scale, but in practice the sensitivity is typically higher.

The current data reduction pipeline extracts position-position-velocity data cubes for 29 "line groups" of up to 100 MHz width that stem from 15 molecular species (i.e., 24 species if including isotopologues). These groups are listed in Table 2. In selected regions we further extract 5 groups from deuterium-bearing molecules at frequencies in the range 72–86 GHz.

The overall philosophy driving the line selection and spectral setup are described in [10]. To give an example, we cover emission lines detected in the nearby M 51 galaxy (e.g., [15]).

It is very likely that the lines extracted by regular pipeline reduction only constitute a small fraction of the transitions that can be meaningfully studied with the LEGO data. We conclude this from an analysis of published line surveys of other targets. Would these objects be covered by LEGO maps of typical sensitivity, we would detect dozens of emission lines (e.g., assuming appropriate frequency coverage and distance, $\geq 92$ transitions in a protostar like NGC 2264 CMM3 [16], $\geq 134$ transition groups in an object like Orion KL [17], and $\geq 79$ transition groups in a Class 0 object like L 483 [18]). Such rich detections of line emission towards compact objects like embedded young stars will only occur towards small beam-sized areas.

### 2.3 Data Processing

The LEGO pipeline builds on data that are calibrated by observatory systems. In addition to the observed spectra, the project also retrieves machine-readable logs from the telescope's data systems. A combination of scripts written in the Python and CLASS languages are used to process the data into science-ready data cubes. Python's superior ability to search and manage data is used to orchestrate the data reduction process, while CLASS is used (via the PyGILDAS interface) to perform low-level data processing steps. Science-ready data cubes in FITS format constitute the final product of the pipeline. In a given target, a cube is extracted for each of the transition groups listed in Table 2. A dedicated server at MIT Haystack Observatory is used to process all of the data in a unified environment.

Jupyter notebooks (https://jupyter.org) are used to document the pipeline, the pipeline execution, and the data products. Specifically, the entire pipeline is written in the form of richly-annotated Jupyter notebooks that can be imported as libraries. One notebook per target is used to control and document the pipeline execution. Additional notebooks are used to document the data quality and the overall features of the targets as revealed by the observations. All of the pipeline, as well as the rich automatically generated documentation on data products, can easily be browsed online via GitHub. All of this material will be part of an upcoming public data release. This includes science-ready data cubes for all targets.

### 2.4 Project Team

The project is headed and managed by J. Kauffmann (MIT Haystack), who also served as PI on the IRAM observing proposals. Observations were conducted, and data processing tools developed and applied, by A. Anderson (U. Hawaii), A.T. Barnes (AIfA Bonn), N. Brinkmann (MPIfR Bonn), A. Broadmeadow (U. Maryland), D. Colombo (MPIfR Bonn), A.E. Guzmán (Joint ALMA Observatory), W.J. Kim (U. Köln), A. Patel (MIT Haystack), L. Szűcs (NRAO), and V. Wakelam (U. Bordeaux). The scientific motivation for LEGO was developed, and the exploration of the data supported, by S. Aalto (Chalmers U.), T. Albertsson (MPIfR Bonn), F. Bigiel (AIfA Bonn), N.J. Evans II (U. Texas), S.C.O. Glover (U. Heidelberg), P.F. Goldsmith (JPL), C. Kramer (IRAM), K. Menten (MPIfR Bonn), Y. Nishimura (U. Tokyo, NAOJ), S. Viti (U. Leiden), Y. Watanabe (Shibaura Institute of Technology), A. Weiss (MPIfR Bonn), M. Wienen (MPIfR Bonn, U. Exeter), H. Wiesemeyer (MPIfR Bonn), and F. Wyrowski (MPIfR Bonn).

## 3 Project Timeline and Expected Impact

LEGO builds on a number of pilot studies that started in 2015. Primary data collection began in fall of 2017, and it was completed in the summer of 2020. Pilot studies of the project were for example presented in [8, 19], while [10] constitutes the first direct output from the LEGO survey. The pandemic required a re-focus of the project: given inefficiencies and hiring restrictions, the project had to concentrate on supporting ongoing student projects. This has

**Table 2.** Summary of transition groups observed by the LEGO survey.

| Species | Transition | Frequency [GHz] |
|---|---|---|
| c–$C_3H_2$ | $J_{K_a,K_c} = 2_{1,2}$–$1_{0,1}$ | 85.3389060 |
| $H^{13}CN$ | 1–0 | 86.3401764 |
| $H^{13}CO^+$ | 1–0 | 86.7542880 |
| SiO | 2–1 | 86.8469950 |
| $HN^{13}C$ | 1–0 | 87.0908590 |
| CCH | $N = 1 - 0, J = 3/2$–$1/2$ | 87.3169250 |
| CCH | $N = 1 - 0, J = 1/2$–$1/2$ | 87.4020040 |
| HNCO | $J_{K_a,K_c} = 4_{0,4}$–$3_{0,3}$ | 87.9252380 |
| HCN | 1–0 | 88.6318473 |
| $HCO^+$ | 1–0 | 89.1885260 |
| HNC | 1–0 | 90.6635640 |
| H | 41$\alpha$ (42–41) | 92.0344340 |
| $N_2H^+$ | 1–0 | 93.1737770 |
| $HC_5N$ | 35–34 | 93.1881250 |
| $HC_5N$ | 36–35 | 95.8503350 |
| $C^{34}S$ | 2–1 | 96.4129500 |
| $CH_3OH$-E | $J_K = 2_{-1}$–$1_{-1}$ | 96.7393630 |
| $CH_3OH$-A | $J_K = 2_0$–$1_0$ | 96.7413770 |
| CS | 2–1 | 97.9809530 |
| SO | $J_K = 3_2$–$2_1$ | 99.2999050 |
| $HC_3N$ | 12–11 | 109.1736380 |
| OCS | 9–8 | 109.4630630 |
| $C^{18}O$ | 1–0 | 109.7821760 |
| HNCO | $J_{K_a,K_c} = 5_{0,5}$–$4_{0,4}$ | 109.9057530 |
| $^{13}CO$ | 1–0 | 110.2013540 |
| $C^{17}O$ | 1–0 | 112.3589880 |
| CN | $N = 1 - 0, J = 1/2$–$1/2$ | 113.1913170 |
| CN | $N = 1 - 0, J = 3/2$–$1/2$ | 113.4909820 |
| CO | 1–0 | 115.2712020 |

resulted in exciting new results. Specifically, studies of IC 5146-W (Anderson et al., in prep.) and G11.11–0.12 (Broadmeadow et al., in prep.) expose additional striking examples of faint HCN emission in dense molecular clouds, just as found in LEGO pilot studies [8].

More generally, the current study of LEGO targets reveals a surprising diversity in line emission properties. Line luminosity ratios $L_i/L_j$ (i.e., for signals integrated over entire clouds) show a strong cloud-to-cloud variation, independent of the specific line pair $i, j$ chosen. At this moment it appears that no line pair can be characterized by a single representative line ratio. Extragalactic line observations typically probe regions of $\gtrsim 100$ pc size that contain several clouds, and so *the line emission properties observed in galaxies constitute complex averages over diverse cloud populations.* To give an example, line ratios are observed to vary within galaxies (e.g., [20]), and the LEGO data imply that *line ratio variations within galaxies might indicate that cloud populations change between galactic environments.*

LEGO is thus likely to provide new and important new information that will ultimately allow us to disentangle the physical substructure of galaxies. By connecting variations in line ratios to physical cloud properties like the gas temperature and density, *LEGO will provide the tools needed to map out the variation of gas properties in galaxies.*

LEGO is now in a push to deliver science-ready cubes and associated documentation for all transitions and targets covered by LEGO. Delivery is planned for the winter of 2022/23.

## 4 Summary

LEGO generates a very detailed picture of how line intensities and ratios vary within and between molecular clouds, and how these variations depend on physical cloud properties like gas temperature and density. The observations reveal strong cloud-to-cloud variations in line ratios. This means that line ratios can be used to uncover important new physical constraints on the physical structure of extragalactic molecular clouds. LEGO is enabled by NSF-AAG grant 1909097.

## References

[1] Y. Gao, P.M. Solomon, The Astrophysical Journal **606**, 271 (2004)

[2] M.Y. Grudić, P.F. Hopkins, C.A. Faucher-Giguère, E. Quataert, N. Murray, D. Kereš, Monthly Notices of the Royal Astronomical Society **475**, 3511 (2018)

[3] D. Colombo, A. Hughes, E. Schinnerer, S.E. Meidt, A.K. Leroy, J. Pety, C.L. Dobbs, S. García-Burillo, G. Dumas, T.A. Thompson et al., The Astrophysical Journal **784**, 3 (2014)

[4] J. Sun, A.K. Leroy, A. Schruba, E. Rosolowsky, A. Hughes, J.M.D. Kruijssen, S. Meidt, E. Schinnerer, G.A. Blanc, F. Bigiel et al., The Astrophysical Journal **860**, 172 (2018)

[5] R. Bachiller, S. Liechti, C. Walmsley, F. Colomer, Astronomy & Astrophysics **295**, L51 (1995)

[6] P. Schilke, C. Walmsley, G. Pineau Des Forets, E. Roueff, D. Flower, S. Guilloteau, Astronomy & Astrophysics **256**, 595 (1992)

[7] R. Meijerink, M. Spaans, F.P. Israel, Astronomy and Astrophysics **461**, 793 (2007)

[8] J. Kauffmann, P.F. Goldsmith, G. Melnick, V. Tolls, A. Guzman, K.M. Menten, Astronomy & Astrophysics **605**, L4 (2017)

[9] J.S. Urquhart, C.C. Figura, T.J.T. Moore, M.G. Hoare, S.L. Lumsden, J.C. Mottram, M.A. Thompson, R.D. Oudmaijer, Monthly Notices of the Royal Astronomical Society **437**, 1791 (2014)

[10] A.T. Barnes, J. Kauffmann, F. Bigiel, N. Brinkmann, D. Colombo, A.E. Guzmán, W.J. Kim, L. Szűcs, V. Wakelam, S. Aalto et al., Monthly Notices of the Royal Astronomical Society **497**, 1972 (2020)

[11] C. Yasui, N. Kobayashi, M. Saito, N. Izumi, The Astronomical Journal **151**, 115 (2016)

[12] A. Mehner, W.J. de Wit, J.H. Groh, R.D. Oudmaijer, D. Baade, T. Rivinius, F. Selman, H.M.J. Boffin, C. Martayan, Astronomy & Astrophysics **585**, A81 (2015)

[13] E. Puga, S. Hony, C. Neiner, A. Lenorzer, A.M. Hubert, L.B.F.M. Waters, F. Cusano, V. Ripepi, Astronomy and Astrophysics **503**, 107 (2009)

[14] M. Wienen, F. Wyrowski, K.M. Menten, J.S. Urquhart, T. Csengeri, C.M. Walmsley, S. Bontemps, D. Russeil, L. Bronfman, B.S. Koribalski et al., Astronomy & Astrophysics **579**, A91 (2015)

[15] Y. Watanabe, N. Sakai, K. Sorai, S. Yamamoto, The Astrophysical Journal **788**, 4 (2014)

[16] Y. Watanabe, N. Sakai, A. López-Sepulcre, R. Furuya, T. Sakai, T. Hirota, S.Y. Liu, Y.N. Su, S. Yamamoto, Astrophysical Journal **809** (2015)

[17] B.E. Turner, The Astrophysical Journal Supplement Series **70**, 539 (1989)

[18] M. Agúndez, N. Marcelino, J. Cernicharo, E. Roueff, M. Tafalla, Astronomy and Astrophysics **625**, A147 (2019)

[19] P.F. Goldsmith, J. Kauffmann, The Astrophysical Journal **841**, 25 (2017)

[20] F. Bigiel, A.K. Leroy, M.J. Jiménez-Donaire, J. Pety, A. Usero, D. Cormier, A. Bolatto, S. Garcia-Burillo, D. Colombo, M. González-García et al., The Astrophysical Journal Letters **822**, L26 (2016)

# The Cygnus Allscale Survey of Chemistry and Dynamical Environments (CASCADE)

## A Max Planck IRAM Observatory Program (MIOP)

*Friedrich* Wyrowski[1,*], *Henrik* Beuther[2], *Karl* Menten[1], and *the CASCADE* Team

[1]MPIfR Bonn
[2]MPIA Heidelberg

**Abstract.** Over the last decades, high-mass star formation research has concentrated on either the large-scale molecular cloud environments or on the small-scale hot dense molecular cores surrounding massive proto-stars/clusters and young stellar objects. However, questions regarding the gas flow from large to small scales, and how the gas is transferred via intermediate-scale filaments and clumps have so far been largely neglected. With "Cygnus Allscale Survey of Chemistry and Dynamical Environments (CASCADE)", a part of the Max Planck IRAM Observatory Program (MIOP), we want to overcome this missing gap via studying one well-known molecular cloud complex with the IRAM NOEMA and 30m facilities from small to large scales, thereby connecting these important physical processes. In the nearby (1.4 kpc) luminous Cygnus X region, recent and on-going star formation combine to present a rich Northern hemisphere laboratory in which star formation and feedback processes can be studied globally as well as locally. Using the new and unique 4 mm capabilities of NOEMA, together with the IRAM 30m telescope, the flow of gas from cloud-to core-scales is being probed with observations of large mosaics covering the ground state lines of many molecules, including their deuterium substituted isotopologs, with the unique large bandwidth only possible with this facility. Here a brief introduction into the program, together with initials results, is presented.

## 1 The nearby Cygnus X region – a star formation powerhouse

The relatively nearby, massive and very luminous Cygnus X region (see, e.g., the spectacular Spitzer imaging by [1]) is an excellent hunting ground to study various phases of massive star formation. Its associated Giant Molecular Cloud complex is harboring a rich collection of various star formation stages, ranging from several OB associations to young and protostellar objects in various evolutionary stages, many of them dense, dusty, hot cores [2]. The latter and the OB association, themselves having different ages, as well as a superbubble driven by the O-stars' stellar winds and the famous Cygnus Loop supernova remnant give testament to intense and widespread star formation happening over (at least) the past few millions of years. Since it is also relatively near to the Sun (at a distance of 1.4 kpc, [3]), many of the above phenomena can be studied here in detail.

*e-mail: wyrowski@mpifr-bonn.mpg.de

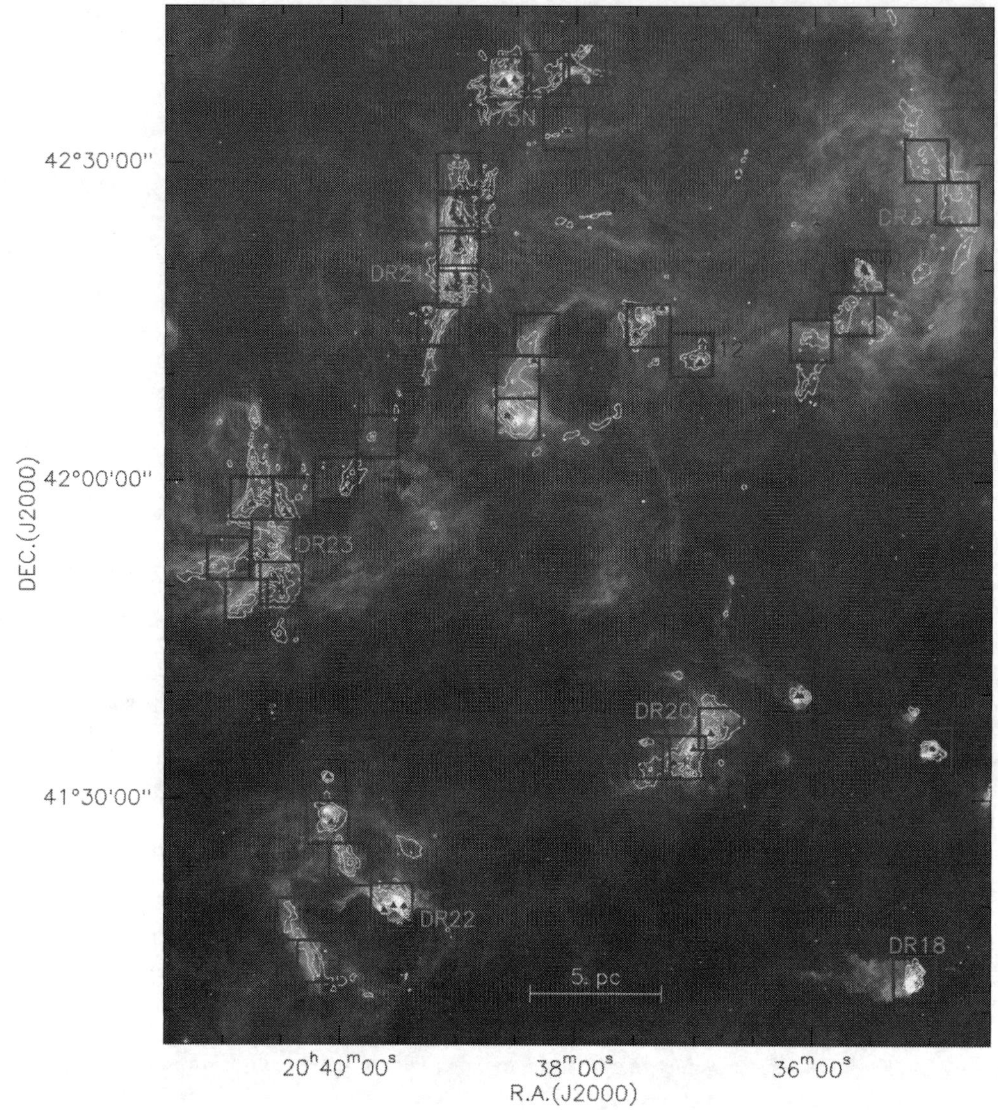


**Figure 1.** Cygnus X as seen in Herschel SPIRE 250 $\mu$m (contours) and Spitzer 8 $\mu$m (heat-scale) emission. The green triangles show the MAMBO clumps revealed by [2]. The blue squares indicate the regions covered by 16 arcmin$^2$ NOEMA mosaics.

The main goal of CASCADE is to connect cloud and core scales with NOEMA C&D array mosaics (3" corresponding to 0.02 pc or 4000 au), combined with complementary short spacings data obtained with the IRAM 30m telecope to probe the larger scales. In total 640 arcmin$^2$ are being observed, covering high column density regions within Cygnus X above $10^{22}$ cm$^{-2}$ (see Fig. 1). The 3 – 4 mm spectral line setup covers crucial ground state lines from, e.g. $N_2H^+$, HCN, HNC, $H_2CO$, including unique lines from deuterated molecules. With a total of 520 hours of NOEMA observing time a continuum sensitivity of 42 $\mu$Jy is reached as well as a line sensitivity of 13.5 mJy (in 0.5km/s channels) corresponding to 0.2 K for the synthesized beam of 3", sufficient to probe even optically thin lines.

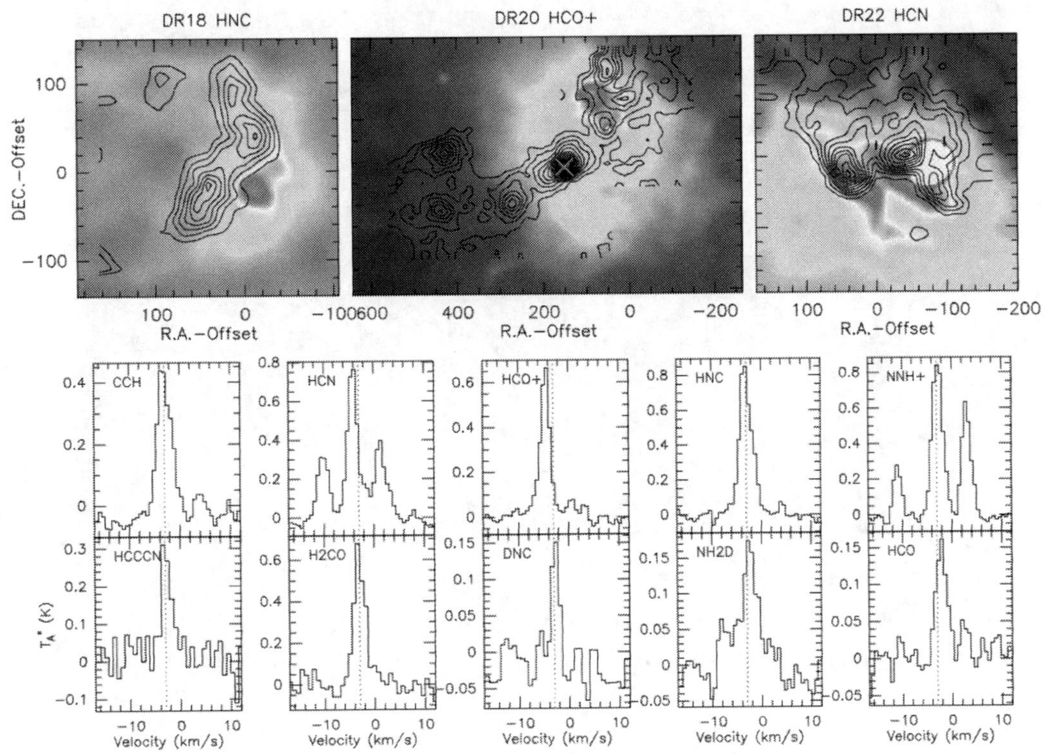


**Figure 2.** First results from the 30 m mapping of NOEMA fields in Cygnus X. The contours show emission in the HNC, $HCO^+$, and HCN $J = 1 - 0$ line, respectively, in some observed sources. The color background shows the 5 cm free-free continuum emission from GLOSTAR, combining VLA and Effelsberg data [4, 5]. The position of a methanol maser discovered in the DR20 field is indicated with a red cross. The individual spectra of selected molecules are from a clump in DR20 close to the (0,0) position. Note the velocity shifts of HCN and $HCO^+$ compared to optically thin lines.

## 2 IRAM 30m results

In 2020, all fields planned for the NOEMA survey were observed with the IRAM 30m telescope. Besides serving as crucial short spacing information to be merged with the high angular resolution NOEMA data, the IRAM 30m results allow chemical and dynamical studies of large scales. In Figure 2, the large scale emission of some of the molecules is shown, together with preliminary evidence for large scale inflow from the blue-skewed profiles of the HCN and $HCO^+$ lines. There is also a strong synergy with the GLOSTAR JVLA/Effelsberg 4 – 8 GHz radio wavelength survey as shown in the Figure 2. A GLOSTAR methanol maser covered in on of the the fields [5] is indicated with a red cross.

## 3 NOEMA results

Three mosaics of the DR20 region were observed in winter 2020/21. All data in the C+D array were taken. For the data processing an imaging pipeline has been setup. Many spectral lines and the continuum emission are well detected in all fields. Figure 3 shows an example of the small scale $HCO^+$ line emission morphology and kinematics, resolving the parcsec scale clumps probed with the IRAM 30m into filaments and cores. In particular in the North-West, close to bright HII regions, the velocity structure of the emission becomes complex. The

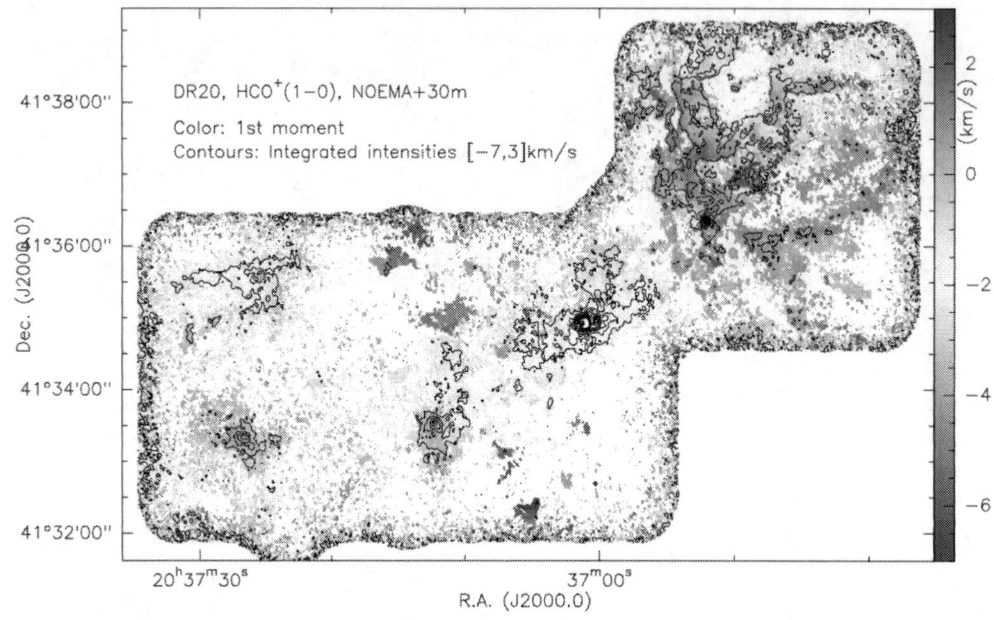


**Figure 3.** Example data for the DR20 region. The entire DR20 region was covered by three mosaics of 78 fields that are combined in the data reduction process. We show the combined NOEMA+30m data for the $HCO^+$ (1–0) line where the colorscale presents the 1st moment map (intensity-weighted peak velocities) and the contours show the integrated emission over the velocity range from -7 to 3 km/s.

eastern part of the DR20 region harbors objects in earlier evolutionary stages associated with more pristine cold cloud structures.

## 4 Outlook

The survey with NOEMA is progressing very well with a potential completion of the survey in 2023. This unique new dataset, in particular through the combination of NOEMA and IRAM 30m data, will allow us to investigate the physical and chemical processes during star formation from large cloud scales down to our spatial resolution limit of $\approx$ 4000 au. Furthermore, covering a broad plethora of spectral lines and continuum emission, we can study a diverse set of phenomena covering accretion flows from large to small scales, cloud collapse, filamentary structures, as well as the physical and chemical properties of the gas associated with this prototypical large star formation complex. A first overview paper about the DR20 region has been submitted [6] and detailed studies of the large scale deuteration and kinematics of the DR21 Main outflow are ongoing.

## References

[1] Kumar et al., MNRAS **374**, 54 (2007)
[2] Motte et al., A&A **476**, 1243 (2007)
[3] Rygl et al., A&A **539**, 79 (2012)
[4] Brunthaler et al., A&A **651**, 85 (2021)
[5] Ortiz-Leon et al., A&A **651**, 870 (2021)
[6] Beuther et al., A&A accepted (2022)

# A complete 3mm line survey of the B1-b and TMC-1 cores

## New discoveries with the IRAM 30m and Yebes 40m telescopes

*Nuria* Marcelino[1,2,*], *José* Cernicharo[3], *Marcelino* Agúndez[3], *Belén* Tercero[1,2], *Carlos* Cabezas[3], and *Pablo* de Vicente[2]

[1]Observatorio Astronómico Nacional, IGN, C/Alfonso XII 3, 28014 Madrid, Spain
[2]Observatorio de Yebes, IGN, Cerro de la Palera s/n, 19141 Yebes, Guadalajara, Spain
[3]Instituto de Física Fundamental, CSIC, C/Serrano 123, 28006 Madrid, Spain

**Abstract.** We present the 3mm spectral line survey performed at the IRAM 30m telescope towards the dense cores B1-b and TMC-1. Within the 46 GHz observed, we have identified more than 500 lines arising from more than 60 molecules. We have also detected tens of unidentified lines, allowing the discovery of new molecular species in space. In this contribution we discuss two examples: the case of $H_2NC$ and $CH_3CO^+$. In the latter, the 30m data was used in combination with the 7mm survey data from the Yebes 40m telescope, which provides lower energy transitions. Our deep 3mm and 7mm spectral surveys reveal a forest of lines at 50-100 mK, showing that dark clouds cannot be considered poor line sources anymore.

## 1 Introduction

Cold ($T_{kin}$ ~10 K) and dense ($n$ ~$10^{4-5}$ cm$^{-3}$) cores are the sites where low-mass protostars and planetary systems like ours will eventually form. These regions are relatively simple since they lack of the physical complexity and energetic events associated with the star formation process. They also present a rather simple chemical scheme, dominated by gas-phase chemical reactions, mainly exothermic ion-molecule and neutral-neutral reactions, due to their low temperatures. Nevertheless, these quiet and cold regions are chemically rich and complex. Spectral line surveys are the best tool to obtain the complete view of the molecular inventory of dense cores. They require sensitive and large bandwidth receivers, but also high spectral resolution is needed in order to resolve the narrow lines arising from these cold sources. In 2011 a major upgrade at the IRAM 30m telescope was completed, with the installation of the EMIR receivers and FFTS spectrometers with a narrow spectral mode, allowing simultaneous observations of 7.2 GHz per frequency setup at high resolution. The extension of the 3mm band limits down to 72 GHz, performed in 2018, also improved the performance of the 30m, by accessing to the fundamental transitions of several deuterated species. With these improvements, the IRAM 30m telescope has become a powerful instrument to perform systematic spectral line surveys.

*e-mail: n.marcelino@oan.es

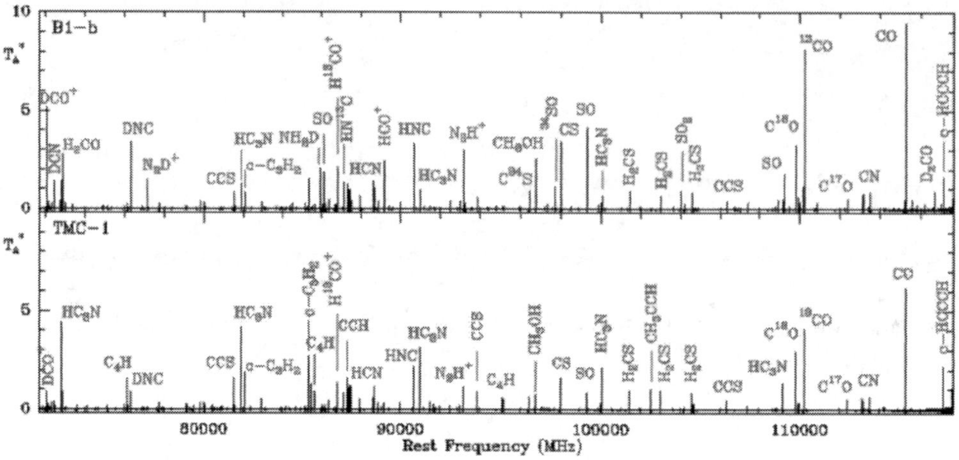


**Figure 1.** Complete 3mm spectral line survey observed towards B1-b and TMC-1.

## 2 The 3mm line survey

Figure 1 shows the full 3mm band observed towards B1-b and TMC-1, with labels indicating the strongest identified lines. In total we have covered 46 GHz, between 71.6 and 117.6 GHz. The final rms obtained in our data ranges between 3-5 mK below 110 GHz, increasing to more than 10 mK at the highest frequencies. Thanks to the high sensitivity reached in our spectra, we are able to confidently detect weak lines of ~20-30 mK (see Fig. 2). Table 1 shows the list of identified molecules in both sources. Different colours indicate those species detected only in B1-b (red) or TMC-1 (blue), and recent new detections are also highlighted. Except for $H_2NC$, all the new species were detected in combination with data from the 7mm spectral survey performed at the Yebes 40m telescope. In total, we have identified 579 lines in B1-b arising from 62 molecular species (127 if we include isotopic substitutions), and 512 lines from 66 molecules (125 including isotopologues) towards TMC-1. As seen in Fig. 1 and Table 1, deuterated substitutions and oxygen-bearing species, including complex molecules typical of hot cores and hot corinos, are mainly detected towards B1-b. On the other hand, TMC-1 is more prolific in carbon chains. These differences could be related to their different physical properties and evolution. While TMC-1 is a starless core, B1-b contains two very young protostars, one of them showing strong emission from COMs as seen by ALMA ([12]). Nevertheless, these objects are not resolved by the 30m beam, and we are essentially observing the cold protostellar envelope. It is also worth noting that although $CH_3OCOH$ and $CH_3OCH_3$ are not detected in our 3mm line survey towards TMC-1, they have been observed through lower energy transitions with the Yebes 40m telescope (see [1]).

## 3 First detection of $H_2NC$

After the extension of our 3mm survey to low frequencies, we detected a series of unidentified lines around 72.2 GHz, which could not be assigned to any molecule. A similar pattern was also observed towards the envelope of the Class 0 protostar L483, which shows stronger and narrower lines than B1-b, and also a larger number of components (see Fig. 2). The observed spectra is typical of a molecule with a complex fine and hyperfine structure. The large splitting betweeen line components suggest that they correspond to a low rotational transition,

**Table 1.** Identified molecules within the 3mm line survey.

| 2 atoms | 3 atoms | | 4 atoms | | 5 atoms | | 6 atoms | 7 atoms | 8 atoms |
|---|---|---|---|---|---|---|---|---|---|
| CN | $C_2H$ | $DCO^+$ | $NH_2D$ | $HOCO^+$ | $CH_3O$ | $C_4H$ | $CH_3OH$ | $CH_3CCH$ | **$CH_2CHCCH$**[j] |
| CO | $C_2D$ | HNO | $NHD_2$ | $DOCO^+$ | l-$C_3H_2$ | $C_4D$ | $CH_2DOH$ | $CH_2DCCH$ | $CH_3OCOH$ |
| CS | HCN | $C_2O$ | $H_2CN$ | $H_2CS$ | l-$C_3HD$ | $HC_3N$ | $CH_3OD$ | $CH_3CCD$ | $CH_2CCHCN$ |
| SiO | DCN | HCS | $H_2NC$[a] | HDCS | c-$C_3H_2$ | $DC_3N$ | $CHD_2OH$ | $CHD_2CCH$ | |
| NS | HNC | HSC | $HCNH^+$ | $D_2CS$ | c-$C_3HD$ | HCCNC | **$CH_2CCH$**[g] | $CH_3CHO$ | |
| $NS^+$ | DNC | $HCS^+$ | $H_2CO$ | $C_3N$ | c-$C_3D_2$ | $HNC_3$ | $CH_3CN$ | $CH_2CHCN$ | |
| SO | $N_2H^+$ | $C_2S$ | $D_2CO$ | **$C_3N^-$**[b] | $H_2CCN$ | $NCCNH^+$ | $CH_2DCN$ | $HC_5N$ | |
| $SO^+$ | $N_2D^+$ | OCS | l-$C_3H$ | $C_3O$ | $H_2CCO$ | **$HC_3O^+$**[e] | **$CH_3CO^+$**[h] | | |
| | HCO | $SO_2$ | c-$C_3H$ | CNCN | HDCCO | **$H_2CCS$**[c] | $CH_3SH$ | | |
| | $HCO^+$ | | c-$C_3D$ | **HCCS**[c] | HCOOH | **HCSCN**[f] | l-$H_2C_4$ | | |
| | | | HCCO | **$HCCS^+$**[d] | | | $HC_3NH^+$ | | 9 atoms |
| | | | HNCO | HNCS | | | **$HCCNCH^+$**[i] | | $CH_2CHCH_3$ |
| | | | HCNO | HSCN | | | HCCCHO | | $CH_3OCH_3$ |
| | | | HOCN | $C_3S$ | | | c-$H_2C_3O$ | | $CH_3C_4H$ |
| | | | DNCO | | | | | | |

**Notes.** Species only detected in B1-b are indicated in red, while those observed only in TMC-1 are indicated in blue. Bold text highlights recent discoveries. **References.** (a): [4] (b): [6] (c): [8] (d): [5] (e): [7] (f): [9] (g): [2] (h): [10] (i): [3] (j): [11]

and our first hypothesis was to consider this as the $N = 1-0$ rotational transition of a linear or quasi-linear species. We requested IRAM 30m time to search for the two immediately higher transitions, in order to confirm this hypothesis. These new observations allowed the detection of a series of lines at the expected frequencies of the $N = 2-1$ and $N = 3-2$ transitions at 144 and 216 GHz, respectively (see [4]). With multiple transitions at hand we investigated the potential carrier, and performed high-level ab initio calculations for all possible candidates. Among them, $H_2NC$, a high energy isomer of $H_2CN$, had the closest value to the observed rotational constant. Furthermore, it is an open-shell species containing three nuclei with non-zero nuclear spin, whose predicted fine and hyperfine splitting is consistent to the observed one. Assuming $H_2NC$ as the carrier of the observed unidentified transitions, we analysed the observed frequencies with the appropriate Hamiltonian in order to obtain the complete set of spectroscopic parameters. The results show an excellent agreement between the theoretical and astronomical values (see [4]).

The observed abundance ratio $H_2NC/H_2CN$ is 1.0±0.3 and 1.25±0.30 for B1-b and L483, respectively ([4]). These two isomers have also been observed in the $z$=0.89 galaxy in front of the quasar PKS 1830-211, where the derived $H_2NC/H_2CN$ ratio is 0.27±0.08. Since the chemical composition of this lensing galaxy is characteristic of diffuse or translucent clouds ([14]), this situation reminds that of the HNC/HCN abundance ratio, which is ~1 in cold dense clouds and <1 in diffuse clouds. Our observations in B1-b and L483 show that the strongest hyperfine component of $H_2NC$ is 3-4 times more intense than the strongest one of $H_2CN$. Therefore, $H_2NC$ should be easily observable in those sources where $H_2CN$ is detected. It is interesting to note that neither $H_2NC$ nor $H_2CN$ are detected towards TMC-1. Ohishi et al. ([13]) reported the discovery of $H_2CN$, although the lines were marginally detected and they should result in antenna temperatures above 20 mK in our 30m spectra. Therefore, the detection of $H_2CN$ in TMC-1 by [13] is questionable.

## 4 Discovery of $CH_3CO^+$

The detection of the acetyl cation, $CH_3CO^+$, is one example of the fruitful combination of the IRAM 30m and Yebes 40m telescopes. Within the unidentified lines in our spectral surveys of TMC-1 at the 3mm and 7mm bands, we found two series of four lines that are harmonically related. The observed spectra (see Fig. 2) reminds that of the $K = 0$ and $K = 1$ components of the rotational transitions of a symmetric rotor. Indeed, if we assume that the carrier is the same for both series of lines and that it has a $C_{3v}$ symmetry, all lines can be fitted with

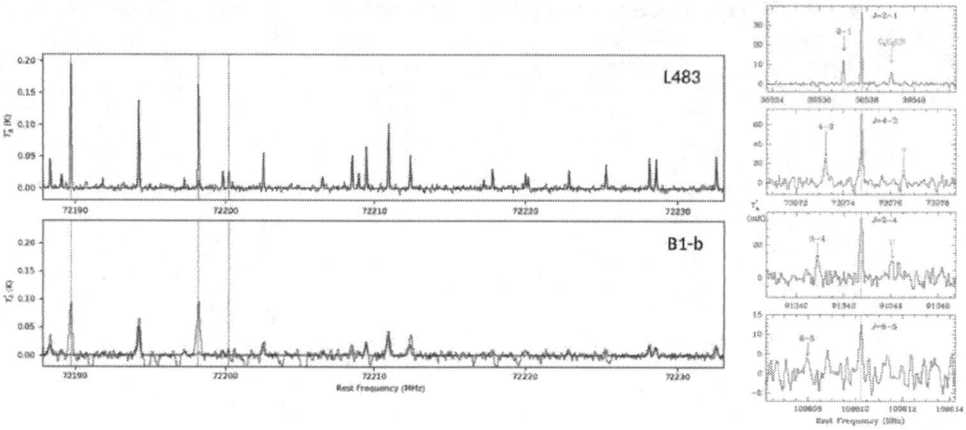


**Figure 2.** *(left):* Series of unidentified lines observed towards L483 (top panel) and B1-b (bottom panel), later assigned to the $1_{0,1} - 0_{0,0}$ transition of $H_2NC$. Lines in blue correspond to hyperfine components of the $N = 1 - 0$ transition of CCD. *(right):* Observed lines of $CH_3CO^+$ towards TMC-1. The blue labels correspond to the series of lines we assign to the $A$ species of $CH_3CO^+$, while the red ones correspond to those of the $E$ species.

a single rotational constant and two distortion constants. The derived rotational constant $B$ from astronomical observations is 9134 MHz, suggesting that the molecule should contain at least three atoms betweeen C, N, and O. Several candidates were proposed based on their similar rotational constants, and we computed their spectroscopic molecular parameters using high-level ab initio calculations (see [10]). The best agreement is found for $CH_3CO^+$, which possible chemical precursor ketene ($H_2CCO$) is one of the most abundant O-bearing species in TMC-1 ([7]). In order to support this possible assignment, we performed microwave laboratory experiments to detect $CH_3CO^+$ ([10]). In total, 79 lines were observed in the laboratory with quantum numbers in the ranges $J$ =10-27 and $K \leq 6$. Thus, the detection in space of the acetyl cation is secure and based on astronomical, theoretical and experimental data.

## References

[1] Agúndez, M., Marcelino, N., Tercero, B., et al., A&A **649**, L4 (2021)
[2] Agúndez, M., Marcelino, N., Cabezas, C., et al., A&A **657**, A96 (2022a)
[3] Agúndez, M., Cabezas, C., Marcelino, N., et al., A&A **659**, L9 (2022b)
[4] Cabezas, C., Agúndez, M., Marcelino, N., et al., A&A **654**, A45 (2021)
[5] Cabezas, C., Agúndez, M., Marcelino, N., et al., A&A **657**, L4 (2022)
[6] Cernicharo, J., Marcelino, N., Pardo, J.R., et al., A&A **641**, L9 (2020a)
[7] Cernicharo, J., Marcelino, N., Agúndez, M., et al., A&A **642**, L17 (2020b)
[8] Cernicharo, J., Cabezas, C., Agúndez, M., et al., A&A **648**, L3 (2021a)
[9] Cernicharo, J., Cabezas, C., Endo, Y., et al., A&A **650**, L14 (2021b)
[10] Cernicharo, J., Cabezas, C., Bailleux, S., et al., A&A **646**, L7 (2021c)
[11] Cernicharo, J., Agúndez, M., Cabezas, C., et al., A&A **647**, L2 (2021d)
[12] Marcelino, N., Gerin, M., Cernicharo, J., et al., A&A **620**, A80 (2018)
[13] Ohishi, M., McGonagle, D., Irvine, W. M., et al., ApJ **427**, L51 (1994)
[14] Tercero, B., Cernicharo, J., Cuadrado, S., et al., A&A **636**, L7 (2020)

# Multi-line characterization of whole molecular clouds using stratified random sampling

*Mario* Tafalla[1,*], *Antonio* Usero[1,**], and *Álvaro* Hacar[2,***]

[1]Observatorio Astronómico Nacional (IGN), Calle Alfonso XII 3, 28014 Madrid, Spain
[2]Department of Astrophysics, University of Vienna, Türkenschanzstrasse 17, 1180 Vienna, Austria

**Abstract.** We have developed a new technique to characterize the multi-line emission from molecular clouds using statistical sampling. Our method uses available extinction maps to select a relatively small sample of cloud positions that cover the full range of column densities in the cloud, and that can be observed with only a modest investment of telescope time. Here we present the first results of applying this technique to the Perseus molecular cloud, a nearby star-forming region that contains a population of both isolated and clustered protostars. We have used the IRAM 30m telescope to observe a sample of 100 random positions that cover two orders of magnitude in $H_2$ column density. These positions have been observed over the full 3mm wavelength band together with selected portions of the 2 and 1mm bands. We find that the emission properties of most species are strongly correlated with the column density over the whole cloud, and that they can be reproduced using a relative simple radiative transfer model.

## 1 Introduction

Characterizing the molecular emission from whole molecular clouds is critical to identifying the physical and chemical processes that act at different spatial scales and lead to the formation of stars. It is also needed to connect spatially-resolved observations of galactic clouds with extragalactic observations that do not resolve the clouds.

The traditional approach to characterize the emission of a cloud is to fully map its area with a telescope. Even when using a time-efficient technique such as on-the-fly mapping, this approach is very time consuming since it requires to fully sample the emission over many square degrees in the sky. Until recently, the mapping of clouds has only been carried out using a very reduced number of tracers (mostly CO), and for selected molecular clouds ([3, 7]). Thanks to the advent of the wide-band receivers, recent years have seen a renewed effort to characterize the emission of molecular clouds using multiple tracers. Examples of this effort are the two IRAM Large Programs ORION-B and LEGO, which have used the IRAM 30m telescope to map large extensions of several target clouds ([1, 6], and related contributions in this volume). Still, the time required to carry out these maps typically exceeds several hundreds of hours, and this strongly limits the ability to extend this approach to multiple clouds.

*e-mail: m.tafalla@oan.es
**e-mail: a.usero@oan.es
***e-mail: alvaro.hacar@univie.ac.at

While mapping a cloud is necessary to fully characterize its detailed emission and complex velocity field, such an effort likely provides more information than is necessary to determine the main cloud emission properties. Molecular clouds are turbulent objects, and much of their appearance is likely highly transient. For this reason, it is tempting to try to characterize the emission of a cloud using a less time-intensive technique, such as random sampling, which could potentially recover the main properties of the emission at a significant fraction of telescope observing time. In this contribution we present our attempt to use this technique to characterize the emission of the nearby Perseus cloud with the IRAM 30m telescope. A full account of this work has been presented in [8].

## 2 Method and observations

Our sampling method uses the $H_2$ column density as a proxy for the molecular emission since principal component analysis has shown that $N(H_2)$ represents the dominant contributor to the emission of most species (e.g., [4]). Sampling the cloud emission therefore requires to first sample its distribution of $H_2$ column densities, and for this, we have used the high quality column density map of Perseus produced by [9] with data from the *Herschel* and *Planck* satellites. This map extends over an area of more than $30 \times 10$ pc$^2$, and covers approximately two orders of magnitude in column density ($10^{21}$-$10^{23}$ cm$^{-2}$). To cover this wide range of column densities, we divided it into 10 logarithmically-spaced bins of 0.2 dex width and chose 10 random cloud positions inside each bin. This approach represents an instance of the so-called *stratified random sampling*, which is often used in polling. Our choice of 10 positions per bin is a compromise between observing enough positions to derive statistically meaningful parameters and accumulating enough integration time per position to detect the emission from the lowest column density bin.

The selected 100 cloud positions were observed with the IRAM 30m telescope using the EMIR receiver followed by the FTS spectrometer configured to provide a velocity resolution of 200 kHz ($\approx$ 0.6 km s$^{-1}$). The observations covered the full available 3mm wavelength band (83.7-115.8 GHz) together with selected portions of the 2 and 1mm bands to include additional transitions of the main 3mm tracers. All data were obtained in frequency switching mode, reduced with the GILDAS software, and converted to the main beam brightness temperature using the recommended telescope efficiencies.

## 3 Results

Fig. 1 summarizes the observational results of our Perseus emission survey and presents the intensity distributions of the different transitions as a function of $H_2$ column density. The top row shows the intensity distribution of the CO isotopologs, the middle row shows the distribution of so-called traditional dense gas tracers (but see, e.g.,[1, 6]), and the bottom row shows the distribution of additional tracers of dense gas. As can be seen, there is a significant correlation between the intensity of each line and $N(H_2)$, a result that supports our initial assumption that the $H_2$ column density represents a good proxy of the molecular line intensity. The typical dispersion of the intensities inside each column density bin is 0.2 dex for both the CO isotopologs and the traditional dense gas tracers, while the additional tracers present a slightly larger dispersion of about 0.3 dex.

An inspection of Fig. 1 reveals a variety of intensity distributions. The main CO isotopologs ($^{12}$CO and $^{13}$CO) present an abrupt drop toward the lowest column density bin, likely due to molecular photo-dissociation by the interstellar UV radiation field at the outer edge of the cloud . Toward the cloud interior, the $^{12}$CO and $^{13}$CO intensities present slopes

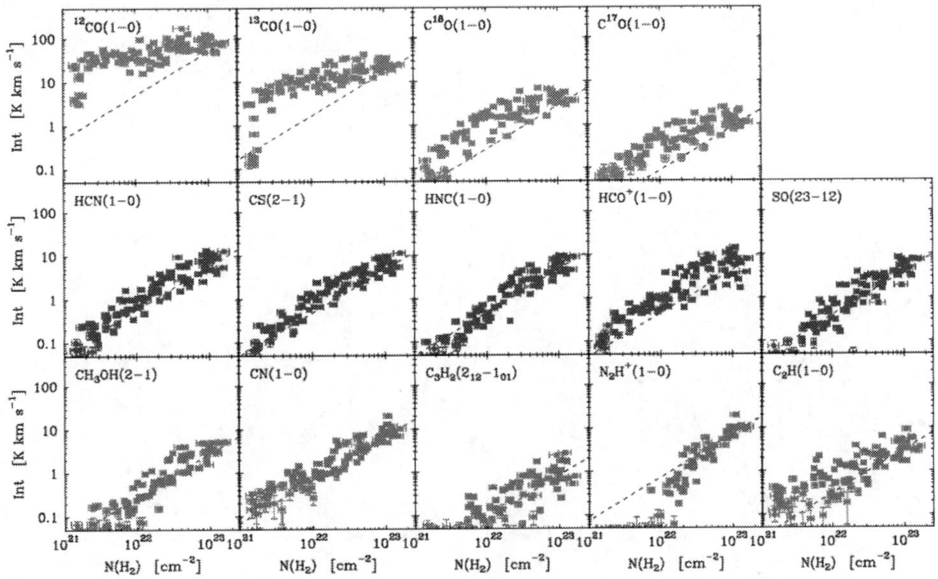


**Figure 1.** Distribution of integrated intensities of the main molecular species detected toward the Perseus cloud. The dashed lines show a linear trend for comparison.

that are significantly flatter than linear (dashed line), and effect that seems to be caused by the combination of thermalization and high optical depth (see below). The $C^{18}O$ and $C^{17}O$ lines are optically thinner, but they still present flatter-than-linear intensity distributions likely caused by freeze out at high column densities.

In contrast with the relatively flat distribution of the CO isotopologs, most dense gas tracers (Fig. 1 middle and bottom rows) present a close-to-linear distribution over the two orders of magnitude range of column densities. A slight flattening at high column densities can be seen in species such as HCN and CS, and is likely due to a combination of optical depth and freeze out. CN and $N_2H^+$, on the other hand, do not flatten out at the highest densities, indicating a stronger resistance to freeze out ([2, 5]). Among all the tracers, however, $N_2H^+$ is the only one that is truly selective of the densest gas component since its emission drops abruptly for $N(H_2)$ values lower than approximately $10^{22}$ cm$^{-2}$. This special behavior likely results from the strong sensitivity of its abundance to the presence of CO in the gas phase ([2]), and makes $N_2H^+$ the tracer of choice to identify the densest gas in a cloud.

The relative simplicity of the intensity distributions seen in Fig. 1 suggest that it may be possible to model the molecular emission of Perseus with a relatively simple model. For this, we have used a large velocity gradient (LVG) radiative transfer code and assumed that the cloud parameters only depend on the $H_2$ column density. Combining a constant temperature of 11 K, a volume density law that is proportional to $N(H_2)^{0.75}$, and abundance profiles for most species that contain both a photo-dissociation edge and freeze out at high column densities (with the obvious exception of $N_2H^+$), it is possible to reproduce the main trends observed in the data (see [8] for further details). The left panels of Fig. 2 illustrate the results of this modeling for the main and rare isotopologs of CO, CS, and HCN. As can be seen, the model fits both main and rare isotopologs of the three species assuming standard isotopic ratios and similar abundance profiles with molecular freeze out at high column densities. Although not shown in the figure, the model also fits the $N_2H^+$ emission assuming that this species survives at high densities and disappears when CO is present in the gas phase.

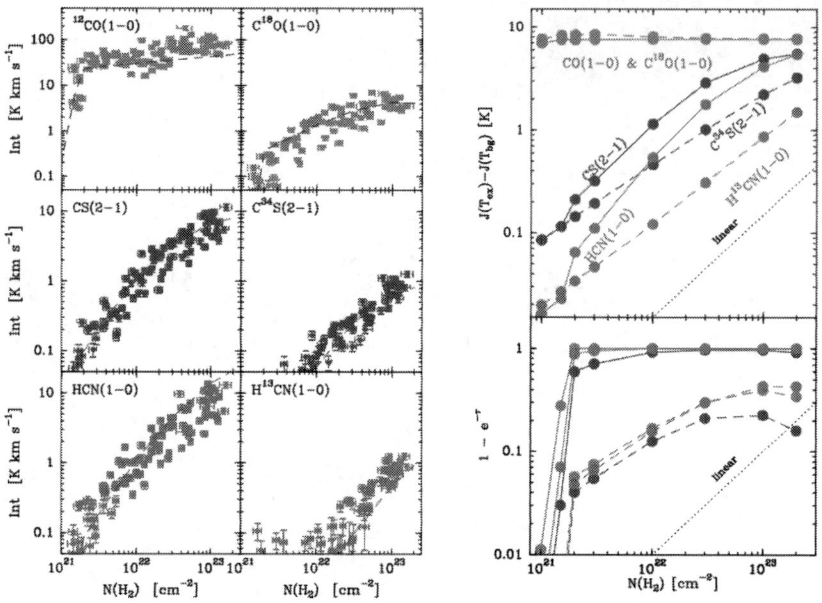


**Figure 2.** *Left:* Integrated intensity distributions of different isotopologs of CO, CS, and HCN together with the results of a simple radiative transfer model. *Right:* Excitation (top) and optical depth (bottom) terms of the equation of radiative transfer predicted by the cloud model. The main difference between the emission of CO and the dense gas tracers CS and HCN arises from their different excitations (top).

Given the success of our radiative transfer model in reproducing the observed intensities, we have used its results to explore how the emission is generated inside the cloud. The right panels of Fig. 2 represent the two terms of the radiative transfer solution, which we designate as the excitation term (top) and the optical depth term (bottom). As can be seen, the excitation of the CO isotopologs is very different from that of CS and HCN. While both CO isotopologs are thermalized through the cloud, the isotopologs of CS and HCN are strongly subthermal, and their excitation depends strongly on $N(H_2)$. The optical depth term (bottom panel), on the other hand, shows that all species follow similar patterns: the emission from the main isotopologs is optically thick, while the rare species are thin. Differences between CO and the dense gas tracers, and the quasi-linear dependence of their intensity, therefore results from the strong increase in the (subthermal) excitation toward the cloud interior.

*Acknowledgements:* We acknowledge support from project PID2019-108765GB-I00.

## References

[1] Barnes, A. T., Kauffmann, J., Bigiel, F., et al., MNRAS, **497**, 1972-2001 (2020)
[2] Bergin, E. A. & Tafalla, M., ARAA, **45**, 339-396 (2007)
[3] Goldsmith, P. F., Heyer, M., Narayanan, G., et al., ApJ **680**, 428-445 (2008)
[4] Gratier, P., Bron, E., Gerin, M., et al., A&A **599**, A100 (2017)
[5] Hily-Blant, P., Walmsley, M., Pineau Des Forêts, G., et al., A&A **480**, L5 (2008)
[6] Pety, J., Guzmán, V. V., Orkisz, J. H., et al., A&A **599**, A98 (2017)
[7] Ridge, N. A., Di Francesco, J., Kirk, H., et al., AJ, **131**, 2921-2933 (2006)
[8] Tafalla, M., Usero, A., Hacar, A., A& A **646**, A97 (2021)
[9] Zari, E., Lombardi, M., Alves, J., et al., A& A **587**, A106 (2016)

# Learning from model grids: Tracers of the ionization fraction in the ISM

*Emeric* Bron[1,*], *Evelyne* Roueff[1], *Maryvonne* Gerin[1], *Jérôme* Pety[2,1], *Pierre* Gratier[3], *Franck* Le Petit[1], *Viviana* Guzman[4], *Jan* Orkisz[5], *Victor* de Souza Magalhaes[2], *Mathilde* Gaudel[1], *Pierre* Palud[6], *Lucas* Einig[2,7], *Sébastien* Bardeau[2], *Maryvonne* Gerin[1], *Pierre* Chainais[6], *Jocelyn* Chanussot[1], *Javier* Goicoechea[7], *Annie* Hughes[7], *Jouni* Kainulainen[7], *David* Languignon[1], *Jacques* Le Bourlot[1], *François* Levrier[7], *Darek* Lis[7], *Harvey* Liszt[7], *Karin* Öberg[7], *Nicolas* Peretto[7], *Antoine* Roueff[7], *Albrecht* Sievers[1], *Pierre-Antoine* Thouvenin[6], and *Pascal* Tremblin[7]

[1]LERMA, Observatoire de Paris, PSL Research University, CNRS, Sorbonne Universités, France
[2]IRAM, 300 rue de la Piscine, 38406 Saint Martin d'Hères, France
[3]Laboratoire d'Astrophysique de Bordeaux, Univ. Bordeaux, CNRS, France
[4]Instituto de Astrofísica, Pontificia Universidad Católica de Chile, Chile
[5]Chalmers University of Technology, Department of Space, Earth and Environment, Sweden
[6]Univ. Lille, CNRS, Centrale Lille, UMR 9189 - CRIStAL, 59651 Villeneuve d'Ascq, France
[7]See https://www.iram.fr/ pety/ORION-B/team.html for the other affiliations

**Abstract.** The ionization fraction in neutral interstellar clouds is a key physical parameter controlling multiple physical and chemical processes, and varying by orders of magnitude from the UV irradiated surface of the cloud to its cosmic-ray dominated central regions. Traditional observational tracers of the ionization fraction, which mostly rely on deuteration ratios of molecules like $HCO^+$, suffer from the fact that the deuterated molecules are only detected in a tiny fraction of a given Giant Molecular Cloud (GMC). In [1], we propose a *machine learning*-based, semi-automatic method to search in a large dataset of astrochemical model results for new tracers of the ionization fraction, and propose several new tracers relevant in different ranges of physical conditions.

## 1 Context

The ionization fraction $x(e^-)$ is a key physical parameter controlling several aspects of the physics and chemistry of neutral interstellar clouds. Fast ion-neutral reactions are one of the main drivers of interstellar chemistry. The ionization fraction also controls the coupling of the gas to the magnetic field, as well as the charge state of grains and thus the efficiency of the photoelectric heating mechanism on grains. Finally, it can affect our interpretation of the observed excitation of high dipole moment molecules such as HCN, which are very sensitive to collisional excitation by electrons. In a typical Giant Molecular Cloud, $x(e^-)$ can vary in the neutral gas from $\sim 10^{-4}$ cm$^{-3}$ at the UV illuminated surface of the cloud, where carbon is fully ionized, to $10^{-6} - 10^{-9}$ cm$^{-3}$ in the central shielded regions, where cosmic rays maintain a residual source of ionization.

*e-mail: emeric.bron@obspm.fr

Usual tracers of the ionization fraction use the molecular deuteration of species such as $HCO^+$ [3] or $N_2H^+$ [4] that indirectly trace the $H_2D^+/H_3^+$ ratio, whose enrichment at low temperature is limited by electronic dissociative recombination and which is thus sensitive to the ionization fraction. However deuteration-based tracers of the ionization fraction suffer from several limitations. They are sensitive to the ortho-para ratio of $H_2$, itself a parameter difficult to estimate observationally. They are also relatively insensitive at very low ionization fractions, yielding estimates with large uncertainties. More importantly, these deuterated species are typically only detected in dense cores, which represent only small fraction of the observable area of a GMC. In the context of the Orion-B IRAM-30m Large Program [7], which has mapped the Orion B GMC over 5 square degrees with an angular resolution of 26" in the full 71 to 116 GHz band, we thus aim at finding new tracers of the ionization fraction to allow a full scale, unbiased view of the ionization fraction in a GMC, covering all the types of environments constituting a GMC.

## 2 Method

Massive datasets of model results produced by running large grids of models exploring wide ranges for multiple parameters and predicting tens or hundreds of observable quantities are now becoming common. These datasets contain a wealth of information on how sensitive the different observable quantities are to different unobservable physical quantities, and can thus help in finding new observable tracers of unobservable quantities such as $x(e^-)$.

The possible relationship between one observable quantity (e.g. an integrated line intensity ratio) and one unobservable quantity of interest (e.g. the ionization fraction) will however be affected by other unobservable parameters (e.g. the gas density, temperature, UV field, etc...). These will act as sources of noise and cause a scatter in the relationship. The predictive power of an observable quantity to predict a given unobservable quantity (here $x(e^-)$) can thus be quantified as the magnitude of this scatter around the mean trend of the relationship. We now describe successively the two components needed for this purpose: a grid of models sampling randomly all of the important unobservable parameters over realistic ranges of values for the environment of interest, and a statistical method to find the mean trend of this relationship and to quantify the scatter around this mean trend.

**Table 1.** Ranges of parameters explored in the two model grids.

| | Translucent medium grid | Cold dense medium grid |
|---|---|---|
| $n_H$ [cm$^{-3}$] | $3\times10^2 - 3\times10^3$ | $10^3 - 10^6$ |
| $T_{gas}$ [K] | $15 - 100$ | $7 - 20$ |
| $G_0$ | $1 - 1000$ | 1 |
| $A_V$ | $2 - 6$ | $5 - 20$ |
| $\zeta$ [cm$^{-1}$] | $10^{-17} - 10^{-15}$ | $10^{-17} - 10^{-16}$ |
| $OPR_{H_2}$ | $0.1 - 3$ | $10^{-4} - 0.1$ |
| depletion factor | 1 | $1 - 10$ |
| [S] | $1.86\times10^{-8} - 1.86\times10^{-5}$ | $1.86\times10^{-8} - 1.86\times10^{-5}$ |

Using the astrochemical model presented in [8], we compute single zone models with fixed density and temperature at stationary state, including a chemical network of 310 species and 8711 chemical reactions. As we expect the best tracers of the ionization fraction $x(e^-)$ to be different in different types of environnement, we run two grids of 5000 models each, exploring ranges of parameter values relevant respectively for translucent medium and for cold dense medium (the ranges of parameters explored are listed in Table 1). We select a list of species based on the frequency range of the Orion-B Large Program : CS, SO, $C_2H$,

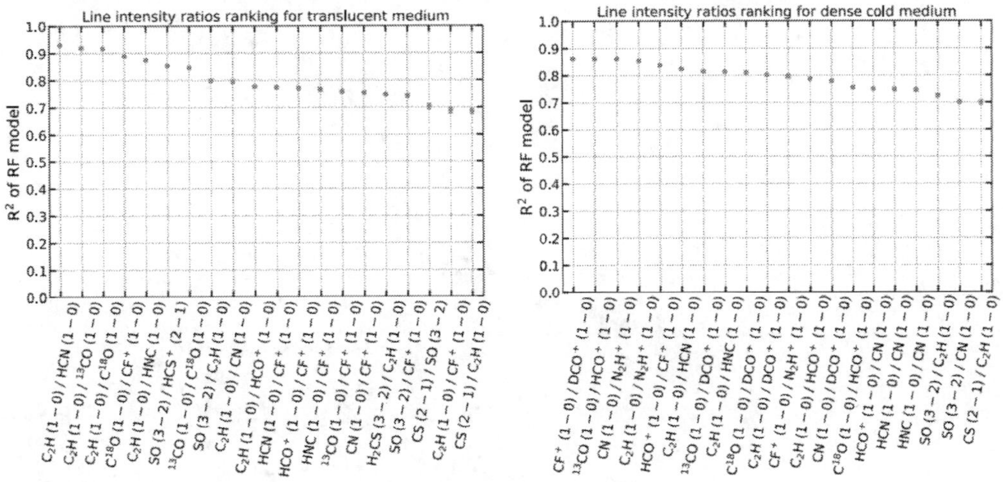


**Figure 1.** $R^2$ value of the line intensity ratios as tracers of the ionization fraction in translucent medium conditions (left panel) and cold dense medium conditions (right panel).

$^{13}CO$, $C^{18}O$, $HCO^+$, HCN, HNC, CN, $H_2CS$, $HCS^+$, $CF^+$ (and $N_2H^+$ and $DCO^+$ for the cold dense medium conditions only). We post-process the results of each model using a non-LTE excitation and radiative transfer model (RADEX, [9]) to compute the integrated line intensity of these species and take as observable quantities all possible ratios between these integrated line intensities. More details on the grids of models used can be found in [1].

In order to capture the mean trend of the relationship between each line ratio and $x(e^-)$, we train a Random Forest statistical model [2] to learn to predict $x(e^-)$ knowing only one given line ratio, based on the dataset of model results. To quantify the predictive power of the given line ratio, we use the $R^2$ value ("out-of-bag" estimate) of the trained predictive model. For a predicted quantity $y$,

$$R^2 = 1 - \frac{\sum_i \left(y_i^{\text{pred}} - y_i^{\text{true}}\right)^2}{\sum_i \left(y_i^{\text{true}} - \overline{y^{\text{true}}}\right)^2},$$

where the sums are over the datapoints, "pred" denotes the prediction while "true" the true value in the dataset, and $\overline{y^{\text{true}}}$ is the mean of $y$ in the dataset. The $R^2$ value thus represents the fraction of the initial variance in the quantity of interest (here $x(e^-)$) in the dataset that can be predicted from the fitted relationship with the single line ratio under consideration. We fit one Random Forest model and compute the corresponding $R^2$ coefficient for each line ratio (separately for each of the two model grids), and then rank the tracers based on the $R^2$ values. More details on how we trained the Random Forest models can be found in [1].

## 3 Results

The rankings of line ratios based on the $R^2$ value are shown on Fig. 1 (left panel: translucent medium conditions, right panel: cold dense medium conditions). Several line ratios can predict a large fraction of the variations of $x(e^-)$ in the model grids. In the translucent case, most of the best ratios involve $C_2H$, which is known to be enhanced in UV-illuminated environments (e.g. [5, 6]). In translucent gas, UV photons are still the main source of ionization, and a tracer of the impact of the UV field such as $C_2H$ can thus also trace accurately the ionization

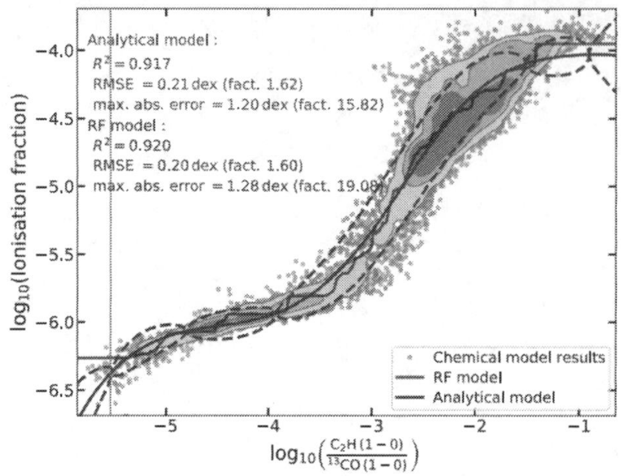


**Figure 2.** Scatterplot of the ionization fraction versus the $C_2H$ (1-0)/$^{13}CO$ (1-0) line intensity ratio for the translucent medium grid of models. Superimposed is the mean trend captured by the Random Forest (red line), with measures of the goodness of fit ($R^2$, mean squared error and maximum absolute error given on the figure. A simpler analytical model described in [1] and fitted to the data is also superimposed (black lines). Credit : Bron, A&A, 645, A28, 2021, reproduced with permission ©ESO.

fraction of the gas. In both types of conditions, we find very good tracers of the ionization fraction, that can give accurate estimates of the ionization fraction despite not knowing the density, temperature, UV illumination, ortho-para ratio of $H_2$, etc. For instance, Fig. 2 shows a scatter plot of the ionization fraction versus one of the best tracers in the translucent case: the $C_2H$ (1-0)/$^{13}CO$ (1-0) line ratio. The scatter induced by the 7 unknown parameters in the model grid remains very limited, as can be expected from the $R^2 > 0.9$ found. An ad hoc analytical fit, more easy to reuse than the Random Forest model, is also provided (black solid line, uncertainties as dashed lines). Complete results can be found in [1].

The method that we propose is very general and can be applied to find good tracers of any physical quantities (e.g., gas density, cosmic ray ionization rate) based on datasets of any types of more complex models (e.g., PDR models, astrochemical models includ-ing surface chemistry). A general implementation of this method is publicly available at http://autorank.ism.obspm.fr/.

## References

[1] Bron E., Roueff E., Gerin M., Pety J., Gratier P., Le Petit F. et al., A&A **645** A28 (2021)
[2] Breiman, L., Mach. Learn. **45** 5 (2001)
[3] Caselli, P., Walmsley, C. M., Terzieva, R., & Herbst, E., ApJ **499** 234 (1998)
[4] Caselli, P., Planet. Space Sci. **50** 1133 (2002)
[5] Gratier, P., Bron, E., Gerin, M., et al., A&A **599** A100 (2017)
[6] Pety, J., Teyssier, D., Fossé, D. et al., A&A **435** 885 (2005)
[7] Pety, J., Guzmán, V. V., Orkisz, J. H., et al., A&A **599** A98 (2017)
[8] Roueff, E., Loison, J. C., & Hickson, K. M., A&A **576** A99 (2015)
[9] van der Tak, F. F. S., Black, J. H., Schöier, F. L., Jansen, D. J. et al., A&A **468** 627 (2007)

# Gas properties in the Early Universe deciphered from spectral line surveys of high-z objects: The Cloverleaf Quasar

*Michel* Guélin[1,*], *Carsten* Kramer[1], *Chentao* Yang[2], *Belén* Tercero[3], and *José* Cernicharo[4]

[1]Institut de RadioAstronomie Millimétrique, 300 rue de la piscine, 38240 St. Martin d'Hères, France
[2]Department of Space, Earth and Environment, Chalmers University of Technology, Onsala Space Observatory, 439 92 Onsala, Sweden
[3]Observatorio Astronomico de Yebes, IGN, Cerro de la Palera, Guadalajara, Spain
[4]Instituto de Fisica Fundamental, CSIC, Serrano 121, Madrid, Spain

**Abstract.** Molecular lines observed in remote quasars arise from exceptionnally strong starburst regions in the vicinity of the QSO. Their study offers a unique opportunity to probe the gas physical and chemical properties at the epoch of peak star formation (z=2-3). We report the preliminary results of an on-going mm/sub-mm spectral line survey of the Cloverleaf Quasar (z=2.6), the archetype of gravitationally lensed quasars. We identify 14 molecular species, including rare isotopologues of CO and $HCO^+$. The high gas density tracers $HCO^+$, HCN, HNC and CN are detected through several rotational transitions, up to J=8-7 ($E_{up}/k \simeq 155$ K). Their emission appears to arise from dense ($> 10^6$ $cm^{-3}$) gas clumps embedded in a circumnuclear disk with a flat rotation curve.

## 1 Introduction and Observations

The Cloverleaf Quasar (H1413+117) is the archetype of gravitationally lensed quasars and one of the most luminous and best studied objects in the distant, hence young Universe. The quasar redshift (z=2.5579, as derived from a Gaussian fit to the CO J=3-2 line of figure 2) implies 11 billion years of light travel time to the Sun and an age of about 2 Gyr after the appearance of the very first stars. Located at the center of a starburst galaxy, its radiation, viewed from the Earth, is magnified by a factor as large as 11 (according to lensing models [8]) by 2 galaxies, with redshifts z≃1 and 2, respectively, that intercept the line of sight and split the quasar image into 4 distinct components (figure 1).

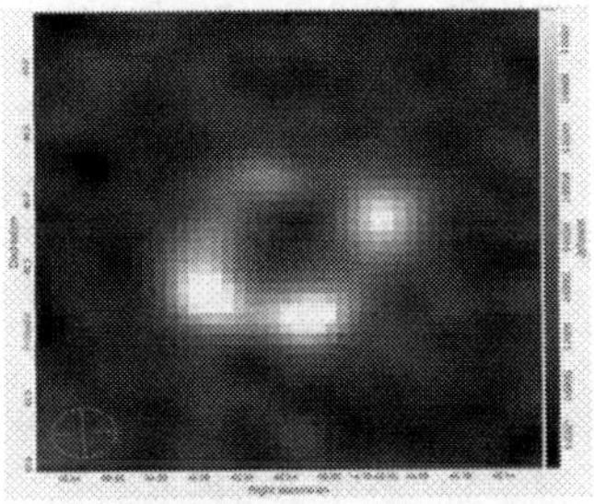

**Figure 1.** The Cloverleaf Quasar continuum emission at 230 GHz observed with ALMA. Image size is 2.5″, its resolution 0.3″ × 0.4″. *Acknowledgement: This paper makes use of the following ALMA data: ADS/JAO.ALMA 2011.0.01234.S and 2017.0.1232.S. ALMA is a partnership of ESO, NSF (USA), and NINS (Japan), together with NRC(Canada), MOST and ASIAA (Taiwan), KASI (Republic of Korea), in cooperation with the Republic of Chile. The Joint ALMA Observatory is operated by ESO, AUI/NRAO and NAOJ.*

The quasar and its host galaxy have been studied in the X-ray, optical and radio domains, and more particularly at sub-millimetre wavelengths. The detection of low-J lines of CO,

*e-mail: guelin@iram.fr

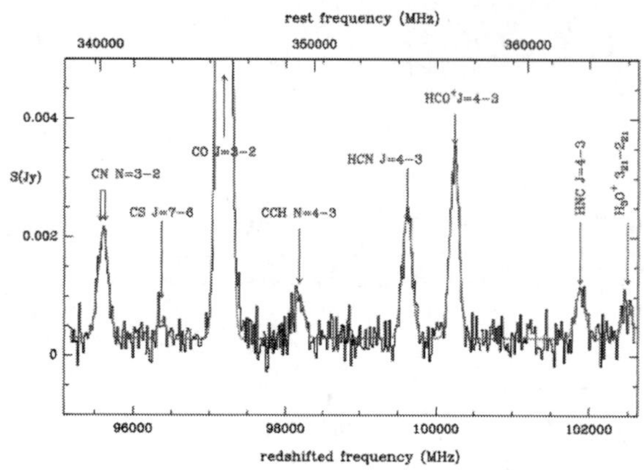


**Figure 2.** ALMA 3-mm spectrum of the Cloverleaf Quasar. Lower scale is redshifted frequency and upper scale rest frequency. Ordinate is line flux density in janskys. The red curve shows Gaussian fits to the detected lines. For the weakest lines, the Gaussian FWHM was set equal to that of the strongest nearby lines – see Fig. 4.

HCN and $HCO^+$ was reported in the early 2000s [1, 2, 5, 7], whereas a low spectral resolution survey of the 680-1100 GHz band (rest frequencies), carried out at the CSO ([3]), allowed to detect the J=6-5 through J=9-8 lines of CO. We report here the results of an on-going spectral line survey of this quasar in the mm-sub/mm domain (from 100 to 800 GHz, rest frequencies) aimed at studying the physical conditions and chemical composition of the gas in the quasar surroundings.

The survey is carried on with NOEMA, ALMA and the IRAM 30-m and Yebes 40-m telescopes. It benefits from a high spectral resolution (few MHz) and a high sensitivity (10 times higher than the pioneering work of Barvainis et al. [2]). So far, it covers the 110-180 GHz (rest frequencies) band, observed with the 40-m telescope, the 340-370 GHz band, observed with the 30-m telescope, NOEMA and ALMA, as well as the 525-625 GHz band and most of the 700-785 GHz band, observed with NOEMA.

Figures 2 and 3 show some of the ALMA and NOEMA spectra. A degree 1 baseline has been subtracted from each spectrum to remove the continuum. The r.m.s. noise per 80 $\mathrm{kms}^{-1}$ channel is 0.3-0.5 mJy. The NOEMA on-source integration time was 8 hours per pair of 8 GHz-wide spectra.

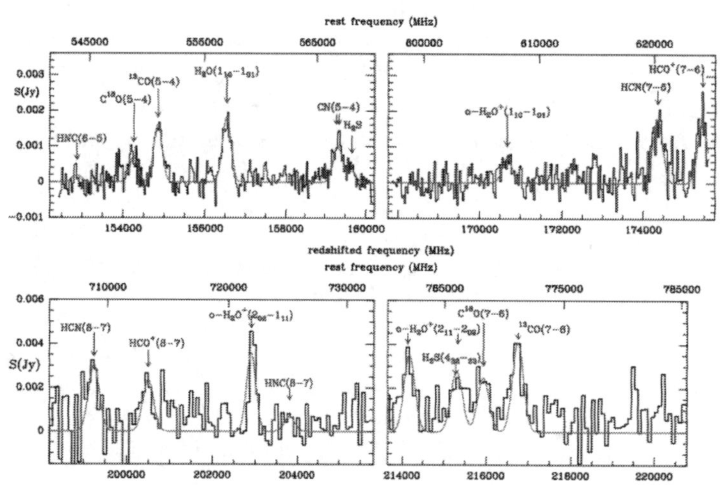


**Figure 3.** NOEMA 2-mm and 1-mm spectra of the Cloverleaf Quasar. As in Fig. 2, the red curves show Gaussian fits to the detected lines, the Gaussian FWHM of the weakest lines being set equal to that of the strongest nearby lines. Note the detections of $H_2O$, $H_2O^+$, $H_2S$ and of the rare isotopologues $C^{18}O$ and $^{13}CO$.

## 2 Results

Fourteen molecular species, including the rare CO isotopologues $^{13}CO$ and $C^{18}O$ and the rare $HCO^+$ isotopologue $H^{13}CO^+$, are detected. Three species, CH, CCH, and $H_2O^+$, are

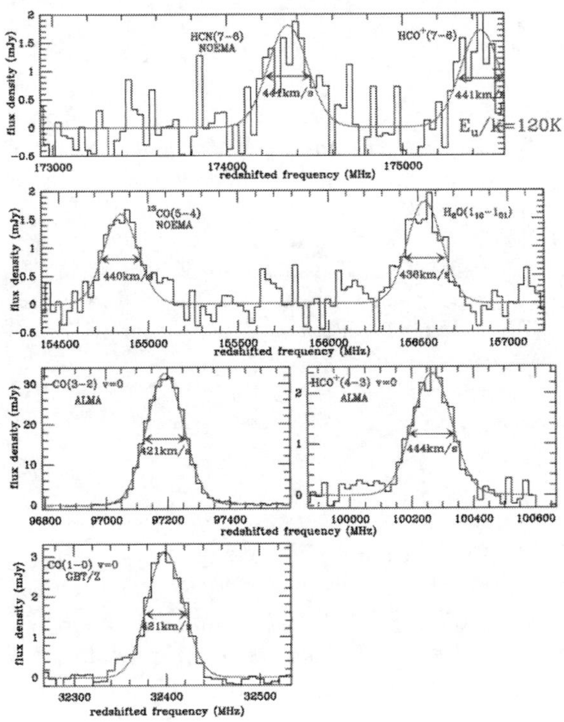


**Figure 4.** Line profiles of the strongest lines detected in the Cloverleaf Quasar. Red curves show the fits by a Gaussian whose FWHM is marked in blue. The upper 6 profiles are extracted from the spectra of Figs. 2 and 3, and the CO(1-0) line profile from GBT observations ([6]).

characteristic of gas irradiated by stellar UV, i.e. of diffuse clouds. The others species, CO, CS, HCN, HNC, $HCO^+$ $H_2O$, $H_3O^+$ and $H_2S$, are more typical of UV-shielded clouds.

The low noise level and the flat baselines provided by the interferometers allow an accurate determination of the shapes and strengths of the strongest lines (figure 4). The CO(3-2) line, which has the highest S/N, exhibits an almost perfect gaussian shape with a full width at half power of $421 \pm 50$ kms$^{-1}$. This is the same width as the CO(1-0) line and, within the errors, as the much weaker, but optically thin, $^{13}$CO(5-4) line ($440 \pm 100$ kms$^{-1}$) — note that the CO(7-6) line reported by [1] was observed with an insufficient bandwidth and that the higher-J CO lines were observed with a too coarse spectral resolution [3].

The lines of the high gas density tracers HCN, $HCO^+$ and $H_2O$ seem to be marginally wider (FWHP $\simeq 440 \pm 100$ kms$^{-1}$), but still compatible with the CO linewidths. The line intensities are far better determined than in previous measurements. The CO(3-2) line luminosity (hence the observed source brightness temperature) is found to be definitely stronger (by 15%) than that of the (1-0) line. This and the near equality of the $^{13}$CO(5-4) and (7-6) line brightness temperatures imply a warmer and/or denser gas than previously assumed (30-50 K, $10^{3-5}$ cm$^{-3}$ according to [6, 9]) and refute the hypothesis that CO lines arise from an homogeneous gaseous disk. As for the dust grains, the SED of the mm/sub-mm continuum [9] is best fitted by a 3-component model with a cold component at 30 K, a lukewarm component at 50 K and a hot component at 115 K, the latter representing only $\simeq 1\%$ of the total dust mass. The detections of several transitions of the high density tracers $HCO^+$, HCN, HNC and CN, as well as that of the rare isotopologues $^{13}$CO and $H^{13}CO^+$ (whose lines have opacities $<1$) allow us to disentangle the relative contributions of the gas kinetic temperature, $H_2$ volume density, infrared pumping and line opacity to the line strengths, hence to better constrain the molecular gas properties.

Figure 5 shows the $HCO^+$, HCN and HNC velocity-integrated line flux density divided by $J^2_{upper}$ (i.e. a quantity that scales with the line luminosity L'), as a function of $J_{upper}$.

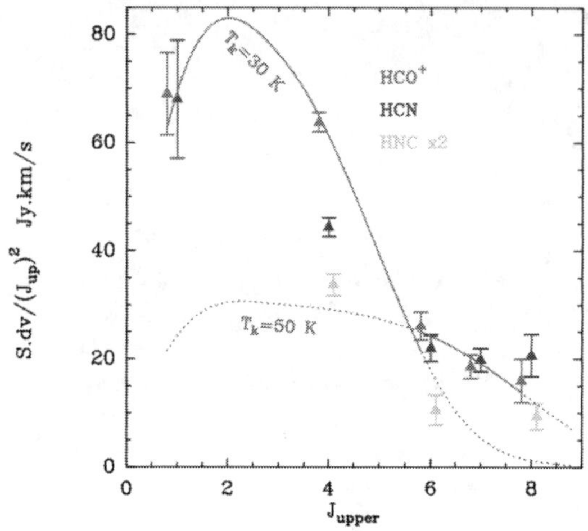


**Figure 5.** The velocity-integrated flux density of the lines of the high density tracers $HCO^+$, HCN and HNC, divided by $J^2_{upper}$. Abscissa is $J_{upper}$. The two red curves show fits to the low-J and high-J $HCO^+$ lines, respectively, with an LVG statistical equilibrium code (MADEX [4]). The fits require 2 different gas temperatures.

Two main results stem from this figure. The first is that fluorescent excitation, via infrared pumping, does not play a major role througout the molecular disk, contrary to what is argued for some superluminous infrared galaxies (e.g. APM08279+5255). The HNC molecule, which has the lowest vibrationally excited state ($\nu 2(1-0) = 21.7\mu$m above the ground state, a wavelength close to the peak of the dust continuum SED), should be the most affected by such a pumping. In contrast, the lowest bending states of HCN ($\nu 2(1-0) = 14.0\mu$m), and $HCO^+$ ($\nu 2(1-0) = 12.1\mu$m), are shifted to wavelengths at which the continuum flux is reduced by factors 6-8. Yet, HCN is the molecules whose luminosity decreases the slowest. In the same vein, despite the lack of bending states for CN, the HCN/CN luminosity ratio remains constant from 350 GHz to 600 GHz, when IR pumping should have strongly increased it.

The second result is that the HCN line luminosity remains almost constant between the J=6-5 and the J=8-7 lines, despite the fact that the latter levels lie twice higher in energy than the former ones. Similarly, the $^{13}$CO(5-4) line luminosity is not significatively lower than that of $^{13}$CO(7-6). This contrasts with the factor of 2 decrease in luminosity between HCN(4-3) and HCN(6-5), and the factor of 1.7 decrease between CO(3-2) and CO(5-4). A probable explanation is that the gas kinetic temperature and the dust temperature increase from the edges of the extended starburst region (whose radius is 700 pc according to [8]) to the quasar immediate surroundings. This is illustrated by the fits to the $HCO^+$ lines in Fig. 5 with an LVG model, which require at least two different temperatures. That the linewidths do not significantly change from line to line suggests that the starburst region is a clumpy gaseous disk rotating around the central quasar with a nearly flat rotation curve.

## References

[1] Alloin, D., Guilloteau, S., Barvainis, R. et al., A&A **321**, 24-28 (1997)
[2] Barvainis, R., Maloney, P., Antonucci, R., Alloin, D., ApJ **484**.695-701 (1997)
[3] Bradford, C.M., Aguirre, J.E., Aikin, R. et al., ApJ **2009**, 112=122 (2009)
[4] Cernicharo, J., Laboratory astrophysics and astrochemistry in the Herschel/ALMA era, ECLA, EAS Pub. Series **58**, 251-261 (2012)
[5] Riechers, D.A., Walter, F., Carilli, C.L. et al., ApJL **645**, L13-L16 (2006)
[6] Riechers, D.A., Carilli, C.L., Maddalena, R.J. et al., A&A **ApJ**, L32-39 (2011)
[7] Solomon, P., Vanden Bout, P., Carilli, C. and Guélin, M., Nature **426**, 636 (2003)
[8] Venturini, S. and Solomon, P.M., ApJ **540**, 740-745 (2003)
[9] Weiss, A. et al., A&A **409**, L41-L45 (2003)

# $z$-GAL – A Comprehensive Redshift Survey of the Brightest Herschel Galaxies

*Pierre* Cox[1,*] (on behalf of the $z$-GAL Team)

[1]Sorbonne Université, UPMC Université Paris 6 and CNRS, UMR 7095, Institut d'Astrophysique de Paris, 98bis boulevard Arago, 75014 Paris, France

**Abstract.** We here introduce $z$-GAL, a Large Program using NOEMA, aimed at a comprehensive 3 and 2-mm spectroscopic redshift survey of a large sample of 137 bright ($S_{500\mu m}$ > 80mJy) *Herschel*-selected high-$z$ dusty star forming galaxies, probing the peak of cosmic evolution. The results highlight the nature of the sources, including lenses and rare hyper-luminous infrared galaxies, as well as, in some cases, their multiplicity and will serve as a foundation for future detailed follow-up observations.

## 1 Introduction

Over the last two decades, surveys in the far-infrared and sub-millimeter wavebands have opened up a new window to our understanding of the formation and evolution of galaxies, revealing a population of massive, dust-enshrouded galaxies forming stars at enormous rates in the early Universe [1–3]. In particular, the extragalactic imaging surveys done with the *Herschel Space Observatory* have increased the number of dust-obscured star-forming galaxies (DSFGs) from hundreds to several hundreds of thousands [4, 5]. Including the all-sky Planck-HFI [6] and the South Pole Telescope (SPT) surveys [7], today we have vast samples of luminous DSFGs that are among the brightest in the Universe, including numerous examples of strongly lensed systems [8] and rare cases of extreme galaxies such as Hyper-Luminous InfraRed Galaxies (HyLIRGS) with $L_{\rm IR} > 10^{13} L_\odot$ [9, 10].

Detailed follow-up studies of these galaxies require robust estimates of their distances to investigate their physical properties. Measuring CO emission lines via sub-millimeter spectroscopy has proved to be the most reliable method to derive redshifts for high-$z$ DSFGs. Recently completed redshift surveys have provided crucial information that enable targeted studies to explore the nature of DSFGs in the early Universe (e.g., [11], [12]).

We here present the NOEMA $z$-GAL project, which is to date the largest of these coordinated efforts (project M18AB - PI: P. Cox, T. Bakx & H. Dannerbauer; see [13] for first results), a comprehensive redshift survey of 137 of the brightest ($S_{500\mu m} \geq$ 80mJy) DSFGs with photometric redshifts $z_{\rm phot} \geq 2$ selected from the *Herschel* surveys in the northern and equatorial fields (H-ATLAS, HerMES and HeRS) [14, 15].

*email: pierre.cox@iap.fr

## 2 Main results

The main goal of the *z*-GAL project has been achieved by measuring for the entire sample of 137 sources robust spectroscopic redshifts based on the detection of at least two emission lines. The targets span the redshift range $0.8 < z < 6.0$ and are centered around the peak of cosmic evolution at $z \sim 2.5$. The sources display an exceptionally wide range of velocity widths, extending to full width half-maximum of $600\,\mathrm{km\,s^{-1}} < \Delta V < 1500\,\mathrm{km\,s^{-1}}$ for ~ 60% of the sources, with many sources revealing double-peaked profiles indicative of merger systems and/or rotating disks (Fig. 1). The sources also reveal a remarkable diversity of morphologies in their dust and molecular gas emission, including evidence for gravitational amplification and binary/multiple systems. This variety is illustrated in Fig. 2 showing a lensed galaxy (HerBS-89a), a binary system (HerBS-70) at $z = 2.3$ which is at the center of a proto-cluster (Bakx et al. in prep.), and a case where two sources were detected at different redshifts in the same field (HerBS-43a at $z = 3.21$ and HerBS-43b at $z = 4.05$).

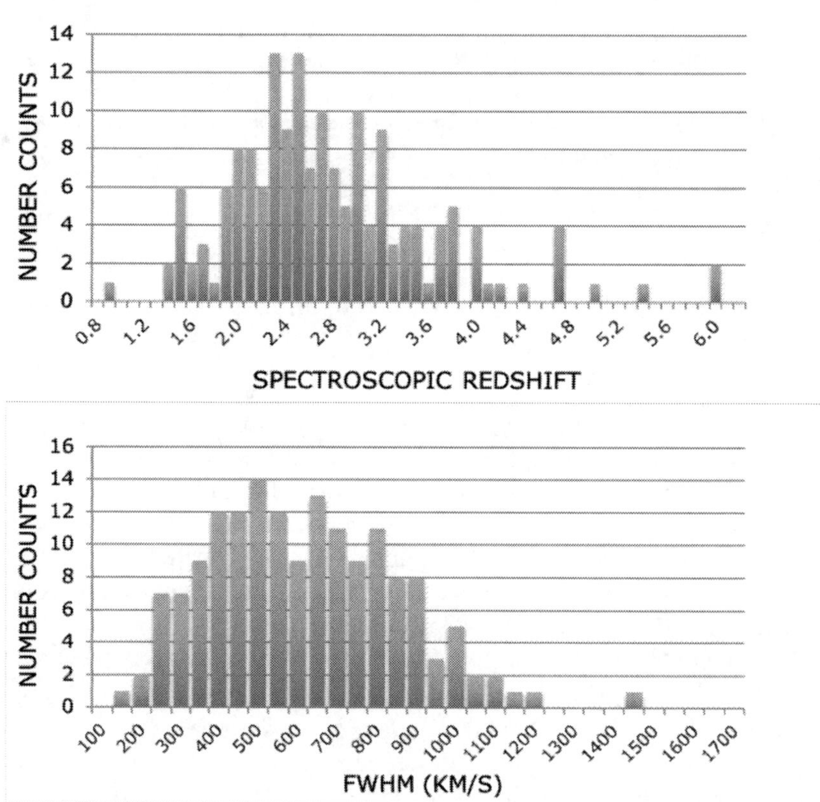


**Figure 1.** *Top panel*: Spectroscopic redshift distribution for the 137 bright *Herschel* sources from the *z*-GAL project. The spectroscopic redshifts are in the range $0.8 < z < 6.0$ with a median value of $z = 2.5 \pm 0.3$. *Lower panel*: Distribution of the full width at half maximum (FWHM) for the entire sample. The linewidths of the sources are large with a median value for the Full Width Half Maximum (FWHM) of $650 \pm 150\,\mathrm{kms^{-1}}$.

In addition to the detailed information obtained on their structure and kinematics (mainly through emission from mid-*J* CO transitions), the large instantaneous bandwidth of NOEMA (~16 GHz) provides, for each source, an exquisite sampling of the underlying dust continuum emission in the 3 and 2 mm bands. Together with the *Herschel* flux densities and, where available, data at 850 $\mu$m (from SCUBA-2) and 1 mm (from NOEMA or ALMA), the dust spectral energy distributions (SEDs) of the $z$-GAL sources are very well constrained.

Based on the successful measurements of $z$-GAL, the following remarks can be made regarding the derivation of robust redshifts: (i) Reliable spectroscopic redshift need the detection of *at least* two emission lines and one emission line is rarely enough; (ii) Photometric redshifts are only indicative and sometimes ambiguous. Their values depend on the adopted photometry and the available SED coverage used in the analysis and are found to be, on average, within 20% of the $z_{spec}$ values; (iii) Most emission lines detected are from CO (from $J$=2-1 to 9-8), with some sources seen in $H_2O$ and [CI] emission lines; for one source at $z = 0.85$, the two detected lines are HCN(3-2) and $HCO^+$(3-2); (iv) Some targets required a third line to lift any ambiguity, i.e., different redshift solutions were compatible with the available photometry. In many of those cases, the relative intensities of the CO emission lines do provide additional useful constraints.

With all the 137 *Herschel*-selected sources detected in the continuum and in at least two emission lines at 3 and 2 mm, the results of the $z$-GAL Large Program demonstrate the ability to efficiently derive redshifts and global properties of high-$z$ galaxies using the new correlator Polyfix and the broad bandwidth receivers on NOEMA. In addition to the redshifts, the NOEMA data provide a richness of information on the sources' morphology, revealing in many cases sources that are complex and multiple, spanning the range from lensed galaxies to genuine binary or triple systems, and providing useful additional information on their properties. For the gravitationally amplified sources, ancillary data from the Hubble Space Telescope and the Keck provide additional information on both the foregound lensing galaxies and, in some cases, the backgound high-$z$ system. Precise measurements of the amplification factors and the derivation of the properties of these sources will require higher resolution follow-up observations in the sub-millimeter and at optical/near-infrared wavelengths to study the characteristics of the foreground amplifying galaxy, as demonstrated in the case of HerBS-89a (see Sect. 3).

## 3 Infall activity in HerBS-89a

HerBS-89a belongs to the sample of 13 *Herschel*-selected galaxies observed by [13]. At $z$=2.94, HerBs-89a is the brightest galaxy of this sample, displaying the broadest CO emission line (FWHM $\sim$ 1100 km $s^{-1}$). HerBS-89a was further investigated using NOEMA and the VLA by [16], revealing a wealth of spectral features, including: two $^{12}$CO line transitions ($J$ =9-8 and 1-0) in addition to those reported by [13] (i.e., $J$=5-4 and 3-2); the three ground-state transitions of the molecular ion $OH^+$ ($1_1-0_1$, $1_2-0_1$, and $1_0-0_1$) and $CH^+$(1−0), all seen in absorption; and transitions of other molecular species including water and tracers of very dense gas (HCN and $NH_2$) - see Fig. 3. The combined continuum measurements available for HerBS-89a provide one of the most exquisite sampling to date of the far-infrared and sub/millimeter SED of a high-$z$ galaxy, with a total of 16 broad-band measurements covering the range from 250 $\mu$m to 3 mm in the observed frame.

This uniquely rich collection of information has allowed a detailed study of HerBS-89a. The NOEMA data reach a spatial resolution of 0.55 × 0.33 arcsec at 254 GHz, revealing a partial Einstein ring (Fig. 3). Thanks to the very high S/N of the continuum emission, a detailed model of the gravitational lens was built, resulting in a picture of the galaxy in the image plane, prior to any beam convolution, and allowing the object reconstruction in the source plane. HerBS-89a is a powerful star forming galaxy, whose flux is magnified by a factor $\sim$ 5 by a foreground spiral galaxy at $z_{phot} = 0.6 - 0.9$. Its intrinsic infrared luminosity is $L_{IR} = (4.6 \pm 0.4) \times 10^{12}\, L_{\odot}$, its rate of star formation SFR > 600 $M_{\odot}$/yr, and its depletion timescale $\tau_{depl} = (3.4 \pm 1.0) \times 10^8$ years. These values potentially position HerBS-89a in the

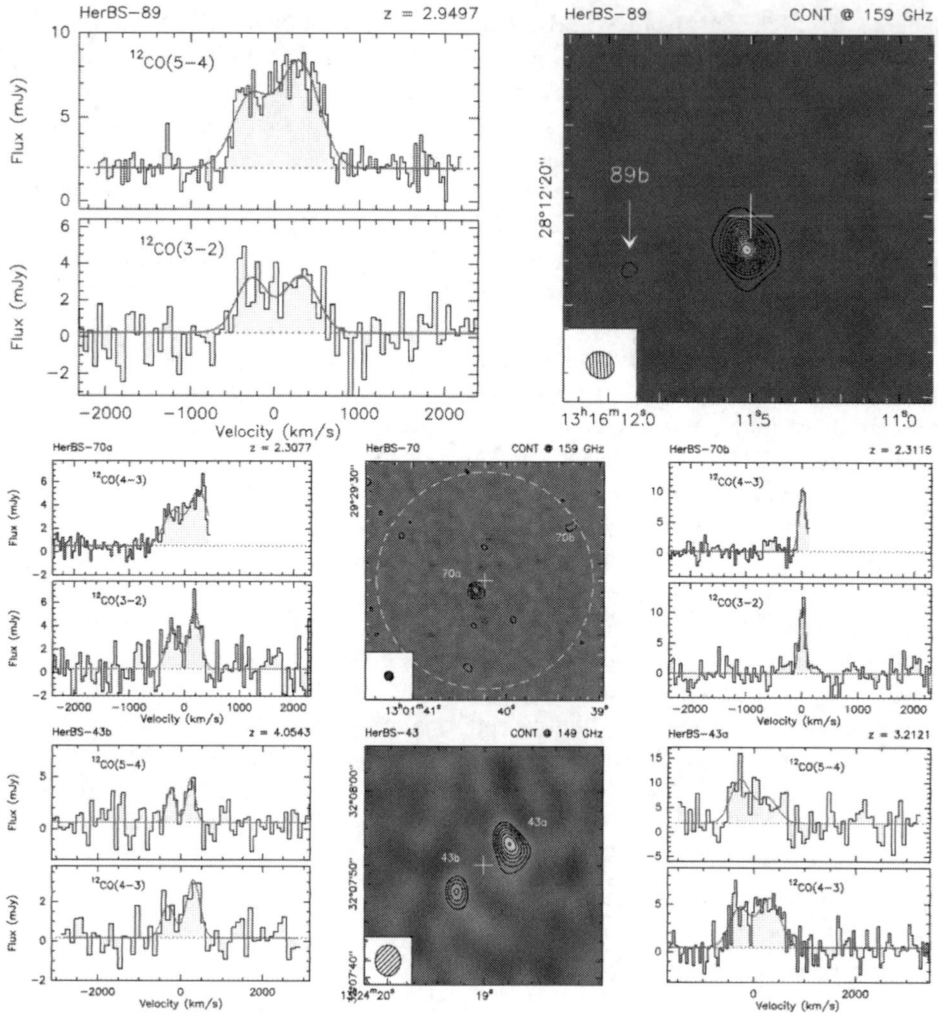


**Figure 2.** Continuum images at 2-mm and spectra from the 2 (top) and 3-mm (bottom) bands for the *Herschel* bright galaxies HerBS-89a, HerBS-70, and HerBS-43, representing, respectively, cases of a gravitationally amplified galaxy, a binary system with projected distance of ~140 kpc, and a field with two galaxies with different redshifts, illustrating the diversity of the $z$-GAL sample. The source name, the continuum frequency, and the derived spectroscopic redshift are indicated along the top of each panel. The emission lines are identified in the upper left corner of each spectrum. The spectra are displayed with the continuum and each emission line is centered at the zero velocity corresponding to its rest frequency. Fits to the continuum and the emission line profiles are shown as dotted and solid red lines, respectively. The primary beam at 50% is shown with a dashed circle. The synthesized beam is shown in the lower left corner of each continuum image. Adapted from [13].

region between the so called "main sequence" of star forming galaxies at $z \sim 3$ and the locus of starburst outliers.

The $^{12}$CO lines detected by NOEMA have a double profile, which can be explained either by a single rotating galaxy or by a system of two interacting objects. In addition to the broad $^{12}$CO emission lines, the NOEMA spectra show deep absorption features of the molecular ions $OH^+$ and $CH^+$ redshifted by $\sim$ 100 km s$^{-1}$ with respect to the galaxy's systemic velocity, revealing a rare example of gas inflow (see Fig. 2).

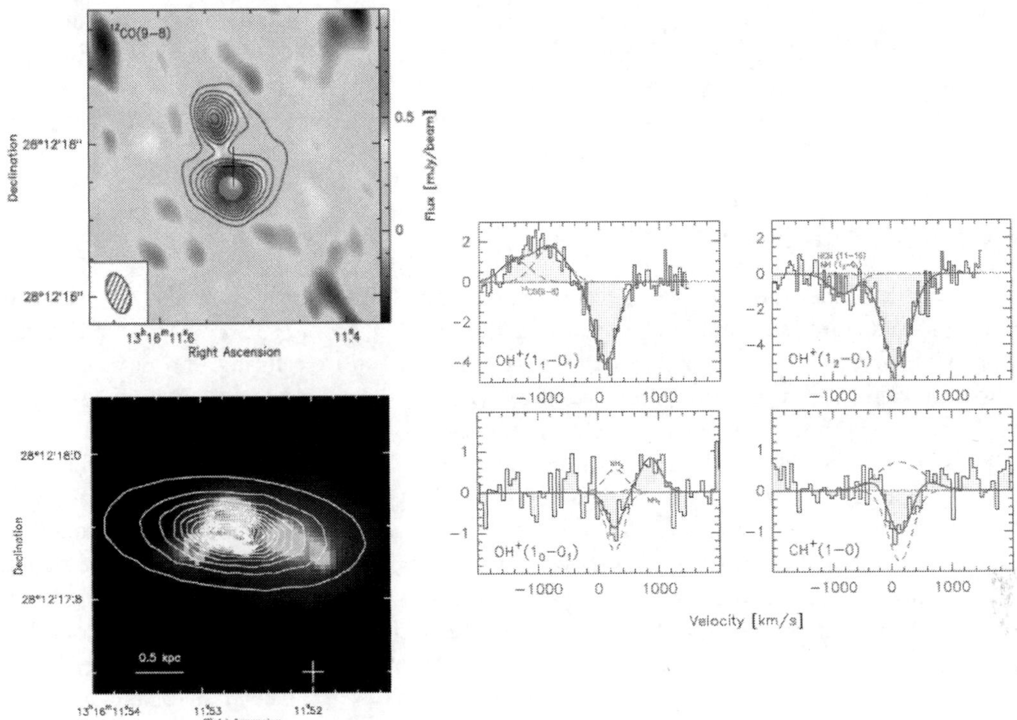


**Figure 3.** *Left panel*: (*Top*) Image of the $^{12}$CO$(9-8)$ emission in HerBS-89a, compared to the underlying continuum. The synthesized beam (0.55"×0.33") is shown in the lower left corner. (*Bottom*) Zeroth moment map of the $^{12}$CO$(9-8)$ emission in the source plane of HerBS-89a, possibly indicating a merging system. The contours show the dust continuum emission. The cross at the lower right corner marks the phase center of the NOEMA observations and the linear scale is indicated in the left corner. *Right panel*: Spectra of the molecular absorption and emission lines detected in HerBS-89a. Each molecular line is identified in the lower corner of each panel (labeled in black) and, in the case of a line overlap, as $^{12}$CO$(9-8)$/OH$^+(1_1-0_1)$, OH$^+(1_2-0_1)$/HCN$(11-10)$ and OH$^+(1_0-0_1)$/o$-$NH$_2(2_{02}-1_{11})(5/2-3/2)$, the second molecular line is labeled in red. The spectra are displayed with the continuum subtracted and, in each panel, the molecular line (labeled in black) is plotted relative to the zero velocity corresponding to its rest frequency. The main red-shifted absorption lines ($\Delta(V) \sim +100$kms$^{-1}$ clearly reveal infall activity in HerBS-89a. Adapted from [16]).

The gravitational lensing model mentioned above was applied to the NOEMA spectral cubes of HerBS-89a, with the aim to reconstruct its velocity field and study the inflow in the source plane. The limited angular resolution and S/N ratio per spectral element of the NOEMA data does not lead to a consensus picture of the kinematics of the system and with the data in hand, it is not possible to be conclusive on whether HerBS-89a is a single rotating galaxy or a merging pair. Consequently, different interpretations of the gas inflow could hold. In the least likely scenario, NOEMA might have detected a flow channeling gas from the intergalactic medium onto the galaxy. While such flow would in principle consist of pristine gas, the detected OH$^+$ and CH$^+$ transitions testify to the presence of enriched material. Enrichment in such kind of flow could happen, for example, in case of an accretion shock as the gas enters the halo of the galaxy. Alternatively, the flow could be part of a galactic fountain, in which a fraction of the gas expelled by the star formation winds falls back onto the galaxy.

In this case, an outflow should also be detected, but no evidence of any outflow feature was found in the NOEMA spectra of HerBS-89a. Finally, if HerBS-89a is in fact a merging pair or a multiple source, a third possibility is that the red-shifted $OH^+$ and $CH^+$ lines originate in a stream of matter between the two components, which would naturally explain the enriched composition of the flow.

## 4 Conclusions and Perspectives

Using NOEMA, the $z$-GAL survey has enabled the first systematic measurements of redshifts of a large and complete sample of high-$z$ galaxies selected from the *Herschel* northern and equatorial surveys. These data provide a unique database for exploring, through dedicated follow-up observations, the properties of these sources, from the gas and dust properties to the feedback and fueling activity, down to linear scales of 100 pc in the case of gravitational amplification. The results of the $z$-GAL comprehensive redshift survey will be reported in a forthcoming first series of papers that will cover the following aspects: (i) Overview of the $z$-GAL survey (Cox et al.); (ii) The dust continuum and the evolution of dust properties with redshift (Ismail et al.); (iii) The molecular gas properties (Berta et al); (iv) The nature of the sources – extended/lensed/companions/HyLIRGs (Bakx et al.). Finally, the $z$-GAL survey will provide, together with other available redshift measurements, a sizeable and homogeneous sample of about ~300 bright *Herschel* selected galaxies with reliable redshifts, increasing significantly the number of known lensed galaxies at the peak of cosmic evolution and providing the largest sample of HyLIRGs and additional rare objects.

**Acknowledgement**: *This work is based on a collaborative work and is presented on behalf of the z-GAL team. We are particularly thankful to the members of the data reduction teams for their relentless efforts and dedication to calibrate, reduce, and, after a critical review, produce the final catalogue, containing all the relevant data and the information on the emission lines and continuum for the entire z-GAL sample.*

## References

[1] Carilli, C.L. & Walter, F., ARA&A, **51**, 105 (2013)
[2] Casey, C.M., Narayanan, D. & Cooray, A., Phys. Rep., **541**, 45 (2014)
[3] Hodge, J.A. & da Cunha, E., Royal Society Open Science **7**, 12 (2020)
[4] Eales, S et al. PASP, 122, 499 (2010)
[5] Oliver, S.J. et al., MNRAS, **424**, 1614 (2012)
[6] Planck Collaboration, Aghanim, N. et al., A&A, **582**, 29 (2015)
[7] Vieira, J.D. et al., ApJ, **719**, 763 (2010)
[8] Negrello, M. et al., Science, **330**, 800 (2010)
[9] Riechers, D. et al., Nature, **496**, 329 (2013)
[10] Ivison, R.J. et al., MNRAS, **489**, 427 (2019)
[11] Reuter, C., Vieira, J.D., Spilker, J.S. et al., ApJ, **902**, 78 (2020)
[12] Urquhart, S., Bendo, G., Serjeant, S. et al., MNRAS, **511**, 3017 (2022)
[13] Neri, R., Cox, P., Omont, A. et al., A&A **635**, 7 (2020)
[14] Bakx, T. et al., MNRAS, **473**, 1751 (2018)
[15] Nayyeri, H. et al., ApJ, **823**, 17 (2016)
[16] Berta, S., Young, A.J., Cox, P. et al., A&A **646**, 122 (2021)

# Molecular richness in protostars: Lessons learnt from spectral observations

*Ana* López-Sepulcre[1,2,*] and *Mathilde* Bouvier[2]

[1]Institut de Radioastronomie Millimétrique (IRAM), 300 rue de la Piscine, 38406 Saint-Martin-d'Hères, France
[2]Univ. Grenoble Alpes, CNRS, IPAG, 38000 Grenoble, France

**Abstract.** The gas associated with the early stages of star formation contains traces of a large variety of molecular species, many of which are organic in nature. Interestingly, we observe a substantial chemical diversity among protostars, with some objects being enriched in what astrochemists label interstellar complex organic molecules (iCOMs), such as methyl formate ($HCOOCH_3$), while others are overabundant in unsaturated carbon chains such as $C_4H$. What is the cause of this diversity? And where should we place the proto-solar-system in this chemical context: was it rich in iCOMs, or in carbon chains, or in both? Thanks to the development of sensitive broadband (sub-)millimetre instrumentation, both in single-dish telescopes and interferometers, we are currently witnessing big steps forward in this area. The present contribution summarises what we have learnt, in the past decade or so, about the molecular contents in solar-mass protostellar sources, and suggests a few guidelines to stimulate progress in the field.

## 1 Introduction: Our chemical origins

Life represents the highest level of chemical complexity that we know of today, yet every single living being is composed of the same basic ingredients: organic molecules. These are essentially carbon-based molecules containing also other elements such as hydrogen, nitrogen, and/or oxygen. The fact that many organic molecules have been detected in comets and other pristine small bodies of the Solar System (e.g. [1, 2]) opens up the appealing possibility that an exogenous delivery of organic matter may have taken place more than 4 billion years ago, when the Earth and the rest of the Solar System had just formed, thus providing substantial quantities of raw material to trigger the origin of life on Earth.

Organic molecules have also been discovered beyond comets, in the interstellar medium (ISM). Most of them have been observed in molecular clouds, which are the birth sites of stellar and planetary systems like our own. This is why star-forming regions (SFRs) are excellent laboratories for astrochemical studies, whose aim is to understand why molecules are present in the ISM, i.e. how they form and how they are destroyed. At the same time, astrochemistry is a powerful tool to understand the process of star formation as a whole.

The present contribution focuses on the organic molecular chemistry present during the early protostellar stages of solar-mass star formation, based primarily on (sub-)millimetre spectral observations of the gas in protostellar cores. Protostars represent the onset of star

*e-mail: lopez@iram.fr

formation, where temperature and density gradients, coupled with infall, outflows, and rotating motions, all leave characteristic molecular imprints on the gas reservoirs surrounding the newly-formed protostellar objects. As we will see in the following sections, protostars are chemically rich sites that can yield precious hints on how our own Solar System chemically looked like during the early phases of its formation.

## 2 Protostellar chemical diversity: A chronological story

What is the typical chemical composition of protostellar cores? The shortest answer to this question is that there is no such thing as *typical*. As evidenced below, if there is one word to describe the chemistry associated with this phase, it is *diversity*.

### 2.1 Hot corinos

In terms of organic contents, hot corinos are the first kind of protostars that were discovered and characterised. They contain relatively high abundances of interstellar Complex Organic Molecules (iCOMs), which astrochemists commonly define as carbon-bearing molecules with 6 or more atoms containing other heavy elements such as oxygen or nitrogen. Examples of iCOMs are methyl formate ($HCOOCH_3$), ethanol ($CH_2CH_3OH$), and formamide ($NH_2CHO$). Hot corinos are dense ($n > 10^7$ cm$^{-3}$), hot ($T > 100$ K), and compact ($< 100$ au) regions of gas surrounding the central protostellar object where the icy mantles formed around dust grains during the pre-stellar phase have sublimated and enriched the gas with all the molecules that were previously trapped in these ices. The release of such molecules into the gas triggers further gas-phase chemical reactions that may lead to more iCOMs. The first hot corino that was discovered and characterised in terms of its iCOM contents is IRAS 16293–2242 [3–5]. Broadband spectral surveys of this and other hot corinos in the (sub-)millimetre wavelength regime, first with single dishes such as the IRAM 30-m telescope, and more recently with interferometers such as ALMA and NOEMA, have played a crucial role in unveiling the outstanding chemical richness of these regions. Notable examples are TIMASSS [6], the IRAM large programs ASAI [7] and SOLIS [8], and the ALMA program PILS [9].

### 2.2 Warm Carbon-Chain Chemistry protostars

It is often assumed that hot corinos are ubiquitous among protostars, which naturally leads to the belief that the Solar System experienced a hot corino phase during its formation. However, it is important to notice that a chemically different type of protostar exists, whose composition differs quite radically from that of hot corinos. Indeed, a few years after the discovery of the iCOM richness associated with hot corinos, the protostar L1527 was thoroughly studied and found to be devoid of iCOM emission. Instead, this source presents a remarkable richness in unsaturated hydrocarbons, carbon chains and rings (hereafter referred to as C-chains for simplicity), such as $C_4H$, $C_4H_2$, and c-$C_3H_2$, whose emission originates in a lukewarm region spanning roughly 1000–3000 au around the central protostellar object [10, 11]. Since then, a few other protostars displaying this kind of chemistry have been discovered [12]. These protostars are known as Warm Carbon-Chain Chemistry (WCCC) objects.

Quantitatively, C-chain abundances are $\sim$10 times higher in WCCC objects than in hot corinos, while their iCOM abundances are about two orders of magnitude lower [13]. Hence, a real chemical diversity exists among protostars in terms of their organic contents. A schematic illustration of a hot corino and a WCCC protostar is shown in Fig. 1 (*left*).

### 2.3 Hybrid protostars

In 2016, when hot corino and WCCC chemistry were still thought to be exclusive occurrences among protostars, the first hybrid, B335, was discovered [14]. Hybrid protostars display both hot corino and WCCC features: a lukewarm region rich in C-chains surrounding a very compact hot corino enriched in iCOMs. Thanks to this discovery, the idea that hot corinos and WCCC objects represent opposite extremes of what is in reality a continuous chemical spectrum began to take shape. The future research on this front is fully open, as only three hybrid protostars have been unambiguously identified so far: B335, L483, and CB68 [14–16]. Furthermore, while their iCOM abundances have been measured to be comparable to those found in pure hot corinos, their WCCC regions have not been as thoroughly characterised as those of pure WCCC protostars. Dedicated searches for more hybrid protostars will allow us to assess how common they are. These searches will require high sensitivity multi-line interferometric observations sampling both compact and more extended spatial scales, a challenging but far from impossible task with the facilities currently available, such as ALMA and NOEMA interferometers, and their planned spectral capability upgrades.

## 3 Protostellar chemical diversity explained... partially

The chemical diversity found among protostars is set by the chemical composition of the ice mantles that coat dust grains during the cold phase preceding the formation of a protostellar object. An ice mantle enriched in methane ($CH_4$) leads to WCCC once it sublimates at a temperature of $\sim$ 25 K, whereas an ice mantle enriched in methanol ($CH_3OH$) will lead to hot corino chemistry once it sublimates at a temperature of $\sim$ 100 K (see [13] for a review). These two molecules are considered parent species of C-chains and iCOMs, respectively. The tough question, still to be fully answered, is what causes such differences in the compositions of dust grain ices. The first explanation that was proposed concerns the duration timescale of the cold pre-stellar phase: shorter timescales ($\sim 10^5$ yr) would favour $CH_4$ ice enrichment and subsequent WCCC activity, while longer timescales ($\sim 10^6$ yr) would favour substantial CO depletion onto grains that, once fully hydrogenated, would lead to $CH_3OH$, triggering hot corino chemistry after ice sublimation. However, recent modelling has challenged this interpretation [17], and various environmental factors present during the cold pre-stellar stage have been proposed as alternative agents that can shape the molecular fate of the future protostar. These include density, temperature, and exposure to either ultraviolet or cosmic ray irradiation. For example, lower density, lower temperature, and higher irradiation favour WCCC (e.g. [17–19]). Even if there has recently been substantial progress in modelling protostellar chemical diversity, some observed features are not fully reproduced yet [17], so model refinement will be needed, with the crucial help of more observational data to provide physical and chemical constraints to such models.

## 4 Unveiling the chemical past of our Solar System: First baby steps

How common are the different types of protostars described in Sect. 2? Recently, a few statistical surveys were carried out with single-dish telescopes having this question in mind (e.g. [20–22]). This is fortunate because, about five years ago, the statistics were seriously poor: less than 10 hot corinos, less than 5 WCCC objects, and 2 hybrid protostars had been clearly identified and characterised. The results of these first surveys suggest that, as suspected, most protostars display chemical features intermediate between pure hot corinos and WCCC objects [22]. However, [23] found that single dish surveys are not optimal for accurately determining the chemical nature of compact protostellar cores, as the molecular emission measured can be dominated by the parsec-scale photo-dissociation region (PDR) at the edge of the parental molecular cloud hosting the protostars. Clearly, high angular resolution observations are needed for a solid assessment of the chemical nature of protostars. This is what interferometric surveys such as CALYPSO [24], which targeted 26 protostars distributed in several SFRs with NOEMA, and PEACHES [25] have accomplished. In particular, PEACHES targeted 50 protostars with ALMA in the Perseus star formation complex, and identified 56% hot corinos, which is quite a large fraction.

Can the results from PEACHES be generalised to other regions? And perhaps more importantly, how representative of the birth environment of the Solar System is Perseus? This last question is pertinent since, as we have seen in the previous section, the environment of stellar formation can determine the chemical nature of protostars. Therefore, it makes sense to study a SFR that resembles what must have been the birth environment of the Sun and its planets if we want to unveil our own chemical past. What does such a region look like? It turns out that Perseus is not exactly the best analog of what must have been the formation environment of the Solar System. Indeed, several pieces of evidence indicate that the Solar System very likely formed in a relatively large clustered environment with high-mass stars ($> 8\,M_{\odot}$) in its vicinity [26, 27]. One should therefore assess the chemical nature of Solar-mass protostars in high-mass SFRs, which is an observationally challenging feat due to the typically large distances of these regions compared to more isolated and/or looser protostellar clusters such as those in Taurus and Perseus. This motivated us to perform the ORANGES survey, a completely analogous survey to PEACHES in terms of sensitivity, spatial resolution and spectral setup, but this time targeting 19 protostars belonging to the Orion Molecular Cloud (OMC; see Fig. 1, *right*), i.e. the nearest high-mass SFR [28]. The survey revealed that, unlike in Perseus, hot corinos are not dominant in Orion. Indeed, ORANGES measured an occurrence of 26%, indicating that environment does matter and that, bizarrely, ORANGES and PEACHES are not the same.

## 5 Final remarks and the bright future ahead of us

Surveys such as PEACHES and ORANGES are providing us with the first precious hints towards understanding chemical diversity and its dependence on the environment. Much more work lies ahead of us to fully comprehend what physical properties have a greater impact on the chemical nature of protostars. This naturally includes surveying other regions to have a larger comparative picture of what variables shape the observed chemical diversity. But it also includes tackling the issue of dust opacity at (sub-)mm wavelengths, which may be responsible for hiding iCOM emission in some (or many) hot corinos. This will require resorting to lower frequency observations, as recently shown by [30].

Additionally, looking for hot corino activity (i.e. hot and compact iCOM emission) is not enough. Indeed, while the number of hot corinos discovered has multiplied by five in the past five years, the number of WCCC objects has barely changed. This is due to an observational bias towards iCOMs, which appear as more appealing due to their potential as precursors of more complex molecules with prebiotic interest. But C-chains also deserve their place, and life as we know it cannot exist without hydrocarbon structures. What would cell membranes be made of otherwise? A complete characterisation of the chemical nature of protostars, also in the context of the origin of life, must include C-chains tracing WCCC activity, and this is something that neither PEACHES nor ORANGES were well designed to do, unfortunately. The good news is, it is never late, and it is perfectly doable with current observatories such as NOEMA. This is our next immediate step as a follow up of ORANGES, which we hope will help bridge the gap between hot corino and WCCC studies.

As a final general note, while the present article has focused on observations, and a little bit of modelling, it is necessary to highlight that astrochemistry is a multi-field research domain, and as such, it must rely on the combined effort of observational astronomers, astrochemical modellers, molecular spectroscopists, and theoretical and experimental chemists. In

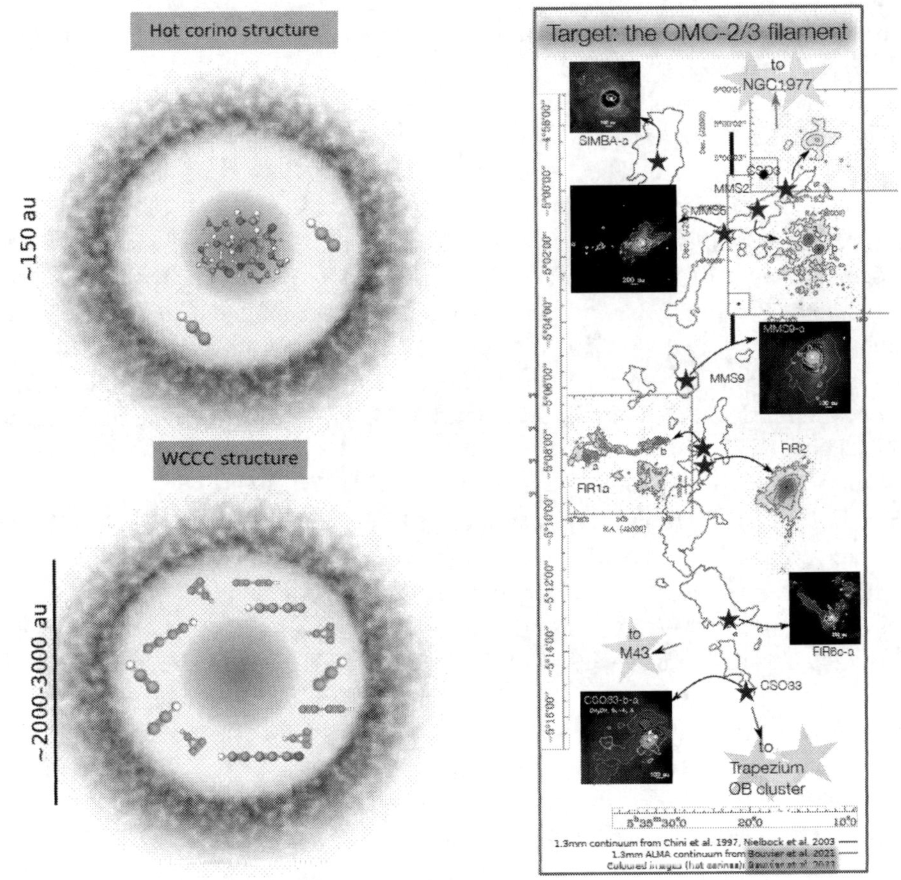


**Figure 1.** *Left*: Schematic illustration of a hot corino (*top*) and a WCCC protostar (*bottom*). *Right*: The 9 fields targeted with ORANGES (black stars), with zoomed-in emission maps resulting in a total of 19 targets due to source multiplicity within the fields. Five new hot corinos were discovered based on hot and compact $CH_3OH$ emission, shown in green contours overlaid on the 1.2 mm continuum emission maps. For the sources where hot corino activity was not detected, the mm continuum emission is displayed in blue contours (adapted from [28, 29]).

this context, interdisciplinary projects such as ACO (Astro-Chemical origins[1]), funded under the European Community, are optimal frameworks to find a common language to communicate, a common goal to pursue, and with this build a solid community that is as beautifully diverse as protostellar chemistry.

## References

[1] E. Bianchi, C. Codella, C. Ceccarelli, F. Vazart, R. Bachiller, N. Balucani, M. Bouvier, M. De Simone, J. Enrique-Romero, C. Kahane et al., MNRAS **483**, 1850 (2019), 1810.11411

[2] M.N. Drozdovskaya, E.F. van Dishoeck, M. Rubin, J.K. Jørgensen, K. Altwegg, MNRAS **490**, 50 (2019), 1908.11290

[3] C. Ceccarelli, L. Loinard, A. Castets, A.G.G.M. Tielens, E. Caux, A&A **357**, L9 (2000)

[1] https://aco-itn.oapd.inaf.it/home

[4] S. Cazaux, A.G.G.M. Tielens, C. Ceccarelli, A. Castets, V. Wakelam, E. Caux, B. Parise, D. Teyssier, ApJL **593**, L51 (2003)
[5] S. Bottinelli, C. Ceccarelli, R. Neri, J.P. Williams, E. Caux, S. Cazaux, B. Lefloch, S. Maret, A.G.G.M. Tielens, ApJL **617**, L69 (2004), astro-ph/0410601
[6] E. Caux, C. Kahane, A. Castets, A. Coutens, C. Ceccarelli, A. Bacmann, S. Bisschop, S. Bottinelli, C. Comito, F.P. Helmich et al., A&A **532**, A23 (2011), 1103.5347
[7] B. Lefloch, R. Bachiller, C. Ceccarelli, J. Cernicharo, C. Codella, A. Fuente, C. Kahane, A. López-Sepulcre, M. Tafalla, C. Vastel et al., MNRAS **477**, 4792 (2018), 1803.10292
[8] C. Ceccarelli, P. Caselli, F. Fontani, R. Neri, A. López-Sepulcre, C. Codella, S. Feng, I. Jiménez-Serra, B. Lefloch, J.E. Pineda et al., ApJ **850**, 176 (2017), 1710.10437
[9] J.K. Jørgensen, M.H.D. van der Wiel, A. Coutens, J.M. Lykke, H.S.P. Müller, E.F. van Dishoeck, H. Calcutt, P. Bjerkeli, T.L. Bourke, M.N. Drozdovskaya et al., A&A **595**, A117 (2016), 1607.08733
[10] N. Sakai, T. Sakai, T. Hirota, S. Yamamoto, ApJ **672**, 371 (2008)
[11] N. Sakai, T. Sakai, T. Hirota, S. Yamamoto, ApJ **722**, 1633 (2010)
[12] N. Sakai, T. Sakai, T. Hirota, M. Burton, S. Yamamoto, ApJ **697**, 769 (2009)
[13] N. Sakai, S. Yamamoto, Chemical Reviews **113**, 8981 (2013)
[14] M. Imai, N. Sakai, Y. Oya, A. López-Sepulcre, Y. Watanabe, C. Ceccarelli, B. Lefloch, E. Caux, C. Vastel, C. Kahane et al., ApJL **830**, L37 (2016), 1610.03942
[15] Y. Oya, N. Sakai, Y. Watanabe, A.E. Higuchi, T. Hirota, A. López-Sepulcre, T. Sakai, Y. Aikawa, C. Ceccarelli, B. Lefloch et al., ApJ **837**, 174 (2017), 1703.03653
[16] M. Imai, Y. Oya, B. Svoboda, F. members, arXiv e-prints arXiv:2206.06603 (2022), 2206.06603
[17] Y. Aikawa, K. Furuya, S. Yamamoto, N. Sakai, ApJ **897**, 110 (2020), 2006.11696
[18] S. Spezzano, L. Bizzocchi, P. Caselli, J. Harju, S. Brünken, A&A **592**, L11 (2016), 1607.03242
[19] J. Kalvāns, ApJ **910**, 54 (2021), 2102.02681
[20] D.M. Graninger, O.H. Wilkins, K.I. Öberg, ApJ **819**, 140 (2016), 1602.01472
[21] J.E. Lindberg, S.B. Charnley, M.A. Cordiner, ApJL **833**, L14 (2016), 1611.06975
[22] A.E. Higuchi, N. Sakai, Y. Watanabe, A. López-Sepulcre, K. Yoshida, Y. Oya, M. Imai, Y. Zhang, C. Ceccarelli, B. Lefloch et al., ApJS **236**, 52 (2018), 1804.08245
[23] M. Bouvier, A. López-Sepulcre, C. Ceccarelli, C. Kahane, M. Imai, N. Sakai, S. Yamamoto, P.J. Dagdigian, A&A **636**, A19 (2020), 2003.06198
[24] A. Belloche, A.J. Maury, S. Maret, S. Anderl, A. Bacmann, P. André, S. Bontemps, S. Cabrit, C. Codella, M. Gaudel et al., A&A **635**, A198 (2020), 2002.00592
[25] Y.L. Yang, N. Sakai, Y. Zhang, N.M. Murillo, Z.E. Zhang, A.E. Higuchi, S. Zeng, A. López-Sepulcre, S. Yamamoto, B. Lefloch et al., ApJ **910**, 20 (2021), 2101.11009
[26] F.C. Adams, ARA&A **48**, 47 (2010), 1001.5444
[27] S. Pfalzner, K. Vincke, ApJ **897**, 60 (2020), 2005.11260
[28] M. Bouvier, C. Ceccarelli, A. López-Sepulcre, N. Sakai, S. Yamamoto, Y.L. Yang, ApJ **929**, 10 (2022), 2202.13835
[29] M. Bouvier, A. López-Sepulcre, C. Ceccarelli, N. Sakai, S. Yamamoto, Y.L. Yang, A&A **653**, A117 (2021), 2107.10743
[30] M. De Simone, C. Ceccarelli, C. Codella, B.E. Svoboda, C. Chandler, M. Bouvier, S. Yamamoto, N. Sakai, P. Caselli, C. Favre et al., ApJL **896**, L3 (2020), 2006.04484

# Feeding a Protostar with 10 000 au Scale Streamers

*Jaime E.* Pineda[1,*], *Dominique* Segura-Cox[1,2], *Paola* Caselli[1], *Nichol* Cunningham[3], *Bo* Zhao[1], *Anika* Schmiedeke[1], *Maria José* Maureira[1], and *Roberto* Neri[3]

[1]Max-Planck-Institut für extraterrestrische Physik (MPE), Gießenbachstr. 1, 85748 Garching, Germany

[2]Department of Astronomy, The University of Texas at Austin, 2500 Speedway, Austin, TX 78712, USA

[3]Institut de Radioastronomie Millimétrique (IRAM), 300 rue de la Piscine, 38406 Saint-Martin d'Hères, France

**Abstract.** Dense cores are the places where stars are formed within the supersonic Molecular Clouds. These dense regions ($n \sim 10^5$ cc) are cold ($T \sim 10$ K) and display subsonic levels of turbulence (Mach $\sim$ 0.5), and represent the initial conditions for both star and disk formation. However, the influence of the parental core properties on the disk formation process is still not well constrained, and it is therefore crucial to study dense cores with interferometers to better understand the dense core and disk connection. We present NOEMA observations of a Class 0 object, which has been suggested to present a disk under gravitational instability (GI) (asymmetrical features in ALMA high resolution dust continuum emission). Our new data reveal a previously unseen large scale ($\sim$10,000 au, or 33″) streamer of fresh gas from the surrounding dense core down to the disk scales. This streamer is almost perpendicular to the outflow, and it contains material with subsonic levels of turbulence, and therefore unperturbed by the outflow. Based on the total mass in the streamer and the free-fall timescale, we estimate infall rates to the disk scales, which clearly show that accretion via streamer can have an important role in the disk formation. Moreover, these results show that previously observed disk asymmetries could also be driven by large scale asymmetric flows instead of GI. This result shows the power and importance of studying dense cores with interferometers to provide a complete and proper picture of star and disk formation.

## 1 Introduction

The classical picture of star formation focuses on the material in an isolated parental dense core that undergoes gravitational collapse [1–3]. In addition, all the material used to form stars and planets must pass through the dense core. In this scenario, star- and disk-formation can be studied in numerical simulations of isolated/closed boxes, allowing the implementation of a range of physical processes [e.g., 4–6].

Non-axisymmetric structures around Class 0 objects were observed with *Spitzer*, and they were interpreted as the result of the collapse of non-equilibrium structures [7]. More

*e-mail: jpineda@mpe.mpg.de

recently, ALMA continuum observations on two young Class 0 objects suggested the presence of accretion streamers [8]. These high angular resolution (better than 150 au, or 0.5″) dust continuum polarisation observations at 870 $\mu$m revealed the presence of two filamentary features aligned with the polarisation vectors reaching down to the disk scales. It is suggested that accretion streamers could explain at least one of these features. Unfortunately, the lack of complementary molecular line observations does not allow for a confirmation of these structures as accretion streams.

Per-emb-2 is a Class 0 object in the Perseus molecular cloud, which was observed with ALMA in dust continuum emission revealing asymmetrical features at disk scales, which are suggestive of a disk under gravitational instability (GI) [9]. Here we present observations carried out with IRAM NOEMA of Per-emb-2 [see also 10], which reveal the presence of a large scale streamer.

## 2 Observations

We observed Per-emb-2 with NOEMA in C and D configurations under project S18AG and using the Band 1 (3-mm) receivers. Thanks to the flexibility available with the PolyFix correlator, we are able to place many high-spectral resolution windows (62.5 kHz) targeting several molecular transitions, including $HC_3N$ (10–9) and (8–7), CCS ($8_7$–$7_6$), $^{13}CS$ (2–1), and $N_2H^+$ (1–0). The observations are calibrated using the standard observatory pipeline in `CLIC`. The imaging is carried out with `MAPPING` using natural weighting, more details can be found in [10]. The typical noise level and beam size are 6 mJy beam$^{-1}$ channel$^{-1}$ and 4″, respectively The integrated intensity maps for $HC_3N$ (10–9), CCS ($8_7$–$7_6$), and $^{13}CS$ (2–1) are shown in Fig. 1.

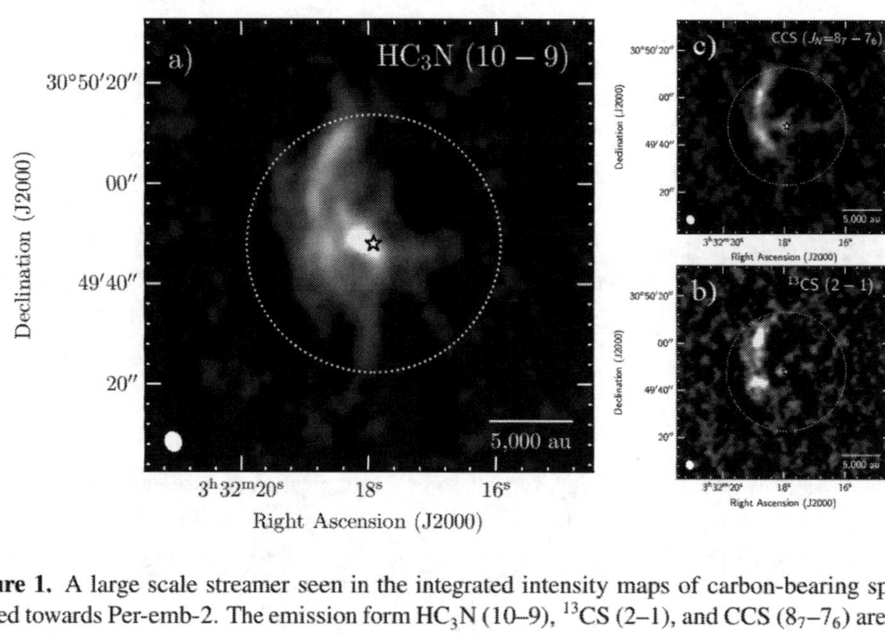


**Figure 1.** A large scale streamer seen in the integrated intensity maps of carbon-bearing species observed towards Per-emb-2. The emission form $HC_3N$ (10–9), $^{13}CS$ (2–1), and CCS ($8_7$–$7_6$) are shown in panels a, b, and c; respectively. The centroid velocity from $HC_3N$ (10–9) is in the background, original data from [10]. The dotted circle shows the Primary Beam at the line frequency. Beam size and scale bar are shown in bottom left and right corner, respectively.

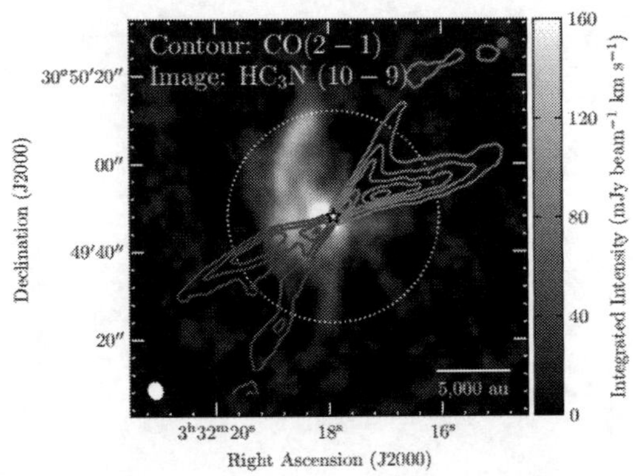


**Figure 2.** The outflow emission as traced by CO (2–1) in red and blue contours, while the background emission is the $HC_3N$ integrated intensity. This shows that the streamer approaches disk-forming scales at an almost perpendicular angle to the outflow direction, and it is unrelated to a possible outflow interaction. The primary beam response for the CO (2–1) observation is shown by the white dotted circle. The CO contours levels are drawn at 5, 15, 25, 35 and 45× rms, where rms is 0.2 Jy beam$^{-1}$ km s$^{-1}$.

## 3 Results

The emission traced by the carbon bearing species reveals a clear narrow and elongated structure, connecting the edge of the primary beam and the disk forming scales, see Fig. 1. These observations detected the streamer in molecular line emission, but not in the (less sensitive) dust continuum emission. The derived velocity map in the streamer is smooth and shows a streamer that begins at $\approx$10 000 au (or 33″) from the central YSO (beyond the classical dense core seen in $NH_3$ and $N_2H^+$).

The streamer emission approaching disk-forming scales is at an almost perpendicular angle to the outflow direction, as traced with the CO (2–1) from the MASSES survey [11], and therefore unrelated to the outflow (see Fig. 2). Moreover, the derived velocity dispersion of the $HC_3N$ emission is subsonic, and therefore the gas is unperturbed by the outflow.

We model the on sky trajectory and velocity along the line of sight with a streamline model [12], which includes rotation on a free-falling parcel of gas. This model matches well the trajectory and velocity of the observations, as seen in Fig. 3. Thanks to the detection of $HC_3N$ (10–9) and (8–7) along the streamer, we estimate an average density of $(4 \pm 2)10^4$ cm$^{-3}$ by comparing the line ratio with LTE models using RADEX [13]. Combing the streamer model and the estimated density, we estimate an average infall rate from the streamer onto the disk forming scales is $10^{-6}$ M$_\odot$ yr$^{-1}$, which is comparable to the current protostellar accretion rate of $7 \times 10^{-7}$ M$_\odot$ yr$^{-1}$ [14, 15]. This suggests that the streamer could modify protostellar accretion by funnelling extra material to the central region.

These results shows how the improved sensitivity and flexibility achieved with NOEMA can play a crucial role in bridging the gap between the large scale molecular cloud studies and the high-angular resolution disk studies. These type of studies are already modifying our picture of star-formation [16], however, it is important to further determine if these streamers are frequent enough and provide sufficient mass to have a decisive role on the disk evolution and planet formation process [e.g., 17, 18].

## References

[1] R.B. Larson, MNRAS **145**, 271 (1969)
[2] F.H. Shu, ApJ **214**, 488 (1977)
[3] S. Terebey, F.H. Shu, P. Cassen, ApJ **286**, 529 (1984)
[4] B. Zhao, P. Caselli, Z.Y. Li, R. Krasnopolsky, MNRAS **473**, 4868 (2018), 1706.06504
[5] M.N. Machida, S. Basu, ApJ **876**, 149 (2019), 1904.04424
[6] P. Marchand, K. Tomida, K.E.I. Tanaka, B. Commerçon, G. Chabrier, ApJ **900**, 180 (2020), 2009.01268

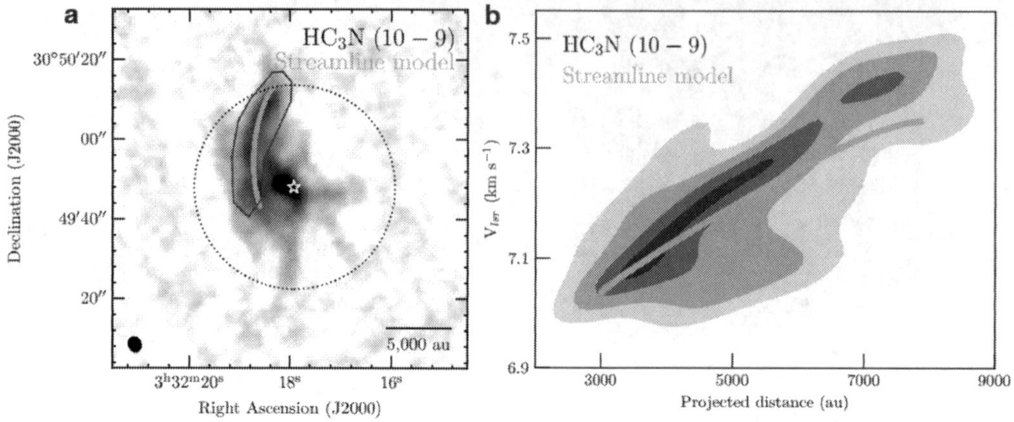


**Figure 3.** Streamline model (shown in orange) reproduces the trajectory and velocity of the streamer. Left panel show the sky trajectory of a streamer overlaid on the integrated intensity map of the $HC_3N$ (10–9) emission, while the blue polygon shows the boundary of the region showing the uncontaminated streamer emission. Right panel shows the kernel density estimate (KDE), for all the points within the blue polygon in the left panel, of the velocity along the line of sight as a function of the projected distance. The dotted circle shows the Primary Beam at the line frequency. Beam size and scale bar are shown in bottom left and right corner, respectively.

[7] J.J. Tobin, L. Hartmann, L.W. Looney, H.F. Chiang, ApJ **712**, 1010 (2010), 1002.2362
[8] V.J.M. Le Gouellec, C.L.H. Hull, A.J. Maury, J.M. Girart, Ł. Tychoniec, L.E. Kristensen, Z.Y. Li, F. Louvet, P.C. Cortes, R. Rao, ApJ **885**, 106 (2019), 1909.00046
[9] J.J. Tobin, L.W. Looney, Z.Y. Li, S.I. Sadavoy, M.M. Dunham, D. Segura-Cox, K. Kratter, C.J. Chandler, C. Melis, R.J. Harris et al., ApJ **867**, 43 (2018), 1809.06434
[10] J.E. Pineda, D. Segura-Cox, P. Caselli, N. Cunningham, B. Zhao, A. Schmiedeke, M.J. Maureira, R. Neri, Nature Astronomy **4**, 1158 (2020), 2007.13430
[11] I.W. Stephens, T.L. Bourke, M.M. Dunham, P.C. Myers, R. Pokhrel, J.J. Tobin, H.G. Arce, S.I. Sadavoy, E.I. Vorobyov, J.E. Pineda et al., ApJS **245**, 21 (2019), 1911.08496
[12] S. Mendoza, E. Tejeda, E. Nagel, MNRAS **393**, 579 (2009), 0803.1020
[13] F.F.S. van der Tak, J.H. Black, F.L. Schöier, D.J. Jansen, E.F. van Dishoeck, A&A **468**, 627 (2007), 0704.0155
[14] S. Frimann, J.K. Jørgensen, M.M. Dunham, T.L. Bourke, L.E. Kristensen, S.S.R. Offner, I.W. Stephens, J.J. Tobin, E.I. Vorobyov, A&A **602**, A120 (2017), 1703.10225
[15] T.H. Hsieh, N.M. Murillo, A. Belloche, N. Hirano, C. Walsh, E.F. van Dishoeck, J.K. Jørgensen, S.P. Lai, ApJ **884**, 149 (2019), 1909.02706
[16] J.E. Pineda, D. Arzoumanian, P. André, R.K. Friesen, A. Zavagno, S.D. Clarke, T. Inoue, C.Y. Chen, Y.N. Lee, J.D. Soler et al., arXiv e-prints arXiv:2205.03935 (2022), 2205.03935
[17] A. Garufi, L. Podio, C. Codella, D. Segura-Cox, M. Vander Donckt, S. Mercimek, F. Bacciotti, D. Fedele, M. Kasper, J.E. Pineda et al., A&A **658**, A104 (2022), 2110.13820
[18] C. Ginski, S. Facchini, J. Huang, M. Benisty, D. Vaendel, L. Stapper, C. Dominik, J. Bae, F. Ménard, G. Muro-Arena et al., ApJL **908**, L25 (2021), 2102.08781

# Interstellar complex organic molecules in the prototypical Class I protostar SVS13-A: From large scales to planet forming disks

*Eleonora* Bianchi[1,*]

[1]Univ. Grenoble Alpes, CNRS, IPAG, 38000 Grenoble, France

**Abstract.** We present a chemical systematic study of the Class I object SVS13-A obtained in the framework of two IRAM Large Programs: ASAI (Astrochemical Survey At IRAM-30m) with the 30m and SOLIS (Seeds Of Life In Space) with NOEMA. Thanks to the ASAI high-sensitivity unbiased spectral survey of the 3, 2 and 1.3mm bands, we detect and analyse several emission lines from deuterated species and interstellar complex organic molecules (iCOMs, e.g. molecules composed by 6 or more atoms and based on carbon). Within SOLIS, we obtain high-sensitivity and high-spatial resolution (~ 180 au) maps of crucial iCOMs. As a follow up, thanks to ALMA we explore the chemistry in the planet forming region (~ 50 au). We image emission lines from methanol ($CH_3OH$), acetaldehyde ($CH_3CHO$), formamide ($NH_2CHO$) and dimethyl ether ($CH_3OCH_3$). The different spatial distributions suggest a chemical differentiation inside the binary system or a different continuum opacity in the two protostellar disks.

## 1 Introduction

Evidence is mounting that planet formation in Sun-like protostars may already start during the early Class I phase (age < $10^5$ years). Indeed, recent observations have shown the presence of substructures such as gaps and rings in very young protostellar disks [e.g. 12]. Moreover, evolved Class II disks are not enough massive to explain the observed exoplanets population, in contrast to early Class 0 and I disks [13]. While close to the protostar a chemical reset is expected, in the outer protostellar disk a consistent part of the molecular complexity is expected to be inherited from the early stages [e.g. 11]. In this respect, observations of protostars in the early Class I stage are crucial to investigate the initial chemical budget of the forming planetary systems.

## 2 SVS13-A: A chemically rich Class I hot corino

SVS13-A is a very well studied Class I protostar, located in the NGC1333 star forming region in Perseus (at a distance of about 300 pc). The source drives a large-scale outflow and the HH7–11 Herbig-Haro chain [8, 9]. From VLA observations SVS13-A is known to be a close binary, with a separation of 0.″3. SVS13-A has been the target of several observational

*e-mail: eleonora.bianchi@univ-grenoble-alpes.fr

large programs dedicated to explore its molecular complexity. The first one was ASAI[1] [10], an unbiased spectral survey performed with the IRAM-30m which disclosed the presence of a hot corino chemistry towards the source with the detection of several iCOMs, such as methanol and its isotopologues ($CH_3OH$, $^{13}CH_3OH$, $CH_2DOH$), acetaldehyde ($CH_3CHO$), methyl formate ($HCOOCH_3$), dimethyl ether ($CH_3OCH_3$), glycolaldehyde ($HOCH_2CHO$), formamide ($NH_2CHO$), ethanol ($CH_3CH_2OH$), and water (HDO) [2, 5].

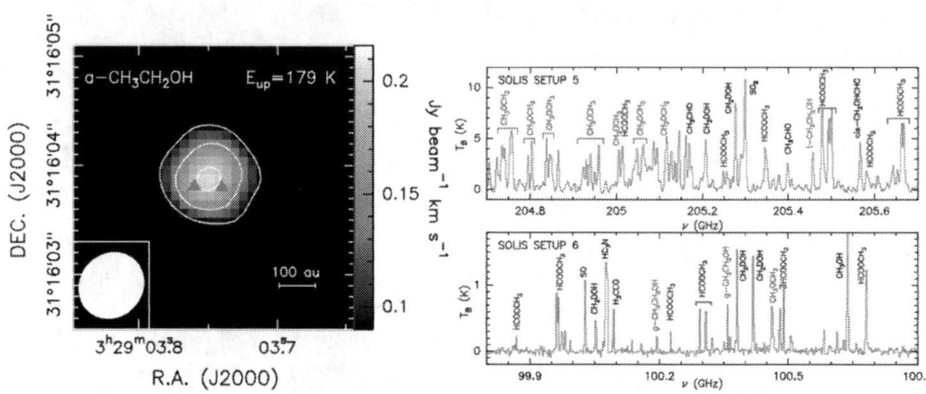


**Figure 1.** Molecular complexity towards SVS13-A, as observed by the NOEMA SOLIS Large Program. On the left, a map of $CH_3CH_2OH$ emission. On the right, spectra extracted towards the source peak position showing emission from iCOMs.

The hot corino has been successively imaged with the NOEMA interferometer in the framework of the CALYPSO [1] and SOLIS[2] (Seeds Of Life in Space; [4]) large programs. Figure 1 shows a map of ethanol emission obtained by SOLIS as well as some spectra extracted towards the source position, showing the rich chemical complexity observed in SVS13-A. More specifically, the broad spectral coverage provided by the IRAM PolyFiX correlator, allowed the detection of more than 100 emission lines from iCOMs. This was crucial to obtain a proper line identification, as well as a precise determination of the gas physical parameters, such as the rotational temperatures and the column densities. Despite the good angular resolution of the NOEMA interferometer ($0\farcs6$, 180 au at 1.4mm), iCOMs emission was unresolved towards this close binary system. The position of the two components is shown by pink triangles in Figure 1. In order to resolve the emission in the planet-forming region (~ 50 au), we performed complementary ALMA Cycle 6 high-angular resolution observations.

## 3 SVS13-A: Exploring the planet-forming region

We observe SVS13-A in ALMA Band 6 with a synthesized beam of $0\farcs2 \times 0\farcs13$ [3], clearly resolving the binary system. The dust emission shows the presence of two possible circumstellar disks as well as several circumbinary structures (see Fig. 2). Performing different beam tapering, we were able to confirm that these structures are real and not related to the beam elongation in the N-S direction. We performe a *uv* plane fitting analysis of the continuum emission and we obtain a beam deconvolved sizes of 180 × 70 au (VLA4A), and 90 × 70 au (VLA4B), for the protostellar disks. From the molecular line emission we derive

[1] https://www.oan.es/asai/
[2] https://solis.osug.fr/

systemic velocities of +7.7 km $s^{-1}$ and +8.5 km $s^{-1}$ for VLA4A and VLA4B, respectively. The moment 1 maps reveal blue-red rotation, roughly perpendicular to the outflow direction.

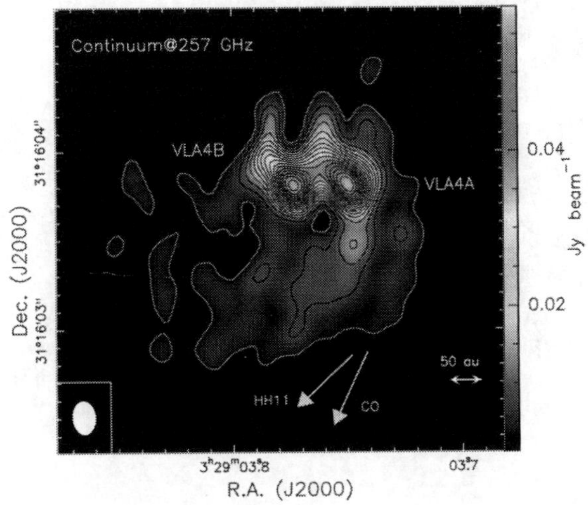


**Figure 2.** The planet-forming region (~ 50 au scale) of the SVS13-A as observed by ALMA (from [3]). The dust continuum emission at 257 GHz shows two protostellar disks and accretion streamers.

We observe several transitions from iCOMs. More specifically, we detected 3 lines of methanol and 5 of $^{13}CH_3OH$, 2 lines of acetaldehyde, 4 lines of dimethyl ether and 3 of formamide, covering a broad range of upper level energies (up to 800 K). The molecular emission is resolved towards the binary system and shows a very complex structure. Figure 3 shows the comparison between one line of $NH_2CHO$ and one line of $CH_3OCH_3$. The maps show that formamide emission is more concentrated towards VLA4A, while dimethyl ether emits towards both the protostars. A similar behaviour is observed for acetaldehyde and methanol. We perform a non-LTE large velocity gradient analysis of the methanol lines and we derive kinetic temperatures of 140 and 170 K for VLA4A and VLA4B, respectively and gas densities larger than $10^7$ $cm^{-3}$ indicating that lines are LTE populated. We assume the derived temperatures to calculate the iCOMs column densities in VLA4A and VLA4B, under LTE conditions. The chemical differentiation between VLA4A and VLA4B is evident when considering the column density ratios which are > 4 for $NH_2CHO$ and between 1–3 for the other O-bearing species. Moreover, the iCOMs emission is partially elongated towards the north, being spatially coincident with the streamers seen in the continuum.

The observed chemical segregation is independent from the transition upper level energy, thus it is very likely not related to different excitation effects. One possibility is that the chemical differentiation is related to O-bearing and N-bearing species. However, the detection of ethylene glycol only emitting from VLA4A was recently reported [7], questioning this hypothesis. A second possibility is that the chemical differentiation is a result of the combined effect of continuum opacity and different iCOMs binding energies. Indeed, continuum opacity can hamper iCOMs detection at mm wavelengths [6]. If we assume that the different molecules are emitted in a larger or more compact region they could be more or less affected by the continuum opacity. To test this scenario we perform new quantum chemistry computations to evaluate the binding energies of formamide and ethylene glycol, which are the two species so far detected only towards one component of the binary system. Their high binding energies suggest that they are emitted in a compact region around the protostar, more compact that the emitting region of methanol or dimethyl ether. This would imply that they

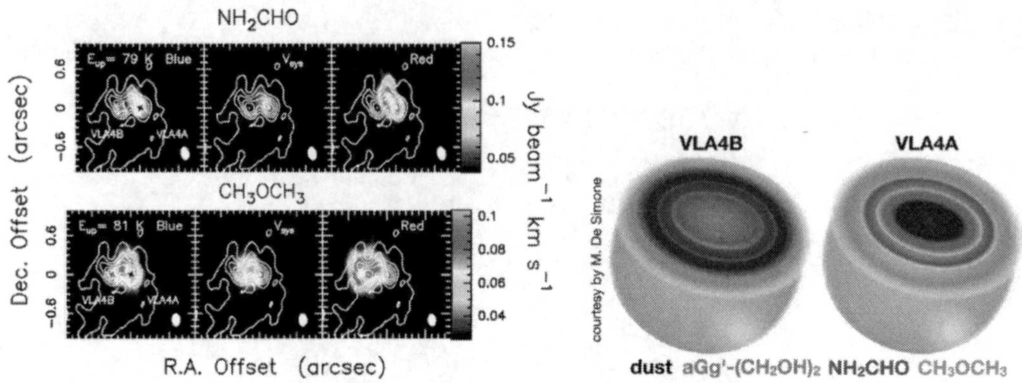


**Figure 3.** *Left panel:* Integrated emission of $NH_2CHO$ and $CH_3OCH_3$, in color scale, superposed on the dust emission in white contours (adapted from [3]). *Right panel:* Sketch illustrating the combined effect of dust opacity and different iCOMs binding energies.

are completely obscured by the dust in VLA4B where the dust is very optically thick while they are still visible in VLA4A, where the dust is less optically thick (see the sketch in Fig. 3, right panel).

## 4 Conclusions

Class I protostellar disks are the birth places of planets and it is crucial to investigate their chemical composition in order to understand the initial conditions of planets formation. In particular, broad spectral surveys and high angular resolution observations are highly complementary to study the chemistry of protostellar disks. Thanks to the synergy between the IRAM and ALMA telescopes, we investigate the chemistry of the Class I source SVS13-A from the large scales to the planet formation region ($\sim$ 50 au). We observe in the continuum two protostellar disks and accretion streamers, in addition to several emission lines from iCOMs. We observe a chemical segregation between the two components of the binary system. We interpret this chemical differentiation as the result of the combined effect of dust opacity and different binding energies.

## References

[1] Belloche, A., et al., A&A, **635**, A198 (2020)
[2] Bianchi, E., et al., MNRAS, **483**, 1850 (2019)
[3] Bianchi, E., et al., ApJL, **928**, L3 (2022)
[4] Ceccarelli, C., et al., ApJ, **850**, 176 (2017)
[5] Codella, C., et al., MNRAS, **462**, L75 (2016)
[6] De Simone, M., et al., ApJL, **896**, L3 (2020)
[7] Diaz-Rodriguez, A. K., et al., ApJ, **930**, 91 (2022)
[8] Lefèvre, C., et al., A&A, **604**, L1 (2017)
[9] Lefloch, B., et al., A&A, **334**, 269 (1998)
[10] Lefloch, B., et al., MNRAS, **477**, 4792 (2018)
[11] Öberg, K. I. & Bergin, E. A., Physics Reports, **893**, 1 (2021)
[12] Sheehan, P. D., & Eisner, J. A., ApJ, **851**, 45 (2017)
[13] Tychoniec, Ł., et al., A&A, **640**, A19 (2020)

# Molecules from evolved stars and their role in the cycle of the ISM

*Marcelino* Agúndez[1,*]

[1]Instituto de Física Fundamental, CSIC, Calle Serrano 123, 28006 Madrid, Spain

**Abstract.** The ejecta of evolved stars are among the most efficient chemical laboratories in the Universe. Many of the chemical compounds synthesized in these environments probably travel along the interstellar medium and may be ultimately delivered to planets. However, we still do not understand many of the chemical processes at work during the evolution from the AGB phase to the PN. Here we review key aspects of our current understanding of the chemistry of AGB and post-AGB envelopes, highlighting the issues that are yet to be understood.

## 1 Introduction

Evolved stars play a key role in the cosmic cycle of matter that connects interstellar clouds with planetary systems hosted by stars. It is well known that at the end of their lives, stars inject an important fraction of their mass into the surrounding interstellar medium. The mass loss process itself is a factory of chemically complex compounds that may survive the whole cycle of matter and end up in planet-forming systems, possibly playing a role on prebiotic chemistry in the early Earth and recently formed planets.

There is evidence that dust grains formed in the atmospheres of Asymptotic Giant Branch (AGB) stars have survived the whole cycle of matter chemically intact since they are identified in primitive meteorites in our Solar System [37].

Large organic molecules are detected in nebulae around evolved stars. The simplest aromatic molecule, benzene, was identified in the pre-planetary nebula (PPN) CRL 618 [8]. The so-called unidentified infrared (UIR) emission features, thought to arise from polycyclic aromatic hydrocarbons (PAHs) [4, 25], are widely observed toward planetary nebulae (PNe) [19], and fullerenes have been also detected mainly toward PNe [6, 18]. These large organics are synthesized around evolved stars and, provided they are large enough to be stable against ultraviolet (UV) radiation, they probably survive all the way along the cycle of matter, either in the gas phase or deposited on dust grains, down to the formation of planets [24]. Moreover, given that $C_{60}^{+}$ has been identified as carrier of various diffuse interstellar bands (DIBs) [7], it is very suggestive to think that DIB carriers are also synthesized during the late evolutionary stages of stars.

It is clear that large organic molecules are synthesized in situ around evolved stars, although it is not yet clear which is the underlying chemical mechanism and whether they are formed during the AGB or post-AGB stage. UIR emission features are not detected toward

*e-mail: marcelino.agundez@csic.es

 https://doi.org/10.1051/epjconf/202226500029

**Table 1.** The 91 molecules detected in IRC +10216 (as of July 2022).

| | | | | | | | |
|---|---|---|---|---|---|---|---|
| **$10^{-3}$** | CO | | | | | | |
| **$10^{-4}$** | | | | | | | |
| | | $C_2H_2$ | | | | | |
| | | | HCN | | | | |
| **$10^{-5}$** | | | | | | | |
| | | $CH_4$ | | | | | |
| | | $C_2H$ | $NH_3$ | | | | |
| | | $C_4H$ | CN | | | | |
| | | $C_2$ | $HC_3N$ | | $SiC_2$ | | |
| **$10^{-6}$** | | $C_3$ | | | SiS | | |
| | | $C_4H_2$ | $C_3N$ | CS | $Si_2C$ | | |
| | | | | | $SiH_4$ | | |
| | | | $HC_5N$ | | SiO | | HCl |
| **$10^{-7}$** | $H_2O$ | $C_5$ | HNC | | | | |
| | | l-$C_3H$ | $CN^-$ | | | | |
| | OH | $C_6H$ | | | SiC | | |
| | | $C_5H$ | $CH_3CN$ | $C_2S$ | | | AlCl |
| | | c-$C_3H_2$ | | | | | |
| | | $CH_3C_2H$ | | | | | |
| | | c-$C_3H$ | $HC_7N$ | | | HCP | |
| | | $C_2H_4$ | | | | | $MgC_4H$ |
| | $H_2CO$ | $H_2C_4$ | | $C_3S$ | $CH_3SiH_3$ | | $MgC_6H$ |
| **$10^{-8}$** | | | | | | CP | HF |
| | | $C_8H$ | $HC_9N$ | $H_2CS$ | SiN | $PH_3$ | MgNC |
| | | | $CH_2CN$ | | | | AlF |
| | | | $HC_2N$ | | | | $MgC_3N$ |
| | | | $C_5N$ | | | | $MgC_5N$ |
| | | $C_7H$ | HCCNC | | | | $MgC_2H$ |
| | | $H_2C_6$ | $C_2H_3CN$ | $H_2S$ | c-$SiC_3$ | | NaCN |
| | | $C_6H^-$ | $C_5N^-$ | | $SiC_4$ | | |
| | $C_3O$ | $C_8H^-$ | $HC_4N$ | | SiCN | | |
| | | $H_2C_3$ | $C_3N^-$ | $C_5S$ | SiNC | PN | NaCl |
| **$10^{-9}$** | | | | | | $C_2P$ | AlNC |
| | $HCO^+$ | | HNCCC | | | | MgCN |
| | | $C_4H^-$ | | | | | KCl |
| | | | | | | | KCN |
| | | | | | | | FeCN |
| | | | | | | | HMgNC |
| **$10^{-10}$** | | | | | | | CaNC |

The 91 molecules detected in IRC +10216 ordered vertically by abundance relative to $H_2$ and grouped in different columns according to their chemical composition. Colors indicate their most probably origin: either chemical equilibrium in the stellar atmosphere (red), non-equilibrium chemistry in the inner envelope (green), and photochemistry in the outer envelope (blue). Molecular hydrogen (not shown) has been also detected in IRC +10216 [17].

AGB stars, although this does not necessarily imply that their carriers are not present at this stage. It may simply reflect the lack of a sufficiently intense UV field, which is necessary to excite PAHs and produce the emission features in the infrared [5]. Attempts to detect DIBs in circumstellar envelopes around AGB stars have resulted inconclusive [23]. In the remaining of this contribution we briefly discuss what do we know about the chemical composition and the chemical processes at work in envelopes around AGB and post-AGB stars.

## 2 Chemistry in the AGB phase

The most important parameter that governs circumstellar chemistry during the AGB phase is the C/O elemental abundance ratio at the stellar surface, which dichotomizes AGB stars

**Table 2.** The 27+7 molecules detected in oxygen-rich AGB envelopes (as of July 2022).

| | | | | | | |
|---|---|---|---|---|---|---|
| **$10^{-4}$** | CO | | | | | |
| | $H_2O$ | | | | | |
| **$10^{-5}$** | OH | | | | | |
| | $CO_2$ | | | | | |
| | | | $SO_2$ | SiO | | |
| | | NO | SO | SiS | | |
| | | | | | PO | |
| **$10^{-6}$** | | | $H_2S$ | | | |
| | | HCN | CS | | | |
| | | $NH_3$ | | | PN | |
| | $H_2CO$ | | OCS | | | |
| **$10^{-7}$** | | CN | | | | TiO |
| | | HNCO | | | | $TiO_2$ |
| | $CH_3OH$ | | NS | | | AlO |
| **$10^{-8}$** | $HCO^+$ | $N_2H^+$ | HNCS | | | AlF |
| | | | | | | AlCl |
| | | HNC | | | | NaCl |
| | | $HC_3N$ | | | | MgS ? |
| **$10^{-9}$** | | | | | | AlOH |

Table shows in color black the 27 molecules detected in oxygen-rich AGB envelopes (from the inventory of molecules in IK Tau, *o* Cet, and R Leo). The molecule MgS, tentatively detected toward OH 26.5+0.6 [21], is included with the abundance predicted by chemical equilibrium [3]. Also included in color magenta are 7 molecules detected in other oxygen-rich evolved stars, such as OH 231.8+4.2 (probably in a post-AGB stage) and the massive evolved stars IRC +10420 and VY CMa.

as carbon-rich (C/O > 1) or oxygen-rich (C/O < 1). Chemical equilibrium indicates that the pressures and temperatures of AGB atmospheres lie in the domain of stability of CO, meaning that this molecule locks most of the limiting element and allow the element in excess to form C-bearing molecules, when C/O > 1, or O-bearing-molecules, when C/O < 1. Carbon-rich envelopes contain a far richer diversity of molecules than oxygen-rich envelopes. As an illustration in Table 1 we show the molecules detected to date in the C-rich envelope IRC +10216 (the nearest high mass loss rate C-rich AGB star), which can be compared with Table 2, where we list the 27 molecules detected in O-rich AGB envelopes together with 7 molecules detected in related environments. That is, the inventory of molecules in C-rich AGB envelopes is about three times larger than in O-rich envelopes, and this is due to the exceptional chemical properties of carbon and its ability to form strong C-C bonds

Chemical equilibrium provides a good overall description of the chemical composition of the inner regions of AGB envelopes, although there are notable failures. Concretely, some molecules are observed to be anomalously abundant compared with the predictions of chemical equilibrium, which points to some non-equilibrium chemical processes at work in the inner layers of AGB envelopes. The most dramatic cases of anomalously abundant molecules are the hydrides $H_2O$, $NH_3$, $SiH_4$, and $PH_3$ in C-rich envelopes, while in O-rich envelopes HCN, $NH_3$, and CS stand out [3]. Two mechanisms have been proposed to account for these anomalies, either shock-induced chemistry associated to the stellar pulsation [13, 14] or photochemistry driven by the penetration of external UV photons through the clumpy structure of the envelope [1, 34]. None of the two mechanisms are fully satisfactory. For example, the shock-induced scenario involve shocks of tens of km $s^{-1}$, which are not evident in the line profiles of IRC +10216 seen at high angular resolution [10], and the abundance of $H_2O$ in C-rich outflows is extremely sensitive to the chemical network adopted [13, 14]. In the case of the photochemical scenario, the resulting molecular abundances are very sensitive to the envelope clumpiness, which is a parameter difficult to constrain from observations.

It is worth noting that chemical equilibrium predicts PAHs to be abundant at a few stellar radii from a C-rich AGB star [11, 33], although models based on chemical kinetics predict a low yield of PAHs in the inner wind of C-rich AGB stars [12]. That is, the presence of abundant PAHs in the inner envelopes of C-rich AGB envelopes is thermodynamically favorable, although their formation would likely occur under non-equilibrium conditions, in close connection with the formation of carbonaceous dust, and models do not predict high yields of PAHs. Recently, the formation of carbonaceous dust in C-rich AGB outflows has been mimicked in the laboratory by exposing carbon atoms to $H_2$ under ultra-high vacuum conditions [27]. In these experiments, in which amorphous carbon nanoparticles were produced, aliphatic carbon clusters were formed but neither aromatic species nor fullerenes were observed.

Understanding how dust is formed in the surroundings of AGB stars is one of the main pending challenges. Chemical equilibrium indicates that the first condensates, graphite or amorphous carbon in carbon-rich stars and alumina ($Al_2O_3$) in oxygen-rich stars, are expected to appear at temperatures below 2000 K [3], while temperatures at the stellar photosphere of AGB stars are typically above 2000 K. That is, the stellar surface should be free of solids and dust formation is therefore expected to occur farther out at the expense of gaseous particles. Identifying these gas-phase precursors of dust is therefore of great interest to shed light on the dust formation process. Some constraint are available from single-dish observations of AGB envelopes, which suggest that $SiC_2$, SiO, and CS (in C-rich objects) while SiO and possibly SO (in O-rich sources) act as gas-phase precursors of dust [28–30]. Interferometric observations of O-rich AGB stars have also allowed to observe two potential gas-phase precursors of alumina dust, AlO and AlOH, in the surroundings of the stellar atmosphere [15, 22]. The abundances derived are well below the elemental abundance of aluminum, which suggests that the remaining aluminum could have already participated in the formation of $Al_2O_3$ dust.

In the outer layers of AGB envelopes the penetration of external UV photons modify the chemical composition of the outflowing material through photochemistry, which in the case of C-rich envelopes drives an interesting chemical processing. Detailed studies of IRC +10216 show that the photodissociation of HCN and $C_2H_2$ result in the formation of polyacetylenic chains of increasing length, such as polyynes $C_{2n+2}H_2$ and cyanopolyynes $HC_{2n+1}N$, with $n = 1, 2, 3, ...$ [2]. Large unsaturated organic molecules are therefore formed under the action of photochemistry in the outer layers of C-rich envelopes, although these are mainly chains. Aromatic rings are not predicted to form through this mechanism.

## 3 Chemistry in the post-AGB phase

As an AGB star evolves from red giant to white dwarf, the chemical composition of the circumstellar material experiences important changes driven by the increasing UV flux from the central star. This energetic radiation causes the ionization and photodissociation of molecules and, as a consequence, the chemical complexity drops drastically. As an example, we show in Table 3 the molecular inventory in the C-rich PN NGC 7027, which can be compared to that of IRC +10216 shown in Table 1. Only a few simple molecules survive the AGB $\rightarrow$ PN transition. This sharp decline in the chemical complexity contrasts with the appearance in PNe spectra of UIR and fullerene emission features, which demonstrate that large and chemically complex organic molecules are present at this stage. Although it is not completely settled whether such molecules are formed during the PPN or PN stage or are inherited from the previous AGB phase, the fact that the simplest aromatic ring, benzene, is detected in the PPN CRL 618 and not in IRC +10216 suggest that aromatic molecules of increasing size are

**Table 3.** The inventory of molecules detected in NGC 7027 (as of July 2022).

| | | | | |
|---|---|---|---|---|
| **$10^{-3}$** | CO | | | |
| **$10^{-4}$** | | | | |
| **$10^{-5}$** | | | | |
| **$10^{-6}$** | | CH | | |
| | OH | | | |
| **$10^{-7}$** | $HCO^+$ | | HCN | |
| | | $CH^+$ | CN | |
| **$10^{-8}$** | $H_2O$ | $C_2H$ | $HC_3N$ | CS |
| | | | HNC | |
| | | $c$-$C_3H_2$ | $N_2H^+$ | |
| **$10^{-9}$** | | | | $HCS^+$ |
| | $CO^+$ | | | |
| **$10^{-10}$** | | | | |

Table constructed from [36] and references therein.

formed during the PPN stage due to photoprocessing, either through a bottom-up mechanism starting from simple hydrocarbons such as $C_2H_2$ [9] or by a top-down scheme involving the photoevaporation of very small carbonaceous grains [31]. Given that DIBs are thought to arise from large organic molecules, it is not unwise to think that they could be synthesized around evolved stars, although attempts to search for DIBs around post-AGB objects have been inconclusive [16, 26].

Insights into the chemical processing taking place after the AGB phase can be provided by the study of objects in a very early post-AGB stage, as it is likely the case of OH 231.8+4.2. Unexpected molecules for an O-rich AGB or post-AGB envelope such as HNCO, HNCS, $HC_3N$, and $CH_3OH$ have been detected in this object [32, 35]. It is yet unclear how are these molecules formed, but shocks emerge as a feasible agent able to drive their synthesis. In particular, $CH_3OH$ emission in OH 231.8+4.2 shows a bipolar outflow morphology [32], similarly to SiO, which strongly suggests that methanol is released from grains by shocks. The most probable formation scenario thus involves the previous condensation of CO onto dust grains, followed by hydrogenation. Although ices are not a typical ingredient around evolved stars, water ice has been observed toward some evolved objects [20]. The detection of methanol implies that CO ice should also be present. The survival of CO ice is more challenging than that of $H_2O$ ice because it requires very low dust temperatures in order to avoid thermal sublimation.

## 4 Conclusions

The ejecta around evolved stars are formidable factories of organics. In particular, planetary nebulae show in their spectra emission features characteristic of very large organic molecules such as aromatic polycycles and fullerenes, the formation of which is still a mystery. In the previous AGB stage, a wealth of organic molecules are formed due to a variety of processes such as photochemistry, but these are mostly carbon chains rather than aromatic cycles. Currently, there is no evidence on the formation of aromatic molecules during the AGB phase from either observations, chemical models, or experiments. The current paradigm is therefore that the aromatization takes place during the post-AGB phase, although this scenario is open for new discoveries.

## References

[1] Agúndez, M., Cernicharo, J., & Guélin, M. 2010, ApJ, 724, L133
[2] Agúndez, M., Cernicharo, J., Quintana-Lacaci, G., et al. 2017, A&A, 601, A4
[3] Agúndez, M., Martínez, J. I., de Andres, P. L., et al. 2020, A&A, 637, A59
[4] Allamandola, L. J., Tielens, A. G. G. M., & Barker, J. R. 1985, ApJ, 290, L25
[5] Allamandola, L. J., Tielens, A. G. G. M., & Barker, J. R. 1989, ApJS, 71, 733
[6] Cami, J., Bernard-Salas, J., Peeters, E., & Malek, S. E. 2010, Science, 329, 1180
[7] Campbell, E. K., Holz, M., Gerlich, D., & Maier, J. P. 2015, Nature, 523, 322
[8] Cernicharo, J., Heras, A. M., Tielens, A. G. G. M., et al. 2001, ApJ, 546, L123
[9] Cernicharo, J. 2004, ApJ, 608, L41
[10] Cernicharo, J., Daniel, F., Castro-Carrizo, A., et al. 2013, ApJ, 778, L25
[11] Cherchneff, I. & Barker, J. R. 1992, ApJ, 394, 703
[12] Cherchneff, I., Barker, J. R., & Tielens, A. G. G. M. 1992, ApJ, 401, 269
[13] Cherchneff, I. 2006, A&A, 456, 1001
[14] Cherchneff, I. 2011, A&A, 526, L11
[15] Decin, L., Richards, A. M. S., Waters, L. B. F. M., et al. 2017, A&A, 608, A55
[16] Díaz-Luis, J. J., García-Hernández, D. A., Kameswara Rao, N., et al. 2015, A&A, 573, A97
[17] Fonfría, J. P., DeWitt, C. N., Montiel, E. J., et al. 2022, ApJ, 927, L33
[18] García-Hernández, D. A., Manchando, A., P. García-Lario, P., et al. 2010, ApJ, 724, L39
[19] Hony, S., Van Kerckhoven, C., Peeters, E., et al. 2001, A&A, 370, 1030
[20] Justtanont, K., Olofsson, G., Dijkstra, C., & Meyer, A. W. 2006, A&A, 450, 1051
[21] Justtanont, K., Muller, S., Barlow, M. J., et al. 2019, in Why Galaxies Care about AGB stars: A Continuing Challenge through Cosmic Time, Proc. IAU Symp. 343, eds. F. Kerschbaum, M. Groenewegen, & H. Olofsson
[22] Kamiński, T., Wong, K. T., Schmidt, M. R., et al. 2016, A&A, 592, A42
[23] Kendall, T. R., Mauron, N., McCombie, J., & Sarre, P. J. 2002, A&A, 387, 624
[24] Kwok, S. 2009, Int. J. Astrobiol., 8, 161
[25] Leger, A. & Puget, J. L. 1984, A&A, 137, L5
[26] Luna, R., Cox, N. L. J., Satorre, M. A., et al. 2008, A&A, 480, 133
[27] Martínez, L., Santoro, G., Merino, P., et al. 2020, Nature Astron., 4, 97
[28] Massalkhi, S., Agúndez, M., Cernicharo, J., et al. 2018, A&A, 611, A29
[29] Massalkhi, S., Agúndez, M., & Cernicharo, J. 2019, A&A, 628, A62
[30] Massalkhi, S., Agúndez, M., Cernicharo, J., & Velilla-Prieto, L. 2020, A&A, 641, A57
[31] Rapacioli, M., Joblin, C., & Boissel, P. 2005, A&A, 429, 193
[32] Sánchez Contreras, C., Alcolea, J., Bujarrabal, V., et al. 2018, A&A, 618, A164
[33] Tejero, J. & Cernicharo, J. 1991, Modelos de Equilibrio Químico Aplicados a Envolturas Circunestelares de Estrellas Evolucionadas (Madrid: IGN).
[34] Van de Sande, M., Sundqvist, J. O., Millar, T. J., et al. 2018, A&A, 616, A106
[35] Velilla-Prieto, L., Sánchez Contreras, C., Cernicharo, J., 2015, A&A, 575, A84
[36] Zhang, Y., Kwok, S., & Dinh-V-Trung 2008, ApJ, 678, 328
[37] Zinner, E. 1998, Annu. Rev. Earth Planet. Sci., 26, 147

# Multi-line diagnostics of the axi-symmetric wind around the MS-type AGB star RS Cancri

*Jan Martin* Winters[1,*], *Thibaut* Le Bertre[2], *Do Thi* Hoai[3], *Ka Tat* Wong[1], *Pham Tuyet* Nhung[3], *Pierre* Lesaffre[4], *Wonju* Kim[5], and *Pierre* Darriulat[3]

[1]IRAM, 300 rue de la Piscine, 38406 St. Martin d'Hères, France
[2]LERMA, Observatoire de Paris, 61 av. de l'Observatoire, 75014 Paris, France
[3]VNSC, VAST, 18 Hoang Quoc Viet, Cau Giay, Ha Noi, Viet Nam
[4]LPENS, 24 rue Lhomond, 75231 Paris, France
[5]I. Physikalisches Institut, Uni Köln, Zülpicher Str. 77, 50937 Köln, Germany

**Abstract.** We present data recently obtained with NOEMA/PolyFiX on the MS-type Asymptotic Giant Branch (AGB) star RS Cnc, allowing to characterize in detail the structure of its wind. These results became possible by the recent advances of the NOEMA observatory, which combine large spectral bandwidth with high sensitivity together with still increasing spatial resolution.

## 1 Introduction

We observed the largely unexplored intermediate type AGB star RS Cnc, a star already mixing Carbon up to its photosphere by the third dredge-up process. It is a relatively nearby semi-regular variable star (d = 150 pc) with a stellar temperature on the upper end for an AGB star ($T_\star$ = 3200 K). It develops a wind with a low terminal outflow speed of 8 $\mathrm{kms}^{-1}$ and a relatively low mass loss rate of the order of $10^{-7} M_\odot \mathrm{yr}^{-1}$. Until our work, only CO, SiO and HI were detected in the mm to radio range in this object (see, e.g., [1, 2]).
We picked RS Cnc because of its peculiar CO line profiles, which are composed of a broad pedestal with a super-imposed narrow central component (Fig. 1, right). An early interpretation of this kind of composite line profiles was given by [3] in terms of successive winds with different mass-loss rate and different outflow velocity, whereas [1] proposed an interpretation for RS Cnc in terms of a bipolar outflow with an equatorial enhanced density structure.

## 2 Results

### 2.1 Molecular lines

A frequency range of 32 GHz in the 1 mm band was covered with two spectral setups using NOEMA's wideband correlator PolyFiX to observe diagnostically useful lines of CO and SiO, but also other interesting lines, e.g., from HCN and $H_2O$. In the end, we detect many more species than expected, including their isotopologs, i.e., SO, $SO_2$, $^{34}SO$, vibrationally excited $H_2O$, $Si^{17}O$, $^{29}SiO$, HCN, $H^{13}CN$, and even PN. In total, we detect 32 lines, emitted by 13 molecules and isotopologs, of which the latter nine species are first detections in RS Cnc in the mm range. All details of the observations are given in [4].

*e-mail: winters@iram.fr

## 2.2 Morpho-kinematics of the wind

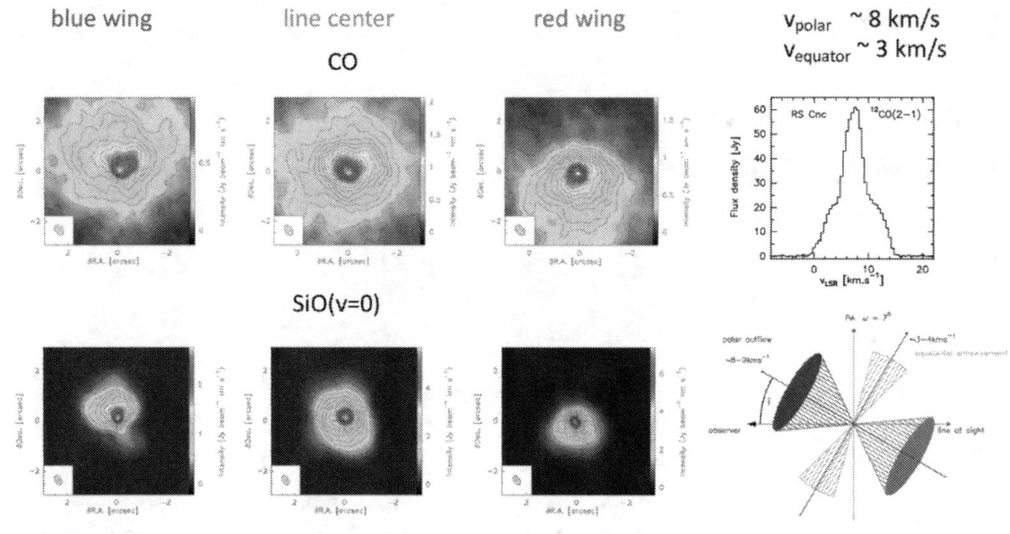


**Figure 1.** Integrated intensity maps in CO(2-1) and SiO(v=0,6-5) (left), CO(2-1) line profile (upper right), and sketch of the outflow structure (lower right).

In particular from CO, but also from SiO, information is obtained on the morpho-kinematic structure of the large-scale outflow. In Fig. 1 we show integrated intensity maps of the blue line wing, the line center, and the red line wing of CO(2-1) on top and of SiO(6-5) on the bottom row. In both lines, our modeling reveals a bipolar structure expanding at ~8 kms$^{-1}$ toward us in the North and receding in the South, plus an equatorial enhanced density structure expanding at a much lower speed of ~3 kms$^{-1}$. We note that SiO particularly probes the inner parts of the circumstellar shell whereas CO extends much farther outward.

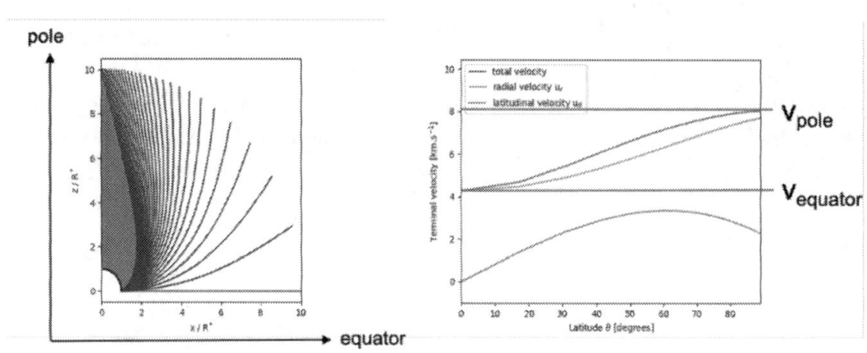


**Figure 2.** Toroidal magnetic field: resulting parcel trajectories in the $x$–$z$ plane (left) and wind velocities versus stellar latitude (right).

A convenient interpretation of this outflow structure would be in terms of a companion. However, no companion has been detected in RS Cnc by now (which does not exclude the presence of one). More importantly, a companion would not easily explain the velocity structure, showing higher velocities toward the poles than toward the equator. We therefore propose an alternative interpretation in terms of a toroidal magnetic field that would funnel material toward the poles and at the same time would naturally provide outflow velocities that are small at the equator and increase toward the polar direction due to acceleration by the Lorentz force. Figure 2 presents a simplified model of a magnetized wind in such a scenario.

### 2.3 Rotating structure

Zooming into smaller scales, as probed by multiple species such as $SO_2$, HCN, SO, and (v=1) vibrationally excited lines of SiO, we find a velocity structure rotated by 90° with respect to the bipolar outflow and with velocities of the order of $\pm 1$ $kms^{-1}$. Figure 4, left plot shows the first-moment map of HCN emission. We interpret this velocity pattern in terms of rotation.

### 2.4 Broad line wings - photospheric shocks?: The innermost region

From SiO we detect several lines in different vibrational states up to v=2 and from several isotopes, a gallery of some of those is shown in Fig. 3. It is worth noting that the SiO(v=1,J=5-4) line was masing in 2017 around light minimum. However, in 2020, about 3.2 cycles later, the maser had disappeared. Temporal variations of SiO masers are quite common for pulsating AGB stars, but that the maser completely disappeared came as a surprise.

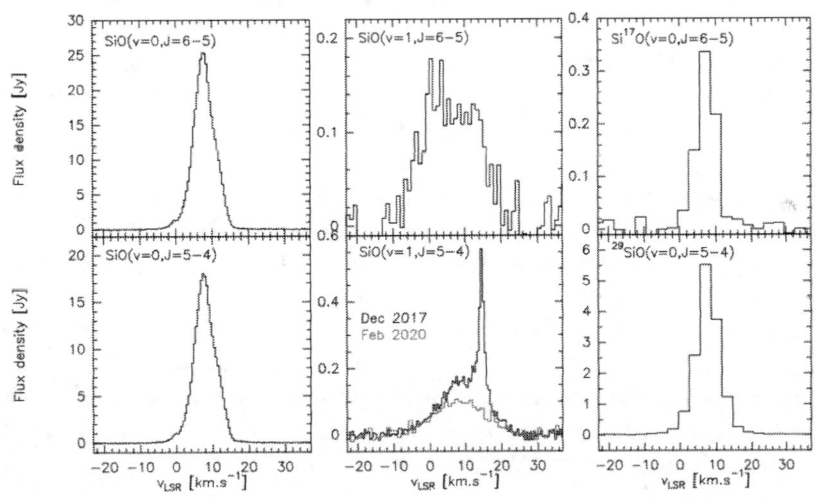


**Figure 3.** Gallery of detected SiO lines.

A striking feature of the SiO lines is their broad wings that indicate velocities almost twice as high as the maximum velocities seen in CO: this has been seen in many other similar stars, like EP Aqr ([5]), R Dor ([6]), and in most of the stars covered by the ALMA large program ATOMIUM on AGB stars ([7]). The map in the right plot of Fig. 4 displays the outflow velocity versus projected distance from the line of sight toward the star, showing that these high velocities only occur close to the central line of sight, i.e. very close to the star (the black line indicates the polar outflow velocity measured on the CO lines). As this type of velocity distribution is seen in many stars, and independent of the orientation of a possible symmetry axis with respect to the line of sight, a common, non-directional mechanism is needed to explain it. Our interpretation is in terms of shocks produced by the stellar pulsation

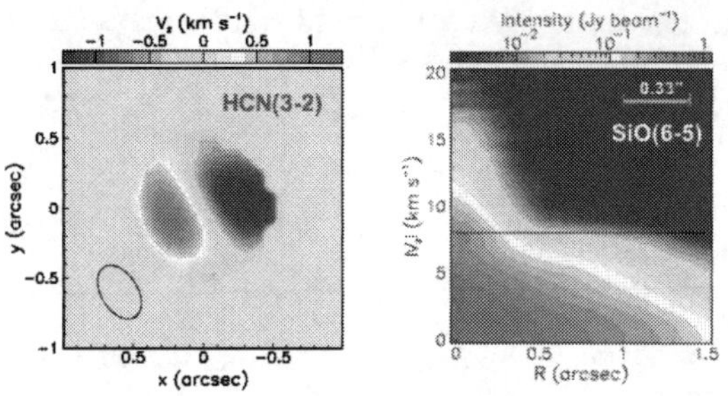


**Figure 4.** Left: Rotating structure as probed in HCN. Right: High velocities close to the line of sight toward the star as probed in SiO.

that provide high velocities close to the star but are damped out when traveling through the stellar atmosphere. These velocities can be higher than the final expansion velocity of the wind, as was already argued by [8] and [9] based on SiO maser observations, and discussed in [10, 11], based on hydrodynamic models.

## 3 Summary

With PolyFiX we detect nine new molecules and isotopologs in RS Cnc (see Sect. 2.1) that serve as detailed diagnostics of the star's environment on very different spatial scales. We essentially confirm the bipolar outflow structure with an equatorial density enhancement. In addition we find a rotating region close to the star, as well as some high-velocity material in the innermost region, which likely indicates shocks traveling through the stellar atmosphere.

In the future we plan to investigate in more detail the role of magnetic fields in shaping the wind structure and to apply a more detailed radiative transfer modeling of the detected lines to better determine the thermodynamic structure of the shell and, in particular, the expected imprint of atmospheric shocks on the line profiles of the different molecular tracers.
In order to investigate, whether the velocity structure seen in HCN and other molecules really indicates rotation, we will propose to carry out observations with higher spatial resolution that may also reveal if the corresponding velocity pattern is Keplerian or not. These observations should become possible by the end of 2022 thanks to the upcoming extended baselines of NOEMA, providing a spatial resolution increased by a factor of two with respect to the current status presented here, and allowing us to resolve a region of about $20\,R_\star$ at RS Cnc's distance. A direct probing of the potentially shocked regions however needs a much higher spatial resolution that may remain a dream for some while.

## References

[1] Y. Libert, J.M. Winters, T. Le Bertre, E. Gérard, L.D. Matthews, A&A **515**, A112 (2010)
[2] P. de Vicente, V. Bujarrabal, A. Díaz-Pulido et al., A&A **589**, A74 (2016)
[3] G.R. Knapp, K. Young, E. Lee, A. Jorissen, ApJS **117**, 209 (1998)
[4] J.M. Winters, D.T. Hoai, K.T. Wong et al., A&A **658**, A135 (2022)
[5] P. Tuan-Anh, D.T. Hoai, P.T. Nhung et al., MNRAS **487**, 622 (2019), 1905.02715
[6] L. Decin, A.M.S. Richards, T. Danilovich, W. Homan, J.A. Nuth, A&A **615**, A28 (2018)
[7] C.A. Gottlieb, L. Decin, A.M.S. Richards et al., A&A **660**, A94 (2022)
[8] J. Cernicharo, J. Alcolea, A. Baudry, E. González-Alfonso, A&A **319**, 607 (1997)
[9] F. Herpin, A. Baudry, J. Alcolea, J. Cernicharo, A&A **334**, 1037 (1998)
[10] J.M. Winters, T. Le Bertre, K.S. Jeong, C. Helling, E. Sedlmayr, A&A **361**, 641 (2000)
[11] J.M. Winters, T. Le Bertre, L.Å. Nyman, A. Omont, K.S. Jeong, A&A **388**, 609 (2002)

# Fragmentation and disk formation in high-mass star formation: The IRAM large program CORE

*Henrik* Beuther[1,*], *Caroline* Gieser[1], *Aida* Ahmadi[2], *Sümeyye* Suri[3], *Jan Martin* Winters[4], *Joe* Mottram[1], and the CORE team[**]

[1]Max Planck Institute for Astronomy, Königstuhl 17, 69117 Heidelberg, Germany
[2]Leiden Observatory, Leiden University, PO Box 9513, 2300 RA Leiden, The Netherlands
[3]Department of Astrophysics, University of Vienna, Türkenschanzstrasse 17, 1180 Vienna, Austria
[4]IRAM, 300 rue de la Piscine, Domaine Universitaire de Grenoble, 38406 St.-Martin-d'Hères, France

**Abstract.** The IRAM CORE large program combines data from NOEMA and the IRAM 30 m telescope to study a diverse set of physical and chemical processes during the formation of high-mass stars. Here, we present a selected compilation of exciting results obtained during the survey.

## 1 Introduction

Some of the central questions in high-mass star formation research relate to the fragmentation and disk formation processes as well as the physical and chemical conditions during the formation of massive stars. To address these questions, we conducted the IRAM large program CORE "Fragmentation and disk formation in high-mass star formation" combining observations with the Northern Extended Millimeter Array (NOEMA) and the IRAM 30 m telescope. An introduction to the survey, its scope and initial results focusing on the 1.3 mm continuum emission can be found in [1]. In addition to the original CORE program targeting 20 high-mass protostellar objects, we extended the survey to earlier evolutionary stages via the CORE-extension project that targets two younger regions at the onset of high-mass star formation [2, 3].

## 2 Selected results

In the following, we will present some selected results corresponding to the important scientific topics studied within the CORE project. In addition to the studies comprising the entire CORE sample [1, 4] and Ahmadi et al. (in prep.), we also conducted a broad range of individual case studies to dissect the processes in individual sources at depth [5–11].

### 2.1 Fragmentation

Figure 1 presents the 1.3 mm continuum data of the entire CORE sample [1]. While the angular resolution varied roughly between 0.3 – 0.4″, here we show all regions converted to

*e-mail: beuther@mpia.de
**http://www.mpia.de/core

the same linear box size of $\sim 4 \times 10^4$ au. Already a visual inspection of the data shows a broad variety in fragmentation properties, from regions dominated by a single massive core to those that harbor up to 20 cores. Analyzing the core separations by means of a minimum-spanning-tree analysis, the CORE regions are consistent with thermal Jeans fragmentation. Nevertheless, the diversity of fragments per region indicates that other processes are likely at play. Variations of the initial density profiles and/or magnetic field could cause the diversity of fragmentation properties. These questions are currently investigated by ongoing follow-up projects.

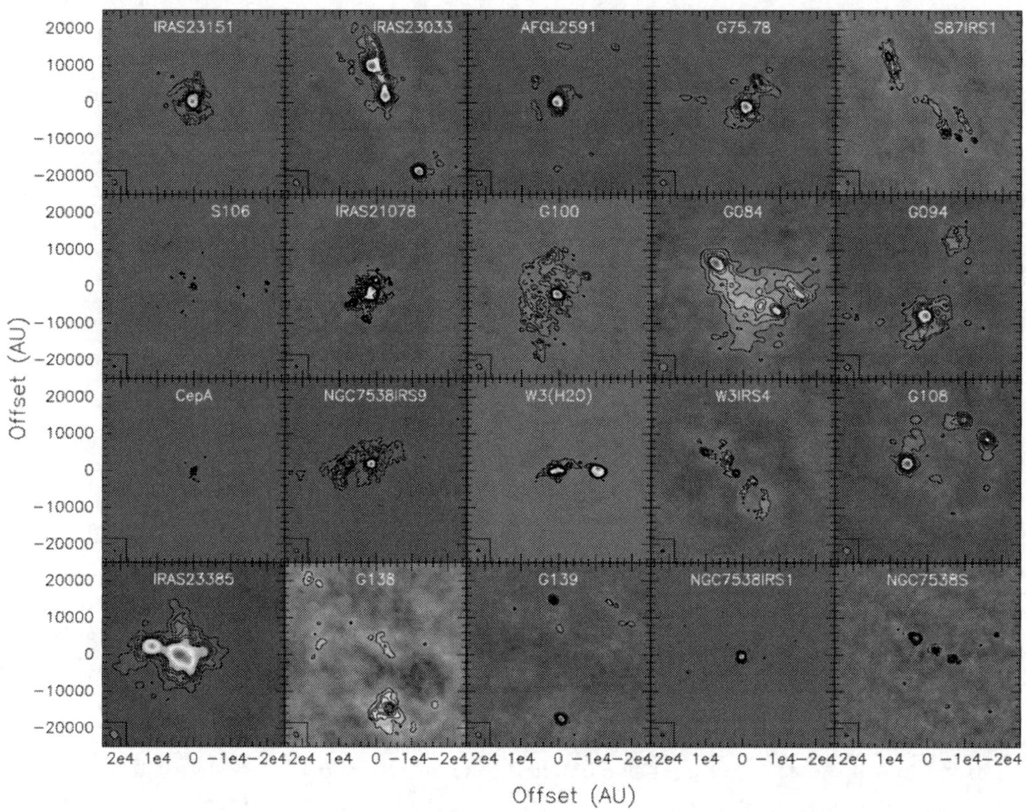


**Figure 1.** Compilation of 1.3 mm continuum images of the CORE sample converted to linear resolution elements [1]. The contouring is in $4\sigma$ steps. The sources are labeled in each panel, and the synthesized beam is shown at the bottom-left of each panel.

## 2.2 Disk formation

While protostellar disks are well studied around low-mass protostars, they are significantly less explored in the high-mass regime. Within the CORE project with linear resolution elements typically on the order of ~1000 au, we usually cannot resolve the innermost disks. But nevertheless, we can study the rotational properties of the larger-scale disk-like structures. From the 20 CORE regions, we identify in 12 of them rotational motions perpendicular to the outflows (Ahmadi et al. in prep.). At high angular resolution, some even fragment into multiple disk-sources like W3($H_2O$) [5]. Fitting the $CH_3CN$ data with XCLASS, we can derive temperature maps of the regions. Employing furthermore the column density information from the 1.3 mm continuum data and assuming Keplerian rotation, we can infer Toomre $Q$ stability maps. Figure 2 (left) presents the corresponding Toomre $Q$ map of the multiple system W3($H_2O$) [5]. While the $Q$ values are high close to the protostars, in the outskirts of the region we find decreasing $Q$ values, in the west coinciding with an additional mm continuum peak. This is a strong indication for fragmentation on disks scales.

For one region, AFGL2591, we have even higher-resolution data at 0.19″ resolution with NOEMA at 843 $\mu$m wavelengths (Fig. 2 right). These data allow us to spatially resolve the fragmenting disk [11]. Combining this disk fragmentation with the thermal Jeans fragmentation investigated in section 2.1, we can differentiate between two fragmentation modes: core fragmentation on cluster scales and disk fragmentation on scales of individual multiple systems.

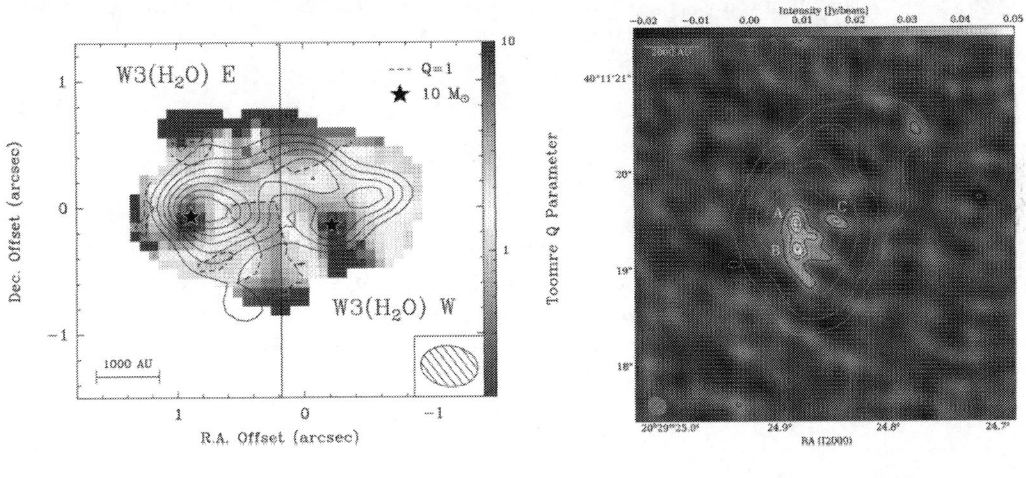


**Figure 2.** *Left:* Toomre $Q$ map of the W3($H_2O$) region [5]. The contours present the 1.3 mm continuum data, and the stars mark the locations of the two assumed 10 $M_\odot$ protostars. *Right:* Resolved disk-structure in AFGL2591 [11]. The color-scale shows the 843 $\mu$m continuum emission at 0.19″ resolution whereas the orange contours present the previous 1.3 mm continuum emission at 0.4″ resolution. The letters mark the three main structures and the red star marks the position of the central protostar assumed to coincide with the 1.3 mm continuum peak.

## 2.3 Physical properties

What are the physical properties, in particular the density and temperature structures of high-mass star-forming regions on small core scales? While density $\rho \propto r^{-p}$ (left) and temperature profiles $T \propto r^{-q}$ have been predicted since decades [12, 13], no statistical analysis of these structural parameters existed on scales of several thousand au around forming high-mass protostars. Within the CORE project, we now derived the temperature as well as the density structures of a large sample of cores.

While temperatures are derived from spectral line fitting, in particular the $H_2CO$ lines in the 1 mm band, the situation is more difficult for the density structures since the derivation relies on the continuum emission. In contrast to the line data where complementary short spacing observations with the 30 m telescope exist, and hence all spatial structures are covered, the continuum data suffer from missing flux. Therefore, the continuum data were fitted directly in the uv-plane. Using the derived slopes of the uv-intensity fits, and adding-in the

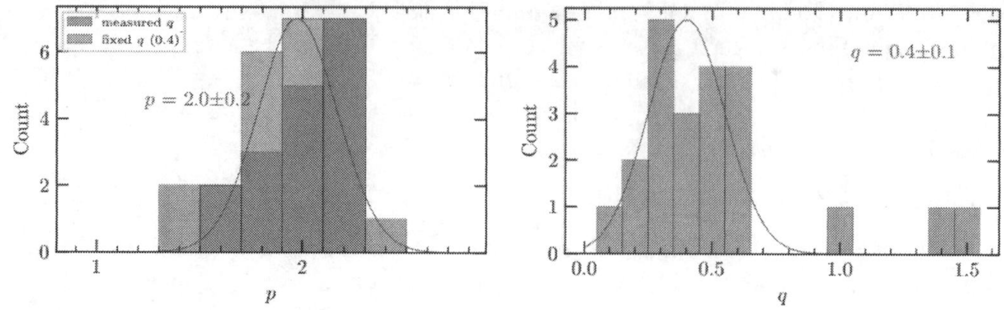


**Figure 3.** Histograms of the power-law distributions of the density $\rho \propto r^{-p}$ (left) and temperature structure $T \propto r^{-q}$ (right) derived on core scales [4]. The green and blue histograms in the left panel correspond to the density coefficients assuming fixed or measured $q$.

temperature information, we can infer the density structures. We apply this analysis to in total 22 fragmented cores. All details can be found in [4].

Figure 3 presents the histograms of all derived power-law indices for the density as well as temperature structures [4]. While the density power-law index ranges around 2, the temperature power-law distribution peaks around 0.4. These mean values agree extremely well with theoretical models of star-forming cores heated from an internal protostar. It is exciting to see that long-standing predictions can finally be confirmed by observations.

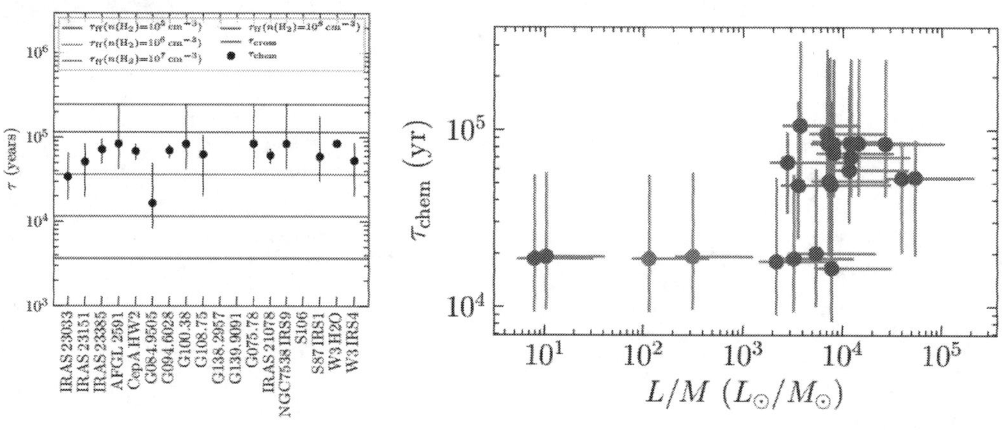


**Figure 4.** Chemical timescales derived from the CORE and CORE-extension projects [3, 4]. The left panel shows the timescales for all sources of the original CORE sample in comparison to crossing and free-fall times. The right panel presents the chemical time-scales of both samples compared to the luminosity-over-mass ratio $L/M$ as independent evolutionary stage tracer. Blue markers are the original CORE sources whereas green and red markers correspond to the CORE-extension sources.

### 2.4 Chemical evolution

The broad bandpass of the IRAM observations allows us to also investigate the chemical properties of the regions in detail. As a first step, Gieser et al. [3, 4] fitted all the spectral lines with XCLASS [14] and derived the gas column densities and abundances. Then, using the physical-chemical modeling tools MUSCLE [15–17], the chemical properties could be investigated in detail [3, 4].

One of the outcomes are chemical timescales of all regions. Figure 4 presents in the left panel the chemical timescales for the original CORE sample, and they are all in the range of several ten-thousand years. These timescales are much shorter than the dynamical timescales of the parental gas clumps, but they are in rough agreement with the free-fall times at densities between $10^5$ and $10^6$ cm$^{-3}$.

Adding in the information derived from a similar analysis for the younger regions of the CORE-extension study, Gieser et al. find evidence that the chemical timescales are indeed correlating with other independent age indicators like the luminosity-over-mass ratio $L/M$ [3]. While chemical timescales are notoriously prone to large error budgets, it is encouraging to see that these chemical times can be well used as independent age indicator.

### 2.5 Dynamical cloud and star formation

Within the CORE-extension program, we targeted two younger regions in larger-scale mosaic observations. In addition to similar analysis as outlined above, these data also allow us to investigate the cloud formation processes. Figure 5 shows the $DCO^+$ 1st moment map and a position-velocity cut through the ISOSS23053 region. Interestingly, we find two well-separated velocity components that meet exactly at the location where most of the dense cores are identified in the 1.3 mm continuum emission. At core positions we measure velocity gradients as large as 50 km s$^{-1}$pc$^{-1}$. These are clear signatures of converging gas flows that appear to be conducive to trigger star formation at their interaction zone. More details can be found in [2, 3].

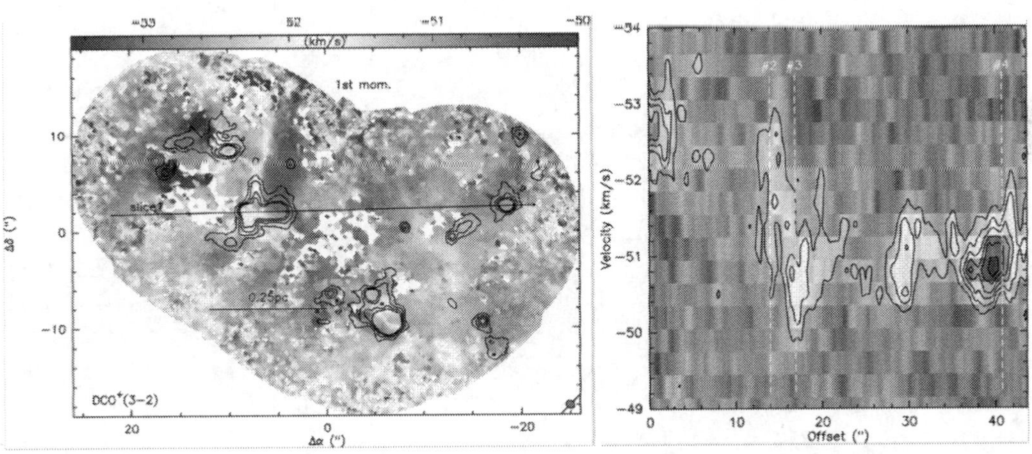


**Figure 5.** $DCO^+$(3–2) data towards ISOSS23053 [2]. The left panel shows in color-scale the $DCO^+$ 1st moment map and in contours the corresponding 1.3 mm continuum data. The right panel presents a position-velocity cut through the $DCO^+$ data cube along the marked slice in the left panel.

## 3 Summary and outlook

As outlined so far, the IRAM CORE program allowed us to derive exciting results for a broad range of physical and chemical questions in the field of star formation. Nevertheless, open questions remain. For example, the spectral setup in the 1 mm band did not cover many important ground-state lines, also from deuterated species. We therefore started a CORE+ study at NOEMA (Gieser et al.), targeting the CORE regions in the 3 mm band. Furthermore, while 20+ regions in the CORE and CORE-extension studies are already a big leap compared to earlier studies, larger statistical studies are now possible. For example the ALMA large program ALMAGAL targets more than 1000 regions in various evolutionary stages in a similar spectral setup. These data will allow us to address many of the here discussed topics on statistical grounds. In addition to that, the advent of JWST will enable studies of the physical and chemical properties of high-mass protostars at mid-infrared wavelengths and will be truly complementary to the CORE program.

## References

[1] H. Beuther, J.C. Mottram, A. Ahmadi, F. Bosco, H. Linz, T. Henning, P. Klaassen, J.M. Winters, L.T. Maud, R. Kuiper et al., A&A**617**, A100 (2018), 1805.01191

[2] H. Beuther, C. Gieser, S. Suri, H. Linz, P. Klaassen, D. Semenov, J.M. Winters, T. Henning, J.D. Soler, J.S. Urquhart et al., A&A**649**, A113 (2021), 2104.02420

[3] C. Gieser, H. Beuther, D. Semenov, S. Suri, J.D. Soler, H. Linz, J. Syed, T. Henning, S. Feng, T. Möller et al., A&A**657**, A3 (2022), 2110.01896

[4] C. Gieser, H. Beuther, D. Semenov, A. Ahmadi, S. Suri, T. Möller, M.T. Beltrán, P. Klaassen, Q. Zhang, J.S. Urquhart et al., A&A**648**, A66 (2021), 2102.11676

[5] A. Ahmadi, H. Beuther, J.C. Mottram, F. Bosco, H. Linz, T. Henning, J.M. Winters, R. Kuiper, R. Pudritz, Á. Sánchez-Monge et al., A&A**618**, A46 (2018), 1808.00472

[6] F. Bosco, H. Beuther, A. Ahmadi, J.C. Mottram, R. Kuiper, H. Linz, L. Maud, J.M. Winters, T. Henning, S. Feng et al., A&A**629**, A10 (2019), 1907.04225

[7] R. Cesaroni, H. Beuther, A. Ahmadi, M.T. Beltrán, T. Csengeri, R. Galván-Madrid, C. Gieser, T. Henning, K.G. Johnston, P.D. Klaassen et al., A&A**627**, A68 (2019), 1905.11257

[8] C. Gieser, D. Semenov, H. Beuther, A. Ahmadi, J.C. Mottram, T. Henning, M. Beltran, L.T. Maud, F. Bosco, S. Leurini et al., A&A**631**, A142 (2019), 1910.05081

[9] J.C. Mottram, H. Beuther, A. Ahmadi, P.D. Klaassen, M.T. Beltrán, T. Csengeri, S. Feng, C. Gieser, T. Henning, K.G. Johnston et al., A&A**636**, A118 (2020)

[10] L. Moscadelli, H. Beuther, A. Ahmadi, C. Gieser, F. Massi, R. Cesaroni, Á. Sánchez-Monge, F. Bacciotti, M.T. Beltrán, T. Csengeri et al., A&A**647**, A114 (2021), 2102.04872

[11] S. Suri, H. Beuther, C. Gieser, A. Ahmadi, Á. Sánchez-Monge, J.M. Winters, H. Linz, T. Henning, M.T. Beltrán, F. Bosco et al., A&A**655**, A84 (2021), 2109.04751

[12] F.H. Shu, ApJ**214**, 488 (1977)

[13] R.B. Larson, MNRAS**145**, 271 (1969)

[14] T. Möller, C. Endres, P. Schilke, A&A**59**, A7 (2017), 1508.04114

[15] D. Semenov, F. Hersant, V. Wakelam, A. Dutrey, E. Chapillon, S. Guilloteau, T. Henning, R. Launhardt, V. Piétu, K. Schreyer, A&A**522**, A42 (2010), 1007.2302

[16] T. Gerner, H. Beuther, D. Semenov, H. Linz, T. Vasyunina, S. Bihr, Y.L. Shirley, T. Henning, A&A**563**, A97 (2014), 1401.6382

[17] T. Gerner, Y.L. Shirley, H. Beuther, D. Semenov, H. Linz, T. Albertsson, T. Henning, A&A**579**, A80 (2015), 1503.06594

# The sulphur depletion problem in molecular clouds: The $H_2S$ case

*David* Navarro-Almaida[1]

[1]Observatorio Astronómico Nacional, IGN-CNIG, Calle Alfonso XII, 3-5, 28014 Madrid, Spain

**Abstract.**
Sulphur is one of the most abundant elements in the Universe and plays a crucial role in biological systems. However, sulphuretted molecules in the ISM are not as abundant as expected and there is no clear answer of where the missing Sulphur is yet. To shed light onto this open question, we focus our attention on the chemistry of $H_2S$, thought to be an important reservoir of Sulphur and formed mainly by grain-phase reactions. To understand the formation of $H_2S$, the growth of ices, and the chemical desorption process, we study the CO, $CH_3OH$, $N_2H^+$, and $H_2S$ abundances towards Barnard 1b, a Sulphur-rich cloud hosting a first Larson core. We look for correlations between gas-phase abundances of $H_2S$ and $CH_3OH$ that better constrain the location of the CO snowline in dark cores. Finally, this provides additional data to benchmark models for a deeper insight on the chemical desorption process and its efficiency.

## 1 Introduction

Despite being one of the most abundant elements in the Universe, with an elemental solar abundance of $S/H \sim 1.35 \times 10^{-5}$, sulphur-bearing molecules are not as abundant as expected. While in diffuse clouds and PDRs the sulphur abundance is found to be close to the cosmic value [1], sulphur is thought to be depleted inside molecular clouds by a factor of 1000 compared to its estimated cosmic abundance [2]. Due to the high hydrogen mobility in the ice matrix, the sulphur atoms depleted in interstellar ice mantles are expected to form $H_2S$ preferentially. Therefore, $H_2S$ is a candidate for the main sulphur reservoir in dark cores [3]. In [4], we found $H_2S$ to be the main sulphur reservoir in TMC 1 and Barnard 1b, released into the gas phase by chemical desorption. The chemical modeling of this molecule allowed us to conclude that it is a tracer of the snow-line in these regions, as the gas-phase abundance at higher extinctions is reduced due to the lower chemical desorption efficiencies. The observed abundances are reproduced with the models using a cosmic elemental Sulphur abundance and allowing an order of magnitude of uncertainty. This work is a follow-up of [4] where we used a subset of the GEMS database [5] and complementary CO and methanol observations to investigate the $H_2S$ and methanol abundances, the interplay of CO, methanol, and the snowline, and the mechanisms and efficiencies of formation of methanol and $H_2S$ in Barnard 1b.

## 2 Observations: Barnard 1b

Barnard 1 is a young, intermediate-mass star forming cloud, embedded in the western sector of the molecular cloud complex Perseus (see Fig. 1). It hosts several dense cores in different

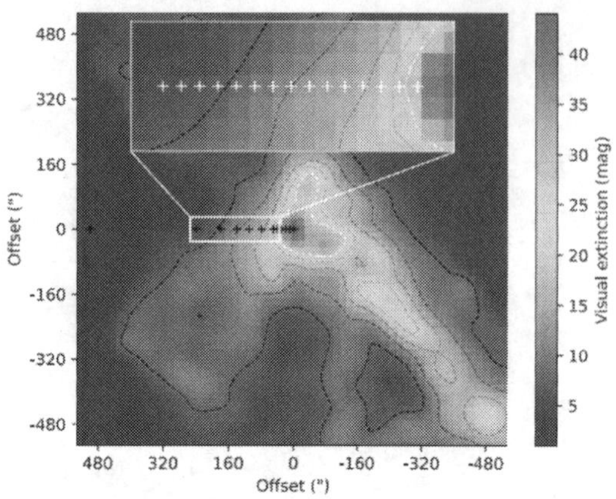


**Figure 1**: Extinction map of Barnard 1b with the observed positions (white crosses) and GEMS positions (black crosses) [4].

**Table 1**: Observed lines

| Molecule | Line | Freq. (GHz) |
|---|---|---|
| $^{13}CO$ | $1 \rightarrow 0$ | 110.201 |
| $C^{18}O$ | $1 \rightarrow 0$ | 109.782 |
| $H_2S$ | $1_{1,0} \rightarrow 1_{0,1}$ | 168.763 |
| $N_2H^+$ | $1 \rightarrow 0$ | 93.174 |
| $CH_3OH$ | E2 $2_{1,2} \rightarrow 1_{1,1}$ | 96.739 |
| $CH_3OH$ | E1 $2_{0,2} \rightarrow 1_{0,1}$ | 96.745 |
| $CH_3OH$ | $A^+$ $2_{0,2} \rightarrow 1_{0,1}$ | 96.741 |
| $CH_3OH$ | $A^+$ $1_{0,1} \rightarrow 0_{0,0}$ | 48.372 |

evolutionary stages, out of which Barnard 1b is the youngest. Barnard 1b has a rich chemistry characterized by large abundances of sulphur-bearing compounds. This work uses a subset of the GEMS database and complementary IRAM 30m observations of the lines shown in Table 1. Additionally, $CH_3OH$ $A^+$ $1_{0,1} \rightarrow 0_{0,0}$ data from the Yebes 40m telescope was observed at the lowest extinctions.

## 3 Analysis and results

For a better understanding of the depletion of species, the presence of the snow-line, and the formation of $H_2S$ and $CH_3OH$, we derived the column densities of the species listed in Table 1, and the physical properties at every position. First, we obtained the hydrogen nuclei number density and the methanol column density using the methanol lines shown in Table 1. To do so, we used the radiative transfer code RADEX with the most up-to-date collisional coefficients for e-methanol and a-methanol [6]. Assuming the gas temperature profile obtained in [4], the intensity ratio W(E2 $2_{1,2} \rightarrow 1_{1,1}$)/W(E1 $2_{0,2} \rightarrow 1_{0,1}$) provides an estimation of the hydrogen nuclei number density that is then used to compute the column density of both methanol species. In positions where at least one of these lines were not detected, we performed additional observations with the Yebes 40m telescope to obtain the density from the W($A^+$ $1_{0,1} \rightarrow 0_{0,0}$)/W($A^+$ $2_{0,2} \rightarrow 1_{0,1}$) intensity ratio. The column density of the rest of the species are derived fitting the integrated intensity of the lines.

The resulting abundances are shown in Fig. 2. We first notice the presence of $H_2S$ in the gas phase at the most external positions where $CH_3OH$ is not detected. This is compatible with an early depletion of Sulphur atoms onto grain surfaces [7], before significant depletion of CO. Since reactions with CO are the main destruction path of $N_2H^+$ [8], when CO depletion occurs, $N_2H^+$ starts increasing its abundance. At the same time, $CH_3OH$ also starts increasing, slightly, its abundance, as it is mainly formed by hydrogenation of depleted CO. The depletion of CO is linked to ice growth which, in turn, reduces the desorption efficiency of molecules formed via grain-phase reactions. This is the case of $H_2S$, whose abundance decreases as $^{13}CO$ starts depleting and methanol and $N_2H^+$ increase their abundances.

Methanol, like $H_2S$, is also produced more efficiently through grain-phase reactions in cold core conditions. However, the gas abundance of methanol does not seem to be affected

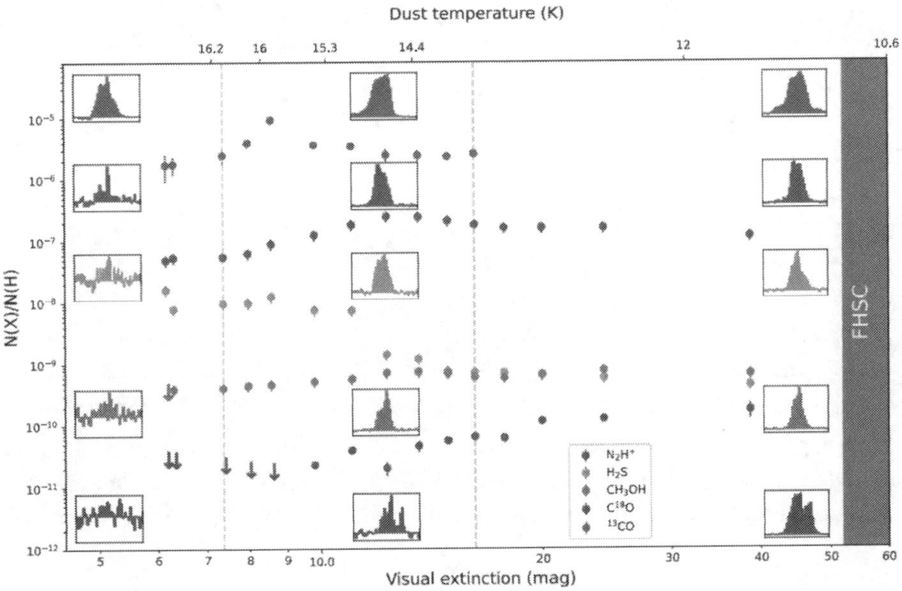


**Figure 2**: Abundances N(X)/N(H) of $^{13}CO$, $C^{18}O$, $H_2S$, $CH_3OH$, and $N_2H^+$ molecules as a function of the visual extinction and dust temperature.

by the decreasing desorption efficiency expected by grain growth. The desorption of COMs like methanol in grains covered by CO ice has been discussed in several works [9]. Current chemical models do not consider a detailed chemical composition of the outermost layers of ices. It is likely that these layers, containing mainly depleted CO, could enhance the chemical desorption of COMs [9] while also reducing the desorption efficiencies of other molecules like $H_2S$. Surface-dependent parameters like the diffusion rate of hydrogen through the ice matrix have also been found to have a great impact in the methanol production [10].

Now we explore the role of ices in the production of $H_2S$ and methanol. The chemical code Nautilus [11], using the densities obtained here and the parameters described in [4], yields the abundances shown in Fig. 3. This model adjusts the desorption efficiency with the ice growth, being at its highest when no ice is present. The efficiency of desorption is reduced as ice grows at higher extinctions. The desorption schemes used in the model are described in [4]. The model results for $H_2S$ (Fig. 3) are in good agreement with the observations, with high desorption efficiencies and low desorption efficiencies best describing low and high extinction areas, respectively, as a consequence of the build-up of ices towards the interior of the core. The model does not however reproduce the observations of methanol, even in a scenario of early chemistry at $10^5$ yr (Fig. 3), since it underestimates its gas-phase abundance. In the following section we discuss several ideas that may be behind this disagreement.

## 4 Discussion and summary

The chemistry of sulphur and COMs are tightly related to grain-phase reactions. The chemical desorption process is crucial for the incorporation of these species into the gas-phase. Even thought $H_2S$ and methanol are mainly formed via chemical desorption, we have revealed great differences between the distribution of their gas-phase abundances.

- Gas-phase abundance of $H_2S$ is anti-correlated with the ice growth expected when CO is being depleted. Our model, which takes into account different desorption schemes as

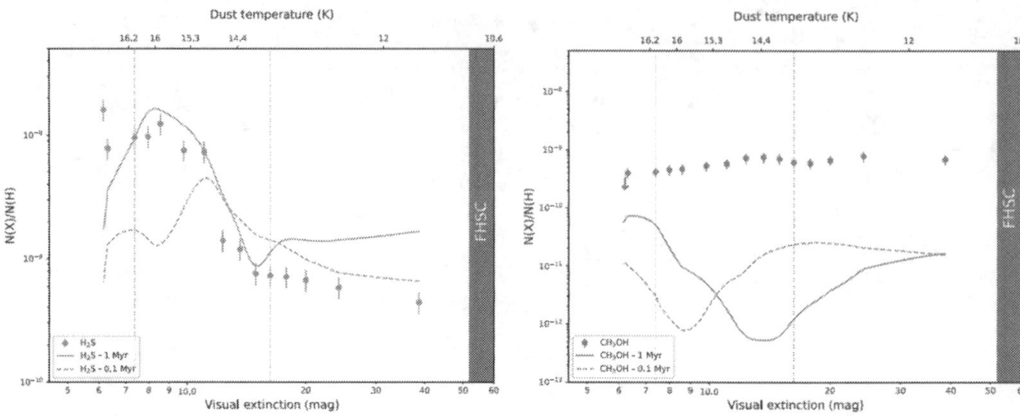


**Figure 3**: *Left:* Observed $H_2S$ abundances (orange dots) and model predictions at two different times: 1 Myr (solid) and 0.1 Myr (dashed). *Right:* Observed methanol abundances (dark green dots) and model predictions at two different times: 1 Myr (solid) and 0.1 Myr (dashed).

ice grows, shows a good agreement with the observations, thus suggesting that the $H_2S$ gas-phase abundance traces the CO snowline. Other possibility that may help explain the observed abundance profile is that $H_2S$, in presence of depleted CO, reacts with it forming other compounds such as OCS and therefore no $H_2S$ is released to the gas phase [12].

- Methanol gas-phase abundance is correlated with CO depletion, in agreement with the formation of methanol by hydrogenation of depleted CO. The measured abundances of methanol and $H_2S$ follow different behaviors despite having similar formation pathways. As discussed previously, CO-ices could enhance the desorption of COMs, although this is yet to be confirmed. Recent results also suggest alternative routes of methanol formation that need to be incorporated into chemical networks [13].
- Finally, uncertainties in the physical structure, grain sizes, and the dynamical behavior may explain the striking differences in methanol and $H_2S$ gas-phase abundances.

## References

[1] D.A. Neufeld, B. Godard, M. Gerin, et al, A&A **577**, A49 (2015), 1502.05710
[2] M. Agúndez, V. Wakelam, Chemical Reviews **113**, 8710 (2013), 1310.3651
[3] T.H.G. Vidal, J.C. Loison, A.Y. Jaziri, et al, MNRAS **469**, 435 (2017), 1704.01404
[4] D. Navarro-Almaida, R. Le Gal, A. Fuente, et al, A&A **637**, A39 (2020), 2004.03475
[5] A. Fuente, D.G. Navarro, P. Caselli, et al, A&A **624**, A105 (2019), 1809.04978
[6] D. Rabli, D.R. Flower, MNRAS **406**, 95 (2010)
[7] S. Cazaux, H. Carrascosa, G.M. Muñoz Caro, et al, A&A **657**, A100 (2022), 2110.04230
[8] P. Caselli, C.M. Walmsley, M. Tafalla, et al, ApJ **523**, L165 (1999)
[9] A.I. Vasyunin, P. Caselli, F. Dulieu, et al, ApJ **842**, 33 (2017), 1705.04747
[10] A. Punanova, A. Vasyunin, P. Caselli, et al, ApJ **927**, 213 (2022), 2112.04538
[11] M. Ruaud, V. Wakelam, F. Hersant, MNRAS **459**, 3756 (2016)
[12] M. el Akel, L.E. Kristensen, R. Le Gal, et al., A&A **659**, A100 (2022)
[13] J.C. Santos, K.J. Chuang, T. Lamberts, et al, arXiv e-prints arXiv:2205.12284 (2022), 2205.12284

# LEGO: A 3 mm molecular line study covering 100 pc of one of the most actively star-forming portions in the Milky Way

*Ashley* Barnes[1,*], *Jens* Kauffmann[2], *Frank* Bigiel[1], and LEGO collaboration

[1]Argelander-Institut für Astronomie, Universität Bonn, Auf dem Hügel 71, 53121 Bonn, Germany
[2]Haystack Observatory, Massachusetts Institute of Technology, Westford, MA 01886, USA

**Abstract.** The current generation of high-sensitivity (sub)mm-telescopes such as the IRAM-30m and NOEMA have allowed the emission from molecular lines to become a major tool for studying the physical, kinematic and chemical properties of extragalactic systems. However, we need to develop a detailed understanding of emission lines in Milky Way molecular clouds before we can exploit extragalactic line emission data to its full potential. With this in mind, the IRAM-30m large project "Line Emission to assess Galaxy Observations" (LEGO) aims at developing a comprehensive picture of how 3mm-band emission lines in Milky Way molecular clouds depend on a variety of environmental factors. In this proceeding, I will focus on our recent detailed investigation of the W49 massive star-forming region published in Barnes et al. 2020 [1].

## 1 Introduction

Investigating the emission properties of molecular gas density tracers – and many other commonly observed molecular line transitions – across a broad range of environments is of current critical importance (e.g. 2–5). This is of particular relevance now more than ever, as more numerous, weaker molecular lines are becoming routinely observable in other galaxies (e.g. 6, 7), where, even at the highest achievable spatial resolution for the closest disc galaxies, each line-of-sight contains a very broad range of environments (e.g. 8–11).

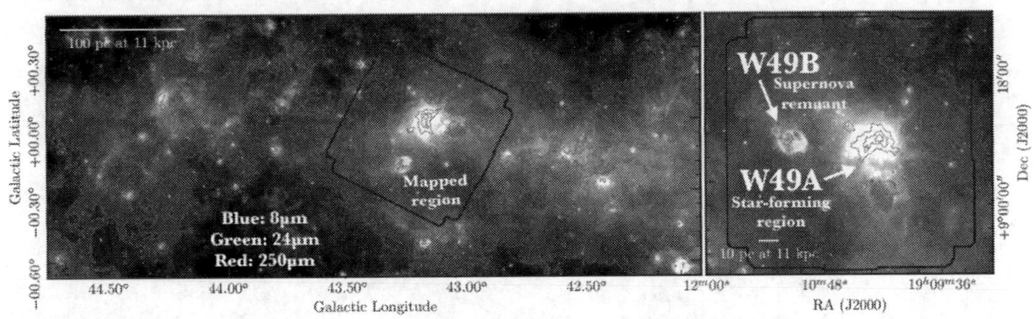


**Figure 1. The region around W49 within the plane of the Milky Way.** The upper panel shows a three-colour image across the Galactic plane, and a zoom-in on the mapped region of W49 indicated by the heavy black line. Shown are the *Spitzer* 8 μm in blue, *Spitzer* 24 m in green, and *Herschel* 250 m in red. Overlaid as red contours is the 250 μm map, which were chosen to best highlight the dense dust/gas distribution. The labels on the zoom-in panel show the positions of the W49A star-forming region and the W49B supernova remnant. Shown in the main and zoom-in panels are scale bars of 100 pc and 10 pc at a distance of 11 kpc, respectively.

*e-mail: ashleybarnes.astro@gmail.com

### 1.1 The LEGO survey

One of the primary aims of the Molecular Line Emission as a Tool for Galaxy Observations (LEGO) project is to further investigate the above issue by obtaining large maps of many commonly observed 3 mm-band molecular lines, towards a selection of 25 Milky Way star-forming regions that span a range of environmental properties (e.g. cloud density, star formation feedback, galactic environment, and metallicity). This survey was conducted using the Eight Mixer Receiver on the 30m telescope of the Instituto de Radioastronomía Milimétrica (IRAM) on Pico Veleta, Spain. Data in the horizontal and vertical polarizations were acquired with the Fast Fourier Transform Spectrometer (FTS) operating in a wide–band mode that provides a spectral resolution of 200 kHz (i.e. $0.6\ \mathrm{km\,s^{-1}}$ at 100GHz). The EMIR receiver was operated in two setups, to cover continuous frequency ranges of 86.10 to 99.89 GHz and 101.78 to 115.57 GHz. The telescope's intrinsic beam size is around 24″ at 100 GHz.

### 1.2 The W49 region and its surroundings

We cover a 100 by 100 pc box covering the bright W49 region (Fig. 1 ), which was chosen to be representative of what can now be observable using interferometric observations of nearby star-forming galaxies. The W49A region is a very well studied actively star-forming region that contains high-mass stars and associated HII regions (total stellar mass of $10^{4-5}\mathrm{M_\odot}$). Parallax distance measurements place it at a distance of $\sim$ 11 kpc (i.e. far-side of MW).

## 2 Molecular line emission across this extragalatic template

Here, we focus on the main results from the analysis of W49. For a comprehensive overview and complete analysis of the molecular line dataset see Barnes et al. 2020 [1].

In Fig. 2, we present integrated intensity maps for a selection of the observed molecular lines, which have been ordered by increasing critical density. This plot was made to be comparable to Fig. 1 of [12], from which we see broadly similar trends to Orion A. We find that $^{12}$CO and $^{13}$CO, have extended emission across all the mapped region, whilst $^{18}$CO and $^{17}$CO emits only towards the centre of W49A. This effect is likely due to the reduced abundance of rarer isotopologues combined with our sensitivity limits.

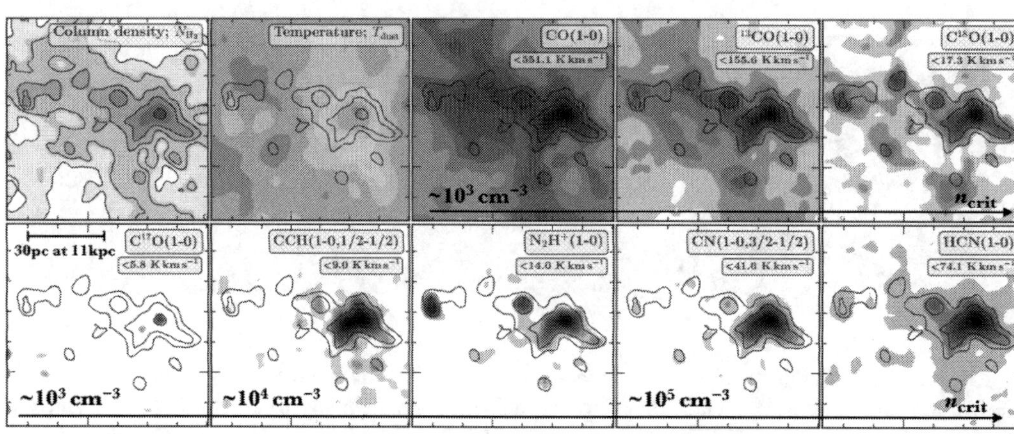


**Figure 2. Maps of the integrated intensity order by increasing critical density.** The upper left two panels show the column density and dust temperature determined from the *Herschel* observations. Overlaid as black contours is the $N_{\mathrm{H2}}$, and the filled contours show the molecular line emission that are drawn from a signal–to–noise ratio of 3. The peak value of integrated intensity maps is labelled in the upper right.

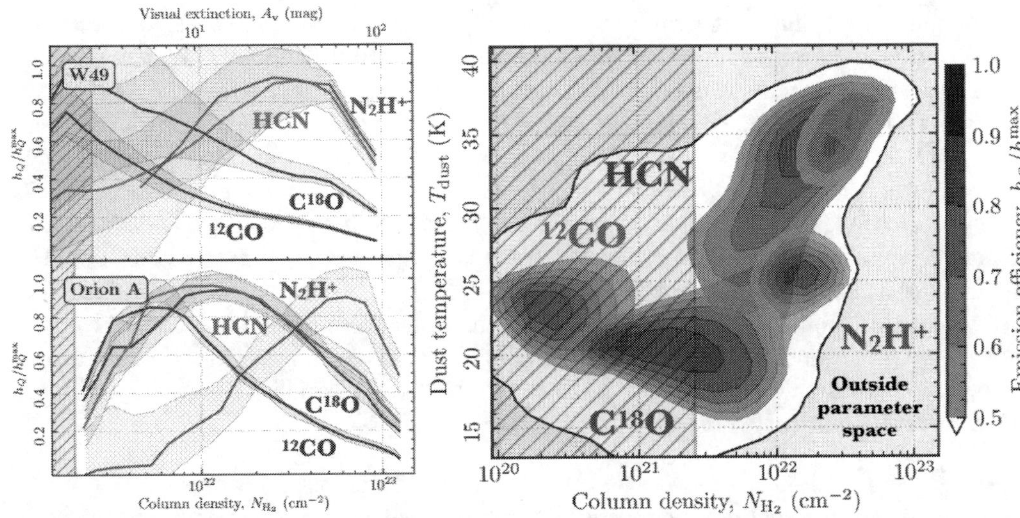


**Figure 3. The ratio of the integrated intensity to the molecular hydrogen column density (or emission efficiency ratio; $h_Q = W(Q)/N_{H2}$) as a function of column density (left) and also dust temperature (right).** We highlight by shading here column densities below the threshold limits for both sources, which are believed to suffer from fore- and background contamination.

Moving beyond the CO lines, we find little correspondence between the extent of the molecular line emission and the increase in critical density. For example, we see comparable spatial morphologies of the CCH, CN and HCN lines, despite them having critical densities differing by almost two orders of magnitude. Moreover, we find that only $N_2H^+$ appears to also trace the column density peak at the east of the mapped region (towards W49B) at comparable brightness to the peak towards the centre of the W49A; potentially highlighting this as a region of differing excitation or chemistry. This clearly shows that the brightness and coverage of a molecular line transition are not completely governed by the critical density; e.g. excitation effects or chemical abundance variations may also be important factors.

In Fig. 3 (left panels), we present the emission efficiency ratio ($h_Q$=$W(Q)/N_{H_2}$; i.e. line brightness per $H_2$ column density) as a function of the column density for a selection of the observed molecular lines for W49 and Orion A [12]. We see that 12CO has a maximum emission efficiency ratio at the lowest column density, whilst $C^{18}O$, HCN and $N_2H^+$ show peak emission efficiency ratio values at increasingly higher column densities. Interestingly, we see that the HCN and $N_2H^+$ peak at a similar column density across the W49 region, which is in contrast to Orion A where $N_2H^+$ peaks much higher relative to HCN.

In Fig. 3 (right panels), We build upon this by assessing how the emission efficiency ratio varies as a function of *both* the column density and dust temperature. We see that CO has an emissivity that peaks towards the lowest column densities, and at moderate dust temperatures. There are several well-studied mechanisms that could cause this effect. Optical depth effects are most likely the major cause of this for the observed range of column densities ($10^{20-23}$ cm$^{-2}$): CO, for example, is thought to be optically thick in all but the lowest density gas within the Galaxy. An additional effect at high values of the column density would be CO depletion. In cold ($T_{dust}$<20 K) and dense ($n_{H_2}$ > $10^4$ cm$^{-3}$) environments, CO molecules can "freeze out" onto the surfaces of dust grains forming icy mantles, such that their observed gas-phase abundances are a factor of several smaller than expected in the case of no depletion.

In addition, we find the distribution of HCN and $N_2H^+$ lines split into two discrete temperature peaks. We find that the $N_2H^+$ peaks towards the lowest observed temperatures at the column density within the observed parameter space (~ 20 K), whereas the HCN peaks towards moderate-to-high temperatures of around ~ 30-25 K. This then highlights the usefulness of the dust temperature as an additional variable to further investigate the emissivity profiles within the regions even at a low spatial resolution (~ 3 pc for these W49 observations). It is not surprising that both HCN and $N_2H^+$ peak towards the densest gas as the abundance of N-bearing species is boosted within dense gas. Moreover, the abundance of $N_2H^+$ is known to be particularly enhanced towards low temperatures, within regions where the CO molecules are frozen-out of the gas phase. This is due to CO being an effective destructor of the precursors in the formation pathway of $N_2H^+$; removing molecules such as $NH^+$ and $H_3^+$. On the other hand, molecules such as HCN (also $HCO^+$, CN, and HNC) are sensitive to the far-UV radiation produced from star formation, where star-forming regions can be identified as having elevated dust temperatures. For example, the HCN$(1-0)$ line is thought to be pumped by mid-infrared photons and/or the effects of X-ray Dominated Regions, which can both be present within actively star-forming regions (i.e. W49).

## 3 Summary

The two main results highlighted here are 1) that the critical density of molecular lines should not be used as an *absolute* proxy for the density regimes from which the emission is originating; 2) variations in both density *and* temperature drive excitation and chemistry which then strongly govern where we detect these molecules within molecular clouds.

## References

[1] A.T. Barnes, J. Kauffmann, F. Bigiel, N. Brinkmann, D. Colombo, A.E. Guzmán, W.J. Kim, L. Szűcs, V. Wakelam, S. Aalto et al., MNRAS**497**, 1972 (2020), 2007.11005

[2] A.T. Barnes, S. Kong, J.C. Tan, J.D. Henshaw, P. Caselli, I. Jiménez-Serra, F. Fontani, MNRAS**458**, 1990 (2016), 1602.06798

[3] A.T. Barnes, J.D. Henshaw, P. Caselli, I. Jiménez-Serra, J.C. Tan, F. Fontani, A. Pon, S. Ragan, MNRAS**475**, 5268 (2018), 1801.06538

[4] A.T. Barnes, S.N. Longmore, A. Avison, Y. Contreras, A. Ginsburg, J.D. Henshaw, J.M. Rathborne, D.L. Walker, J. Alves, J. Bally et al., MNRAS**486**, 283 (2019), 1903.06158

[5] A.T. Barnes, J.D. Henshaw, F. Fontani, J.E. Pineda, G. Cosentino, J.C. Tan, P. Caselli, I. Jiménez-Serra, C.Y. Law, A. Avison et al., MNRAS**503**, 4601 (2021), 2103.09122

[6] F. Bigiel, A.K. Leroy, M.J. Jiménez-Donaire, J. Pety, A. Usero, D. Cormier, A. Bolatto, S. Garcia-Burillo, D. Colombo, M. González-García et al., ApJL**822**, L26 (2016), 1604.05316

[7] J.S. den Brok, F. Bigiel, K. Sliwa, T. Saito, A. Usero, E. Schinnerer, A.K. Leroy, M.J. Jiménez-Donaire, E. Rosolowsky, A.T. Barnes et al., arXiv e-prints arXiv:2201.05165 (2022), 2201.05165

[8] M.J. Jiménez-Donaire, F. Bigiel, A.K. Leroy, A. Usero, D. Cormier, J. Puschnig, M. Gallagher, A. Kepley, A. Bolatto, S. García-Burillo et al., ApJ**880**, 127 (2019), 1906.08779

[9] I. Bešlić, A.T. Barnes, F. Bigiel, J. Puschnig, J. Pety, C. Herrera Contreras, A.K. Leroy, A. Usero, E. Schinnerer, S.E. Meidt et al., MNRAS**506**, 963 (2021), 2106.09742

[10] C. Eibensteiner, A.T. Barnes, F. Bigiel, E. Schinnerer, D. Liu, D.S. Meier, A. Usero, A.K. Leroy, E. Rosolowsky, J. Puschnig et al., A&A**659**, A173 (2022), 2201.02209

[11] J.S. den Brok, D. Chatzigiannakis, F. Bigiel, J. Puschnig, A.T. Barnes, A.K. Leroy, M.J. Jiménez-Donaire, A. Usero, E. Schinnerer, E. Rosolowsky et al., MNRAS**504**, 3221 (2021), 2103.10442

[12] J. Kauffmann, P.F. Goldsmith, G. Melnick, V. Tolls, A. Guzman, K.M. Menten, A&A**605**, L5 (2017), 1707.05352

# Nitrogen isotopes in the interstellar medium: A chemical journey across the Galaxy

*Laura* Colzi[1,*], *Francesco* Fontani[2], *Víctor M.* Rivilla[1], and *Paola* Caselli[3]

[1]Centro de Astrobiología (CAB), CSIC-INTA, Ctra. de Ajalvir Km. 4, 28850 Torrejón de Ardoz, Madrid, Spain

[2]INAF-Osservatorio Astrofisico di Arcetri, Largo E. Fermi 5, 50125 Florence, Italy

[3]Max-Planck-Institut für extraterrestrische Physik (MPE), Giessenbachstrasse 1, 85748 Garching, Germany

**Abstract.** One of the most important tools to investigate the chemical history of our Galaxy and our own Solar System is to measure the isotopic fractionation of chemical elements. This is the process that distributes the less abundant stable isotopes of an element in different molecules. The isotopic ratios are governed by two main processes: 1. chemical evolution of the whole Galaxy due to stellar nucleosynthesis; 2. local fractionation effects. In this Proceeding we report some results highlighting both processes towards massive star-forming regions.

## 1 Introduction

One of the main tools to understand the heritage that has been received by the Solar System is to study the isotopic variation. In fact, pristine Solar System materials, like comets and carbonaceous chondrites, present enrichment in $^{15}$N and D, with respect to the proto-Solar Nebula (PSN) value (see e.g. the left panel of Fig. 1). This means that there was an alteration of isotopic ratios (e.g. the $^{14}$N/$^{15}$N ratio) during the Solar formation process, whose causes are still not well understood.

From an observational point of view the $^{14}$N/$^{15}$N ratio has been studied towards different type of molecular clouds, such as low-mass pre-stellar cores and protostellar objects, infrared dark clouds, massive star-forming regions, and protoplanetary discs (e.g. [2]; [3]; [4]; [5]; [6]; [7]; [8]; [9]), as shown in the right panel of Fig. 1. Most of the observed $^{14}$N/$^{15}$N ratios lie in between the cometary value (144±3, [10]) and the PSN value (441±6, [11]). However, there are also some exceptions, as the $^{14}$N/$^{15}$N ratios measured from $N_2H^+$ towards pre-stellar cores and the ratios in some massive cores are even higher than the PSN value (e.g. [5]; [12]).

All of these different $^{14}$N/$^{15}$N ratios depending on the physical conditions of the sources and on the observed molecular species, raising interest also from a theoretical point of view. First, the so-called low-temperature isotopic-exchange reactions in chemical models have been invoked to explain the observed ratios (e.g. [13]; [14]; [15]), as:

$$N^{15}N + N_2H^+ \rightarrow N^{15}NH^+ + N_2 + 10.3\ K, \quad (1)$$

which exchange one of the $^{14}$N of $N_2H^+$ with $^{15}$N. These reactions are very efficient in a low-temperature environment (e.g. 5–10 K for pre-stellar cores, [16]) because of their exothermicities. Moreover, such reactions are also very successful in reproducing the D/H ratios

*e-mail: lcolzi@cab.inta-csic.es

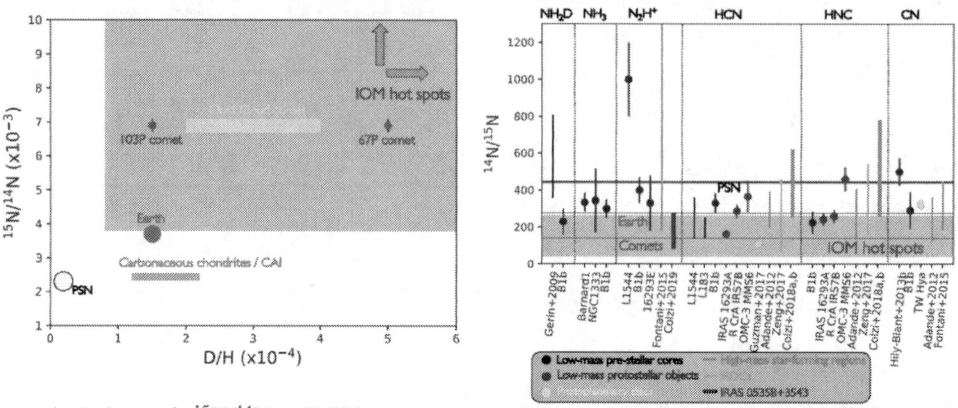


**Figure 1.** *Left panel:* $^{15}N/^{14}N$ vs D/H in comets, chondrites, hot spots in the Insoluble Organic Matter (IOM) of meteorites, Earth and PSN. *Right panel:* $^{14}N/^{15}N$ ratios obtained for different molecules. Black points represent low-mass pre-stellar cores, the blue points are low-mass protostellar objects, and the cyan points indicate protoplanetary discs. The horizontal yellow and red solid lines represent the PSN value of 441 and the terrestrial atmosphere value of 272, respectively. The green horizontal line denotes the average value measured in comets. The pink area represents measurements in carbonaceous chondrites, where the lower ones are the so-called "hot-spots". Adapted from [1].

in molecular clouds (e.g. [17]). However, this is not so for nitrogen fractionation. This indicates that there are other chemical processes that need to be taken into account. Other mechanisms, such as different rates for the dissociative recombination of $N_2H^+$ (e.g. [15]), or the isotope-selective photodissociation of $N_2$ (e.g. [18]) have been also proposed.

## 2 The role of the Milky Way chemical evolution

The $^{14}N/^{15}N$ ratios are also governed by the chemical evolution of the whole Galaxy due to stellar nucleosynthesis (e.g. [19]). In this respect, we have analysed a sample of 87 massive star-forming regions observed with the IRAM 30m radiotelescope[1]. The sources span galactocentric distances, $D_{GC}$, from 2 up to 12 kpc allowing to study the $^{14}N/^{15}N$ ratio of HCN and HNC as a function of $D_{GC}$ (Fig. 2). For the two datasets we have performed a linear regression fit (light blue solid lines in Fig. 2) and a parabolic analysis (red parabolas in Fig. 2). From the linear trends we have derived a new local ISM $^{14}N/^{15}N$ value of 375±75 ([9]). Then, the observational results have been compared with the predictions of the Galactic Chemical Evolution (GCE) model [19] (magenta lines in Fig. 2). They predict a linear positive trend up to 8 kpc because of novae outbursts as the main way to produce the $^{15}N$, and a flattening trend above this distance because of the low-level star formation and gas infall on long timescales (inside-out galaxy formation) that reduce the nova effect (see e.g. [19, 20]).

GCE models can reproduce the observational trend but not the absolute values. However, the shift on the y-axis of the models is defined by the ejected mass in the form of $^{15}N$, $M_{ejec}^{15N}$, which is an assumed parameter considering valid ranges taken from hydrodynamic simulations (see [19] for more details). More recent GCE models [20], which take into account different rotation velocities of low-metallicity massive stars, updated $M_{ejec}^{15N}$ and could better reproduce also the absolute value of observations.

We are now extending this trend towards the outer part of our Galaxy ($D_{GC}$ >12 kpc), also updating GCE models with new stellar yields and approaching distances where CNO isotopes

[1] For more information on the observations and the analysis done see [8, 9].

https://doi.org/10.1051/epjconf/202226500034

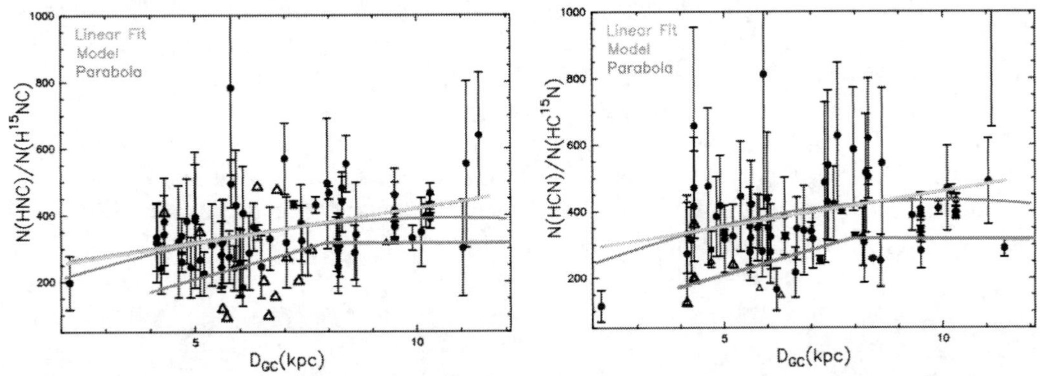


**Figure 2.** $^{14}N/^{15}N$ ratios for HNC (*left panel*) and HCN (*right panel*) as a function of the galactocentric distance ($D_{GC}$). The cyan solid line is the linear regression fit computed for the two data sets, the red parabola is the results from the parabolic analysis described in detail in [9], and the magenta solid line represents the GCE model of [19].

have never been studied (Colzi et al. in prep). In fact, this zone of the Galaxy present lower metallicities, a factor of four lower at $D_{GC}$=19 kpc with respect to the inner Galaxy, which could also affect the behaviour of the elemental $^{14}N/^{15}N$ ratio.

## 3 Local chemical fractionation effects

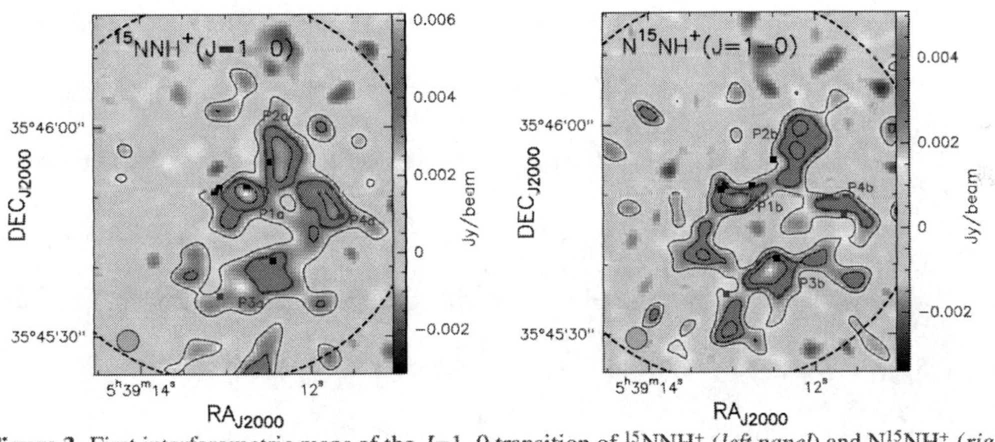


**Figure 3.** First interferometric maps of the $J$=1–0 transition of $^{15}NNH^+$ (*left panel*) and $N^{15}NH^+$ (*right panel*). The black contour levels are 3, 5 and 7 times the 1$\sigma$ of the maps (0.5 and 0.72 mJy beam$^{-1}$, respectively). The blue contours correspond to the 5$\sigma$ of the map, from which the $^{14}N/^{15}N$ ratios have been derived. The black and blue squares indicate the positions of the continuum sources and other $N_2H^+$ peak positions, respectively. The dashed circle represents the NOEMA field of view and the synthesized beam is the ellipse indicated in the lower left corner.

To investigate the local effect of chemistry in individual star-forming regions we have studied for the first time N-fractionation of $N_2H^+$ at high angular resolution (~3", i.e. ~6200 au at a distance of 1.8 kpc). In particular, we have used IRAM NOEMA observations towards the massive star-formig region IRAS 05358+3543 ([21]).

Figure 3 shows the first interferometric maps of the $^{15}$N-isotopologues of $N_2H^+$ at core scales (about 0.03 pc). From this work it appears that the $^{14}N/^{15}N$ ratio of $N_2H^+$ is lower in the inner denser cores (∼100–200) with respect to the more diffuse gas (>250). These results highlight the importance of local fractionation effects, and in particular they demonstrates that isotope-selective photodissociation of $N_2$ should be introduced in chemical models to explain $^{15}$N-enrichment in star-forming regions (e.g. [18]; [21]; [22]).

We have also obtained new NOEMA observations to study N-fractionation of the nitriles HCN and HNC towards the same source and with the same spatial resolution. The comparison of the $^{14}$N/$^{15}$N ratio of $N_2H^+$ with that of nitriles will allow a deeper insight of the local chemical processes affecting isotopic ratios.

## References

[1] L. Colzi, *Isotopic fractionation study towards massive star-forming regions across the Galaxy, PhD Thesis, Premio Tesi di Dottorato Series, Florence University Press* (2020)

[2] F. Daniel, M. Gérin, E. Roueff, J. Cernicharo, N. Marcelino, F. Lique, D.C. Lis, D. Teyssier, N. Biver, D. Bockelée-Morvan, **560**, A3 (2013), 1309.5782

[3] P. Hily-Blant, L. Bonal, A. Faure, E. Quirico, **223**, 582 (2013), 1302.6318

[4] S.F. Wampfler, J.K. Jørgensen, M. Bizzarro, S.E. Bisschop, **572**, A24 (2014), 1408.0285

[5] F. Fontani, P. Caselli, A. Palau, L. Bizzocchi, C. Ceccarelli, **808**, L46 (2015), 1506.05180

[6] S. Zeng, I. Jiménez-Serra, G. Cosentino, S. Viti, A.T. Barnes, J.D. Henshaw, P. Caselli, F. Fontani, P. Hily-Blant, **603**, A22 (2017), 1705.04082

[7] V.V. Guzmán, K.I. Öberg, J. Huang, R. Loomis, C. Qi, **836**, 30 (2017), 1701.07510

[8] L. Colzi, F. Fontani, P. Caselli, C. Ceccarelli, P. Hily-Blant, L. Bizzocchi, **609**, A129 (2018), 1709.04237

[9] L. Colzi, F. Fontani, V.M. Rivilla, A. Sánchez-Monge, L. Testi, M.T. Beltrán, P. Caselli, **478**, 3693 (2018), 1804.05717

[10] P. Hily-Blant, V. Magalhaes, J. Kastner, A. Faure, T. Forveille, C. Qi, **603**, L6 (2017), 1706.10095

[11] B. Marty, L. Zimmermann, P.G. Burnard, R. Wieler, V.S. Heber, D.L. Burnett, R.C. Wiens, P. Bochsler, **74**, 340 (2010)

[12] E. Redaelli, L. Bizzocchi, P. Caselli, J. Harju, A. Chacón-Tanarro, E. Leonardo, L. Dore, **617**, A7 (2018), 1806.01088

[13] E.S. Wirström, S.B. Charnley, M.A. Cordiner, S.N. Milam, **757**, L11 (2012), 1208.0192

[14] E. Roueff, J.C. Loison, K.M. Hickson, **576**, A99 (2015), 1501.01141

[15] J.C. Loison, V. Wakelam, P. Gratier, K.M. Hickson, **484**, 2747 (2019)

[16] A. Crapsi, P. Caselli, M.C. Walmsley, M. Tafalla, **470**, 221 (2007), 0705.0471

[17] P. Caselli, C. Ceccarelli, **20**, 56 (2012), 1210.6368

[18] K. Furuya, Y. Aikawa, **857**, 105 (2018), 1803.06787

[19] D. Romano, F. Matteucci, Z.Y. Zhang, P.P. Papadopoulos, R.J. Ivison, Monthly Notices of the Royal Astronomical Society **470**, 401 (2017), 1704.06701

[20] D. Romano, F. Matteucci, Z.Y. Zhang, R.J. Ivison, P. Ventura, arXiv e-prints arXiv:1907.09476 (2019), 1907.09476

[21] L. Colzi, F. Fontani, P. Caselli, S. Leurini, L. Bizzocchi, G. Quaia, **485**, 5543 (2019), 1903.06567

[22] A.N. Heays, R. Visser, R. Gredel, W. Ubachs, B.R. Lewis, S.T. Gibson, E.F. van Dishoeck, **562**, A61 (2014), 1401.1630

# Feedback from young stars, the molecular signature of shocks and outflows

*Antoine* Gusdorf[1,2,*]

[1]Laboratoire de Physique de l'École Normale Supérieure, ENS, Université PSL, CNRS, Sorbonne Université, Université Paris Cité, 75005 Paris, France
[2]Observatoire de Paris, PSL University, Sorbonne Université, LERMA, 75014 Paris, France

**Abstract.** Why do we study shocks ? Because they are there. Shocks are ubiquitous in the interstellar medium (ISM), where they constitute a major source of energy injection, together with photons and cosmic rays (CRs). Galactic shocks, and converging flows at the basis of the formation of molecular clouds and filaments, are examples of interstellar shocks. Shock waves are also generated during the birth, life and death of stars in the form of jets and protostellar outflows, stellar winds and supernovae and supernova remnants (SNRs). Hence, they are a major route of feedback of stars on galaxies. As such, they are a proficient tool to better understand the cycle of matter and energy in galaxies, but also the formation of stars. In this review, I will describe the recent advances on the study of shocks that can be observed and characterized with the IRAM instruments, with emphasis on the study of protostellar jets and outflows.

## 1 Introduction

In this review, I first focus on the direct questions in relation with shock phenomenology: what are the processes at work and how we progressively overcame our limitations on their understanding (section 2). I will then show how we measure energetic and chemical impacts of interstellar shocks (section 3). And finally, beyond the phenomenology of shocks, I will describe the general properties of protostellar jets and outflows and subsequent star formation that we can infer from their observations (section 4).

## 2 Mechanisms at work

**Physical description.** Interstellar shock waves are best described by numerical codes that self-consistently calculate the thermodynamical, physical, dynamical and chemical state of a one-dimensional layer of gas in stationary conditions. A reference for such code is the Paris-Durham model, that also allows to approximate the propagation of a non stationary shock wave ([1] and references therein). Depending on the pre-shock parameters (density, ionization, magnetic field...) and shock characteristics (velocity), it allows to simulate the propagation of various kinds of shocks. Combined with a radiative transfer it has been used to intperpret observations in a number of protostellar outflows and SNRs (e.g., [2], [3], [4], [5], [6]). In some cases, more simple, parametric approaches have been used (e.g., [7], [8]).

*e-mail: antoine.gusdorf@phys.ens.fr

**Limitations: geometry.** An important limitation of such otherwise sophisticated models is the way they treat the geometry of the shock structure. Getting observationally close to resolving a one-dimensional shock structure has been claimed in [9] but remains challenging. Using the spectral resolution of observational infrastructures such as SOFIA has been proposed by [10] and [11] but this raises the question of modelling line profiles, that necessitates at least an approximate 3D modelling. So-called pseudo 2D or 3D solutions have recently been implemented, where collections of 1D structures are 'stitched' onto a 2D or 3D structure to model the optically thin emission of $H_2$ maps ([12], [13]), or line profiles and excitation diagrams ([14]).

**Limitations: irradiation.** Another limitation until very recently came from the lack of models enabling to simultaneously treat the effects of shocks and energetic UV photons. This effect was exposed when e.g. [15] or [16] tried to interpret fine-structure emission lines observed with *Herschel* by the same shock models that were so successful in accounting for $H_2$, CO or SiO emission. The energetic photons, in these cases come either from an external source of radiation (like a massive protostar in a protostellar outflow), or from the shock itself (for high velocity shocks, above 30 km $s^{-1}$). Both requirements have been addressed more or less recently: by [17] for the former, and [18], [19] and [20] for the latter.

**The next steps.** The (ongoing) next steps in terms of modelling consist in developing pseudo-3D solutions for optically thick lines, and providing user-friendly and standard, extensive grid of models covering as many input parameters as necessary. Observationnally, these new tools will allow to interpret observations of shocks closer to the protostar ([21], [22], [23], [24], [25],...) including accretion shocks (where the material falls on the disk - [26], [27], [28],...). They will also provide support to interpret observations from high-mass protostellar regions ([29], [30], [31], [32],...), and also from later feedback stages ([33], [34], [35],...). These new shock models are fed with more input parameters, and require more observational constraints. Multi-wavelengths studies will be needed. Instruments such as the IRAM30m, NOEMA, Yebes40m, APEX or ALMA ones, sometimes aimed at new lines (like recombination lines, e.g. [36] with APEX, [37] with the IRAM30m, and [38] with the Yebes40m), will have to be combined with JWST, Spitzer or ISO (to observe $H_2$, but also OH and $H_2O$), IRTF (to observe $CH^+$, [39]), and SOFIA, or *Herschel* ([40], [41], [42], [43], [44], [45]...). Fine-structure, but also meta-stable lines in the optical and UV ranges, as well as Lyman $\alpha$ and $\beta$ lines will also bring new constraints (observed with the VLT for instance, see [46], [47], [48]...).

## 3 Impact of shocks

**Energetic impacts: isolated outflows.** Measurements of the energetic impacts of protostellar shocks have been performed since the observation capabilities have reached the necessary spatial and spectral resolutions [49]. Interpreting observations by means of shock models is a sophisticated way to measure the energetic impacts of shocks at one position [5]. Since it requires both numerous observations and an easy access to grids of shock models, this method has only been performed over shock positions, and not generalized. In fact, observational studies can provide a more direct access to e.g., the momentum, the mechanical force, the kinetic energy, and the associated luminosity of a shock, not only on one position, but over entire maps, i.e., potentially over several spatial scales (see e.g. [6], where an analysis of CO lines observed by NOEMA, the IRAM-30m, and other telescopes led to the classification of Cep E as a jet-driven outflow, see also [50]).

**Energetic impacts: perspectives.** IRAM instruments have a key role to play in the evaluation of energetic impacts of shocks in more and more interstellar regions. With multiple beam receivers combined with wide-band receivers, the required observations of various lines of various isotopologues of CO will be much faster than it is now, enabling us to map large portions of the interstellar medium. This perspective is particularly interesting to map entire filaments possibly originating from cloud-cloud collisions, or evolved SNRs interacting with the ISM over tens of pc (see e.g. the CO observations presented in [51], and the energetic parameters in [52]). Interferometers will also open new perspectives in this direction: if short-spacings observations are performed and allow to recover the flux missing from these generally bright lines, then a next step could be to quantify the energetic impact of clusters of protostellar outflows, whether these are low-mass (like NGC1333, e.g. [53, 54]) or high-mass (like W43-MM1, see [55, 56]).

**Chemical impacts: SiO.** SiO has been considered the ultimate shock tracer for a long time. The classical intepretation for its systematic association to shock regions was that it is necessary to have a rather violent (velocity above 25 km $s^{-1}$) shocks to release the silicon almost entirely locked in grain cores in the quiescent medium. This necessary sputtering was made possible by the existence of a drift velocity between neutrals and charged grains, a requirement met in C-type shocks (see [57], [58], [59], with reference observational papers such as [60]). On the other hand, the interpretation of the SiO emission from young shock regions (nearby the protostar in jets and outflows, [61], [2], [6]), or the inclusion of grain-grain interactions in shock models ([62], [63], [64]), and more recently the interpretation of extended SiO emission in low-velocity shock regions (e.g. [65], [66], [67], [68], [69],...) made it necessary to adopt another model where a small fraction (1 to 10%. of the elemental solar neighborhood value) of silicon is free in the pre-shock phase (in a form depending on the conditions in the pre-shock region). In such conditions, SiO formation can be achieved above 3 km $s^{-1}$ (chemical route) or ~12 km $s^{-1}$ (mantle sputtering route; [55]). One must then explain one SiO is not detected/observed elsewhere: in the diffuse medium, it could be due to photodissociation and ionization into Si and $Si^{+}$. In the dense medium, it could be due to freeze-out on grains or to conversion into $SiO_2$.

**Chemical impacts: systematic studies.** Beyond SiO, the IRAM instruments have played a pivotal role in the study of chemical impacts of shocks so far, and should continue to do so. Among the large programs that have brought several breakthroughs, one can cite SOLIS (Seeds Of Life in Space) with NOEMA ([70]), ASAI (Astrochemical Surveys At IRAM) with the IRAM-30m telescope ([71]), and CALYPSO (Continuum And Lines in Young Protostellar Objects) with both telescopes ([72], [73]). A source like L1157-B1, for instance, has been extensively observed to study the chemistry of silicon (SiS, [74], and SiO, [75]), sulphur ($H_2S$, [76]; CCS, $H_2CS$, OCS, SO, $SO_2$, [77], [78]), but also nitrogen (formamide, [79], NO, [80], DCN and $H^{13}CN$, [81], HCN/HNC, [82], $HC_3N$ and $HC_5N$, [83], $H^{13}CN$ and $HC^{15}N$, [84]), and phosphorus (PO and PN, [85]). The production of complex organic molecules in this region was also studied by [86], [87], and [73], each time involving IRAM instruments. Interestingly, recent observations of the Cep E protostellar outflow by [88] might serve as a basis for irradiated shock studies.

## 4 Jets and outflows

**General properties.** Protostellar jets and outflows can also be studied for themselves rather than for shock-related goals. In this case, observations by IRAM or IRAM-like telescopes usually shed light on the process of star formation. The recent CALYPSO program was the first survey dedicated to globally characterize properties (presence of wide angle winds and high-velocity jets, jet mass loss rates, total jet power...) of outflows from young Sun-like stars with NOEMA ([89], with an equivalent ALMA program, see [90]).

**Episodicity.** The recent advents of interferometry also enabled to measure the episodicity of jets and outflows driven by low-mass ([91]), intermediate-mass ([92]), and high-mass protostars embedded in a protostellar cluster (W43-MM1, see [56]), with an episodicity in the ejection process occurring at dynamical timescales of tens to hundreds years, confirming the episodic nature of star formation processes.

**Precession.** The precession of jets and outflows was also more and more studied, specially around low-mass protostars (e.g., L1157, [93], SVS 13, [94], and IRAS 15398, [95]). Recently the precession of the intermediate-mass Cep E system was studied by [92]. The authors usually use models by [50] or [96] to describe the precession caused by either the orbital model of a binary system (when the precession has a W shape), or by the precession of the inner disk induced by a companion on inclined orbit (see also [97]). In the case of Cep E, the role of the B protostellar companion reported in [98] was excluded, but the authors were able to place constraints on a tighter possible companion (located at 10 to 40 AU from the protostar driving the Cep E outflow).

**Launching mechanisms.** At very high angular resolution, the recourse of interferometry has enabled to test theories of launching mechanisms for jets and outflows. The theoretical, magneto-hydrodynamical (MHD) disk wind model first developed by [99] was progressively moved closer to the observations by [100, 101]. It has now been tested in the HH212 system by [21, 22, 25, 102, 103]. Although very high angular resolution is required, these studies offer interesting perspectives for our understanding of a key process occurring almost systematically during the formation of stars.

## 5 Perspectives

Observations and models have progressed in parallel these past years, and more interaction is probably needed between these too distinct communities. Confrontations must systematized between observations and models. From this respect, observations must be as multi-wavelengths as possible, and include constraints on magnetic field properties. Models need to include radiative transfer, and also pseudo 3D effects. In this necessary dialog, the IRAM telescopes will play an instrumental role in the coming years, towards a more complete understanding of the formation of stars and better measurements of their feedback on the ISM.

## References

[1] Flower, D.R. & Pineau des Forêts, G., A&A, **578**, A63 (2015)
[2] Gusdorf, A., Giannini, T., Flower, D. R., et al., A&A, **532**, A53 (2011)
[3] Gusdorf, A., Anderl, S., Güsten, R., et al., A&A, **542**, L19 (2012)
[4] Anderl, S., Gusdorf, A., Güsten, R., A&A, **569**, A81 (2014)
[5] Gusdorf, A., Riquelme, D., Anderl, S., et al., A&A, **575**, A98 (2015)
[6] Lefloch, B., Gusdorf, A., Codella, C., et al., A&A, **581**, A4 (2015)
[7] Giannini, T., Nisini, B., Neufeld, D. A., et al., ApJ, **738**, 80 (2011)
[8] Neufeld, D. A., Gusdorf, A., Güsten, R., et al., ApJ, **781**, 102 (2014)
[9] Cosentino, G., Jiménez-Serra, I., Caselli, P., et al., ApJL, **881**, L42 (2019)
[10] Neufeld, D. A., De Witt, C., Lesaffre, P., et al., ApJL, **878**, L18 (2019)
[11] Reach, W. T., Tram, L. N., Richter, M., et al., ApJ, **884**, 81 (2019)
[12] Kristensen, L. E., Ravkilde, T. L., Pineau des Forêts, G., et al., A&A, **477**, 203 (2008)
[13] Gustafsson, M., Ravkilde, T. L., Kristensen, L. E., et al. A&A, **513**, A5 (2010)

[14] Tram, L. N., Lesaffre, P., Cabrit, S., et al., MNRAS, **473**, 1472 (2018)
[15] Benedettini, M., Gusdorf, A., Nisini, B., et al., A&A, **598**, A14 (2017)
[16] Gusdorf, A., Anderl, S., Lefloch, B., et al., A&A, **602**, A8 (2017)
[17] Godard, B., Pineau des Forêts, G., Lesaffre, P., et al., A&A, **622**, A100 (2019)
[18] Hollenbach, D., & McKee, C. F., ApJ, **342**, 306 (1989)
[19] Lehmann, A., Godard, B., Pineau des Forêts, G., et al., A&A, **643**, A101 (2020)
[20] Lehmann, A., Godard, B., Pineau des Forêts, G., et al., A&A, **658**, A165 (2022)
[21] Tabone, B., Cabrit, S., Bianchi, E., et al., A&A, **607**, L6 (2017)
[22] Tabone, B., Raga, A., Cabrit, S., et al., A&A, **614**, A119 (2018)
[23] Tabone, B., Godard, B., Pineau des Forêtes, G., et al., A&A, **636**, A60 (2020)
[24] Tychoniec, L., van Dishoeck, E. F., van't Hoff, M. L. R., et al., A&A, **655**, A65 (2021)
[25] Lee, C. F., Tabone, B., Cabrit, S., et al., ApJL, **907**, L41 (2021)
[26] Codella, C., Bianchi, E., Tabone, B., et al., A&A, **617**, A10 (2018)
[27] Csengeri, T., Belloche, A., Bontemps, S., et al., A&A, **632**, A57 (2019)
[28] van Gelder, M. L., Tabone, B., van Dishoeck, E. F., et al., A&A, **653**, A159 (2021)
[29] Leurini, S., Wyrowski, F., Wiesemeyer, H., et al., A&A, **584**, A70 (2015)
[30] Gusdorf, A., Güsten, R., Menten, K. M., et al., A&A, **585**, A45 (2016)
[31] Maud, L. T., Cesaroni, R., Kumar, M. S. N., et al., A&A, **620**, A31 (2018)
[32] Gieser, C., Beuther, H., Semenov, D., et al., A&A, **648**, A66 (2021)
[33] Schneider, N., Bontemps, S., Motte, F., et al., A&A, **591**, A40 (2016)
[34] Luisi, M., Anderson, L. D., Schneider, N., et al., Science Advances, **7**, 15 (2021)
[35] Tiwari, M., Karim, R., Pound, M/ W., et al., ApJ, **914**, 2, 117 (2021)
[36] Kim, W.-J., Urquhart, J. S., Wyrowski, F., et al., A&A, **616**, A107 (2018)
[37] Cuadrado, S., Salas, P., Goicoechea, J., et al., A&A, **625**, L3 (2019)
[38] Goicoechea, J. R., & Cuadrado, S., A&A, **647**, L7 (2021)
[39] Neufeld, D. A., Godard, B., Bryan Changala, P., et al., ApJ, **917**, 15 (2021)
[40] Nisini, B., Santangelo, G., Giannini, T., et al., ApJ, **801**, 121 (2015)
[41] Kristensen, L. E., Gusdorf, A., Mottram, J. C., et al., A&A, **601**, L4 (2017)
[42] Kristensen, L. E., van Dishoeck, E. F., Mottram, J. C., et al., A&A, **605**, 93 (2017)
[43] Sperling, T., Eislöffel, J., Fischer, C., et al., A&A, **642**, A216 (2020)
[44] Sperling, T., Eislöffel, J., Fischer, C., et al., A&A, **650**, A173 (2021)
[45] Yang, Y.-L., Evans, II., N. J., Karska, A., et al., ApJ, **925**, 93 (2022)
[46] Nisini, B., Giannini, T., Antoniucci, S., et al., A&A, **595**, A76 (2016)
[47] Nisini, B., Antoniucci, S., Alcalá, J. M., et al., A&A, **609**, A87 (2018)
[48] Alcalá, J. M., Cupani, G., Evans, C. J., et al., Experimental Astronomy, special issue: Science with the Cassegrain U-Band Efficient Spectrograph CUBES (2022)
[49] Bontemps, S., André, P., Terebey, S., et al., A&A, **311**, 858 (1996)
[50] Raga, A. C., Cantó, & S. Biro, MNRAS, **260**, 163 (1993)
[51] Dell'Ova, P., Gusdorf, A., Gerin, M., et al., **644**, A64 (2020)
[52] Dell'Ova, P., PhD thesis (2022)
[53] Plunkett, A. L., Arce, H. G., Corder, S. A., et al., ApJ, **774**, 22 (2013)
[54] Dionatos, O., & Güdel, M., A&A, **597**, A64 (2017)
[55] Louvet, F., Motte, F., Gusdorf, A., et al., A&A, **595**, A122 (2016)
[56] Nony, T., Motte, F., Louvet, F., et al., A&A, **636**, A38 (2020)
[57] Caselli, P., Hartquist, T. W., & Havnes, O., A&A, **322**, 296 (1997)
[58] Schilke, P., Walmsley, C. M., Pineau des Forêts, G., et al., A&A, **321**, 293 (1997)

[59] Gusdorf, A., Cabrit, S., Flower, D. R., et al., A&A, **482**, 809 (2008)
[60] Nisini, B., Codella, C., Giannini, T., et al., A&A, **462**, 163 (2007)
[61] Gusdorf, A., Pineau des Forêts, G., Cabrit, et al., A&A, **490**, 695 (2008)
[62] Guillet, V., Pineau des Forêts, G., & Jones, A. P., A&A, **527**, A123 (2011)
[63] Anderl, A., Guillet, V., Pineau des Forêts, G., et al., A&A, **556**, A69 (2013)
[64] Leurini, S., Codella, C., López-Sepulcre, A., et al., A&A, **570**, A49 (2014)
[65] Jiménez-Serra, I., Caselli, P., Tan, J. C., et al., MNRAS, **406**, 187 (2010)
[66] Kauffmann, J., Pillai, T., & Zhang, Q., ApJL, **765**, L35 (2013)
[67] Sanhueza, P., Jackson, J. M., Foster, J. B., et al., ApJ, **773**, 123 (2013)
[68] Nguyen-Lu'o'ng, Q., Motte, F., Carlhoff, P., et al., ApJ, **775**, 88 (2013)
[69] Duarte-Cabral, A., Bontemps, S., Motte, F., et al., A&A, **570**, A1 (2014)
[70] Ceccarelli, C., Caselli, P., Fontani, F., et al., ApJ, **850**, 176 (2017)
[71] Lefloch, B., Bachiller, R., Ceccarelli, C., et al., MNRAS, **477**, 4792 (2018)
[72] Maury, A., André, P., Testi, L., et al., A&A, **621**, A76 (2019)
[73] Belloche, A., Maury, A. J., Maret, S., et al., A&A, **635**, A198 (2020)
[74] Podio, L., Codella, C., Lefloch, B., et al., MNRAS, **470**, L16 (2017)
[75] Spezzano, S., Codella, C., Podio, L., et al., A&A, **640**, A74 (2020)
[76] Holdship, J., Viti, S., Jimenez-Serra, I., et al., MNRAS, **463**, 802 (2016)
[77] Holdship, J., Jimenez-Serra, I., Viti, S., et al., ApJ, **878**, 64 (2019)
[78] Feng, S., Codella, C., Ceccarelli, C., et al., ApJ, **896**, 37 (2020)
[79] Codella, C., Ceccarelli, C., Caselli, P., et al., A&A, **605**, L3 (2017)
[80] Codella, C., Viti, S., Lefloch, B., et al., MNRAS, **474**, 5694 (2018)
[81] Busquet, G., Fontani, F., Viti, S., et al., A&A, **604**, A20 (2017)
[82] Lefloch, B., Busquet, G., Viti, S., et al., MNRAS, **507**, 1034 (2021)
[83] Mendoza, E., Lefloch, B., Ceccarelli, C., et al., MNRAS, **475**, 5501 (2018)
[84] Benedettini, M., Viti, S., Codella, C., et al., A&A, **645**, A91 (2021)
[85] Lefloch, B., Vastel, C., Viti, S., et al., MNRAS, **462**, 3937 (2016)
[86] Lefloch, B., Ceccarelli, C., Codella, C., et al., MNRAS, **469**, L73 (2017)
[87] Codella, C., Ceccarelli, C., Bianchi, E., et al., A&A, **635**, A17 (2020)
[88] Ospina-Zamudio, J., Lefloch, B., Favre, C., et al., MNRAS, **490**, 2679 (2019)
[89] Podio, L., Tabone, B., Codella, C., et al., A&A, **648**, 45 (2021)
[90] Tychoniec, L., PhD thesis (2021)
[91] Plunkett, A. L., Arce, H. G., Mardones, D., et al., Nature, **527**, 70 (2015)
[92] Schutzer, A. de A., Rivera-Ortiz, P. R., Lefloch, B., et al., A&A, in press (2022)
[93] Podio, L., Codella, L., Gueth, F., et al., A&A, **593**, L4 (2016)
[94] Lefèvre, C., Cabrit, S., Maury, A. J., et al., A&A, **604**, L1 (2017)
[95] Okoda, Y., Oya, Y., Logan, F., et al., ApJ, **910**, 11 (2021)
[96] Masciadri, E., & Raga, A. C., ApJ, **568**, 733 (2002)
[97] Terquem, C., Eislöffel, Papaloizou, J. C. B., et al., **512**, L131 (1999)
[98] Ospina-Zamudio, J., Lefloch, B., Ceccarelli, C., et al., A&A, **618**, A145 (2018)
[99] Ferreira, J., Dougados, C., & Cabrit, S., A&A, **453**, 785 (2006)
[100] Panoglou, D., Cabrit, S., Pineau des Forêts, G., et al., A&A, **538**, A2 (2012)
[101] Yvart, W., Cabrit, S., Pineau des Forêts, G., et al., A&A, **585**, A74 (2016)
[102] Lee, C.-F., Li Z.-Y., Codella, C., et al., ApJ, **856**, 14 (2018)
[103] Tabone, B., Cabrit, S., Pineau des Forêts, G., et al., A&A, **640**, A82 (2020)

# Shocks in the surroundings of the NGC 1333 IRAS 4 system

*Marta* De Simone[1,2,*]

[1]Université Grenoble Alpes, IPAG, 38000 Grenoble, France
[2]INAF, Osservatorio Astrofisico di Arcetri, Largo E. Fermi 5, 50125 Firenze, Italy

**Abstract.** The IRAS 4A system is part of the southern filament of the Perseus NGC 1333 molecular cloud. As most of the star forming regions, NGC 1333 is known to be heavily shaped by external triggers, such as shock fronts around OB stars or supernovae remnants, cloud-cloud collisions, and instabilities. Recently, it has been hypothesized that the entire southwest region of NGC 1333, encompassing the filament where IRAS 4A lie, is due to a colliding "turbulent" cell, a clash that triggered the birth of the protostars on the filaments. However, no specific signatures of a clash have been reported so far, leaving unanswered how and where the energy of this clash, if real, is dispersed. To answer this question, we analyzed new high spatial resolution (~600 au) observations of $CH_3OH$ and SiO, known shock tracers, obtained in the context of the Large Program IRAM/NOEMA SOLIS searching for specific signature of the clash event. We detected three parallel elongated structures, called fingers, with narrow line profiles (~1.5 km $s^{-1}$), peaked at the systemic velocity of the cloud, tracing gas with high density (5-20×$10^5$ $cm^{-3}$) and high temperature (80-160 K). They are chemically different with the northern finger traced by both SiO and $CH_3OH$ and the other two only by SiO. Among various possibilities, a train of three consecutive shocks, due to an expanding bubble coming behind NGC 1333 and from the southwest, can reproduce the observations. Finally, we propose a solution for the two-decades long debate on the nature and the origin of the widespread narrow SiO emission observed in the south part of NGC 1333, namely unresolved trains of shocks.

## 1 Introduction

The Perseus molecular cloud is a nearby [orders of 300-400 pc; 1]) giant molecular cloud characterized by a large population of low-mass young stellar objects, of which the majority is associated with the two major clusters IC 348 and NGC 1333. In particular, the latter is one of the most active sites of on-going star formation [e.g., 2].

Like most of the star forming regions, NGC 1333 is characterized by filaments and cavities that, together with a clear velocity gradient (see Figure 1), suggest that it has been shaped by external triggers (such as, ionization/shock fronts by stellar winds or supernovae remnants, cloud–cloud collisions, and instabilities). In particular, Dhabal et al. [3] hypothesized that the entire blue-shifted southwest region of NGC 1333 is due to a colliding "turbulent cell", a clash that created the filament and triggered the birth of the protostars on it. However, no specific signatures of a clash, namely shocks, have been reported so far, leaving unanswered how and where the energy of this clash, if real, is dispersed.

*e-mail: marta.desimone@univ-grenoble-alpes.fr

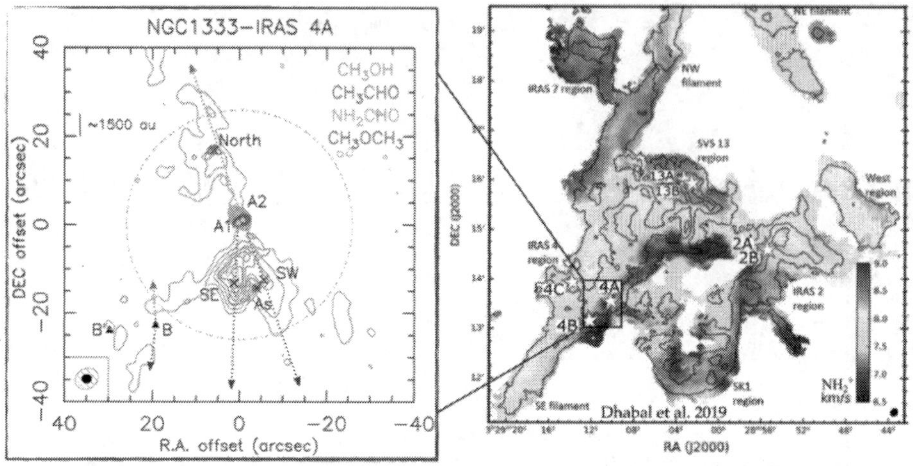


**Figure 1.** *Right*: $N_2H^+$ (1-0) velocity map of NGC 1333 south [3]. The main protostellar cores are marked with white stars. *Left*: iCOMs contour emission in the IRAS 4A outflows observed with NOEMA/SOLIS Observations [4]. The dashed blue and red arrows indicate the direction of outflows.

## 2 NGC 1333 IRAS 4: A region full of surprises

One of the protostellar systems that lies along the NGC 1333 southern filaments is IRAS 4 [distant ~300 pc; 1]. It is characterized by the Class 0 protostar IRAS 4B and the protostellar binary system 4A composed of 4A1, the brightest in mm continuum, and 4A2 separated of 1.″8, driving a spectacular large bipolar outflow [e.g., 5–8].

### 2.1 The molecular complexity of IRAS 4A outflows

As part of the NOEMA SOLIS[1] Large Program, we analyzed the molecular emission of the IRAS 4A outflows. The high angular resolution and high sensitivity observations allowed us to detect and image, for the first time, several interstellar Complex Organic Molecules (iCOMs) in the IRAS 4A outflow (see Figure 1): the methanol emission is extended (~ 1′) and covers the lobes of the two outflows, the acetaldehyde emission traces intermediate scales (~ 15″) and it is bright toward the southern-east lobe, the dimethyl ether and formamide emissions are compact and not resolved in two position along the outflows [4].

Analyzing the emission in different positions of the outflows, and focusing on methanol and acetaldehyde, we confirmed the presence of chemical differentiation, with the southeast lobe (driven by 4A1) brighter in iCOMs than the southwest-north (driven by 4A2). As explained in full details in De Simone et al. [4], comparing the observations with astrochemical model predictions we could test the formation path of the acetaldehyde suggesting that it is synthesized in the gas phase by the reaction of atomic oxygen with ethyl radical.

### 2.2 Elongated structures around IRAS 4A

While analyzing the methanol emission in the IRAS 4A outflows, we noticed the presence of an elongated structure almost perpendicular to the outflow with a narrow spectral profile (FWHM~1.5 km s$^{-1}$) peaking at the systemic velocity of the cloud. Similar structures were

[1] Seed Of Life in Space; https://solis.osug.fr

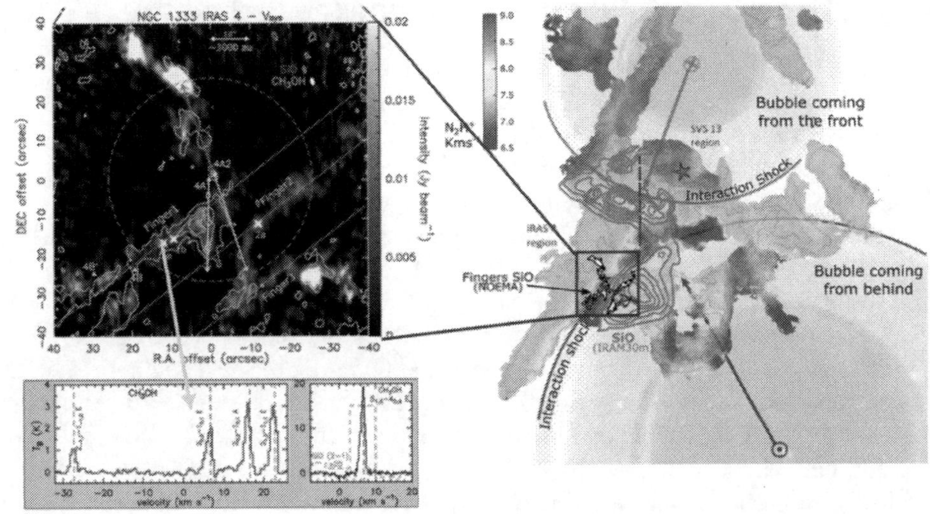


**Figure 2.** *Upper left*: emission of the three fingers, marked by orange lines, in SiO (color scale) and $CH_3OH$ (contours) observed with NOEMA SOLIS [12]. *Bottom left* $CH_3OH$ (black) and SiO (red, magnified by a factor of 20) spectra extracted at the Finger1a position. *Right:* Large-scale view of NGC 1333 south overlapping the $N_2H^+$ velocity map by Dhabal et al. [3] with our SOLIS SiO fingers, and the IRAM 30m widespread narrow SiO emission by Lefloch et al. [9]. The red and blue curves represent the interaction of the expanding bubbles: The former is expanding from the north-west towards SVS13 from the front, and the latter is expanding from the south-west towards IRAS 4A from behind.

observed in SiO by Choi [5] with the VLA in the southwest part of the region, and by Lefloch et al. [9] and Codella et al. [10] with IRAM 30m observations as a widespread SiO emission connecting the IRAS 4A region with the SVS13 one.

In order to understand if these features were related to each other we analyzed our NOEMA/SOLIS observations of $CH_3OH$ and SiO (see Figure 2). We identified three parallel fingers separated by about 10″ (~3000 au): i) Finger1, traced by $CH_3OH$ and SiO, is the northern one; ii) Finger2, traced only by SiO, is 10″ south to Finger1; iii) Finger3, traced only by SiO, is an additional 10″ south. While Finger2 and Finger3, detected only in SiO (2-1), are perfectly in agreement with the SiO (1-0) VLA observations by Choi (2005), Finger1 was never detected before and it is traced by both $CH_3OH$ and SiO. Additionally the fingers are located in the large-scale blueshifted southern region of NGC 1333 (see Figure 2).

Analyzing the methanol and SiO emission using a non-LTE LVG analysis [11] as described in details in [12], we derived the gas properties and the SiO/$CH_3OH$ abundance ratio in Finger1 and Finger2. We confirmed the chemical diversity observed in the fingers, and we found that they are tracing material at high density (> $10^5$ $cm^{-3}$) and high temperature (> 80 K). However, what is their origin?

We investigated two main possibilities: the fingers are tracing i) Kelvin Helmholtz instabilities or ii) a train of shocks. However, even if the former could reproduce the quasi periodicity of the fingers, it could not reproduce the observed gas density and temperature. Therefore, the only plausible possibility was a train of shocks, well justified by the presence of both SiO and methanol, and by the derived gas physical condition.

## 3 Train of shocks as a signature of an unexpected clash

Given that methanol is efficiently formed on the grain mantles through the CO hydrogenation in the prestellar phase, the gaseous methanol observed in the fingers must have been extracted from the grain mantles and injected into the gas phase. The same applies to the observed SiO. The observed chemical differentiation in the fingers can be explained in two ways: the two species are injected from the grains with i) different abundance ratio in the different fingers or ii) the same abundance ratio that changes in time because of chemical processes occurring in the gas phase.

To investigate that, we compared the observed $CH_3OH$/SiO abundance ratio with astrochemical model predictions putting constraints on when the shocks occurred: the northern finger is the youngest one and it happened at least 5000 yr after the next southern one. Therefore, we suggest that the three fingers are the result of a successive train of shocks created by the clash of an expanding bubble coming from south-east and moving towards the IRAS 4A main cloud from behind [12].

Finally, we suggest that the widespread narrow SiO emission observed towards the NGC 1333 IRAS 4 and SVS 13 region with single-dish observations in the late 1990s [9, 10] is due to unresolved trains of shocks like the SOLIS fingers. These shocks would be the signature of the interaction of the bubble giving rise to the IRAS 4A fingers in the south and of another bubble pushing from the north towards SVS 13.

## Acknowledgment

MDS thanks all the co-authors of the works presented in this proceeding [4, 12]. This project has received funding within the European Union's Horizon 2020 research and innovation programme from the European Research Council (ERC) for the project "The Dawn of Organic Chemistry" (DOC), grant agreement No 741002.

## References

[1] C. Zucker, E.F. Schlafly, J.S. Speagle, G.M. Green, S.K.N. Portillo, D.P. Finkbeiner, A.A. Goodman, ApJ **869**, 83 (2018)

[2] J.K. Jørgensen, D. Johnstone, H. Kirk, P.C. Myers, L.E. Allen, Y.L. Shirley, ApJ **683**, 822 (2008)

[3] A. Dhabal, L.G. Mundy, C.y. Chen, P. Teuben, S. Storm, ApJ **876**, 108 (2019)

[4] M. De Simone, C. Codella, C. Ceccarelli, A. López-Sepulcre, A. Witzel, R. Neri, N. Balucani, P. Caselli, C. Favre, F. Fontani et al., A&A **640**, A75 (2020)

[5] M. Choi, ApJ **630**, 976 (2005)

[6] G. Santangelo, C. Codella, S. Cabrit, A.J. Maury, F. Gueth, S. Maret, B. Lefloch, A. Belloche, P. André, P. Hennebelle et al., A&A **584**, A126 (2015)

[7] A. López-Sepulcre, N. Sakai, R. Neri, M. Imai, Y. Oya, C. Ceccarelli, A.E. Higuchi, Y. Aikawa, S. Bottinelli, E. Caux et al., A&A **606**, A121 (2017)

[8] V. Taquet, C. Codella, M. De Simone, A. López-Sepulcre, J.E. Pineda, D. Segura-Cox, C. Ceccarelli, P. Caselli, A. Gusdorf, M.V. Persson et al., A&A **637**, A63 (2020)

[9] B. Lefloch, A. Castets, J. Cernicharo, L. Loinard, ApJ **504**, L109 (1998)

[10] C. Codella, R. Bachiller, B. Reipurth, A&A **343**, 585 (1999)

[11] C. Ceccarelli, S. Maret, A.G.G.M. Tielens, A. Castets, E. Caux, A&A **410**, 587 (2003)

[12] M. De Simone, C. Codella, C. Ceccarelli, A. López-Sepulcre, R. Neri, P.R. Rivera-Ortiz, G. Busquet, P. Caselli, E. Bianchi, F. Fontani et al., MNRAS **512**, 5214 (2022)

# Protostellar jets: A statistical view with the CALYPSO IRAM-PdBI survey

*Linda* Podio[1,*], *Benoit* Tabone[2], *Claudio* Codella[1], and *the* CALYPSO team

[1]INAF - Osservatorio Astrofisico di Arcetri, Largo E. Fermi 5, 50125 Firenze, Italy

[2]Leiden Observatory, Leiden University, PO Box 9513, 2300 RA Leiden, The Netherlands

**Abstract.** In the context of the CALYPSO IRAM-PdBI Large Program we performed the first statistical survey of protostellar jets by analysing molecular emission in a sample of 21 protostars covering a broad range of internal luminosities ($L_{int}$ from 0.035 $L_{\odot}$ to 47 $L_{\odot}$). We find that the outflow phenomenon is ubiquitous in our sample of protostars, with wide-angle outflows detected in CO $(2-1)$ in all sources, and high-velocity collimated jets detected in SiO $(5-4)$ in 80% of the sources with $L_{int} > 1$ $L_{\odot}$. The protostellar flows have an onion-like structure, with the SiO jet (opening angle, $\alpha \sim 10°$) nested into a wider angle SO ($\alpha \sim 15°$) and CO ($\alpha \sim 25°$) outflows. Interestingly, protostellar jets show several properties in common with the atomic jets associated with more evolved sources ($10^6$ yr), e.g. one third of the jets show velocity asymmetry of $\sim 1.3-2$ between the two lobes, and the mass-loss rates are $\sim 1\% - 50\%$ of the mass accretion rates. This suggests that the same launching mechanism is at work and that the correlation between mass ejection and mass accretion holds along the star-formation process from $10^4$ yr up to a few Myr.

## 1 Introduction

Supersonic collimated jets are believed to be launched from the disk due to a magneto-centrifugal mechanism. Theoretical studies suggest that they are a key ingredient for the formation of Sun-like stars, as they can remove the excess angular momentum from the disk thus allowing accretion onto the central star [see, e.g., 1].

Despite their importance, their formation mechanism is still very debated, in particular it is not clear if they are launched from the inner star-disk region [e.g., 2], or instead from a larger range of disk radii out to tens of au [e.g., 3]. While there have been a number of detailed studies of a few prototypical protostellar jets, such as, e.g., HH 212 [4, 5], there has been a lack of statistical studies devoted to characterize the protostellar jet properties on a large sample. Observational surveys of CO low-J transitions at intermediate resolution ($3'' - 4''$) [e.g., 6] have shown that large-scale outflow emission is commonly associated with protostellar objects. Are these outflows accelerated by high-velocity collimated jets launched from the disk?

*e-mail: linda.podio@inaf.it

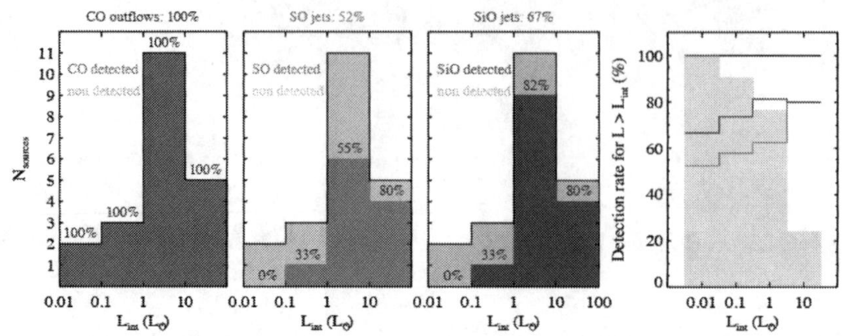


**Figure 1.** Number of sources associated with outflows and jets as traced by CO (2 – 1) (black, *left*), SO ($5_6 - 4_5$) (red, *middle-left*), and SiO (5 – 4) (blue, *middle-right*) as a function of the internal luminosity, $L_{\rm int}$, of the 21 Class 0 protostars in the CALYPSO sample. The detection rate for each $L_{\rm int}$ bin (in logarithmic scale) is reported above the histogram, while the detection rate for the whole sample is labeled on top of each panel. The detection rate of jets and outflows for $L > L_{\rm int}$ is shown in the right panel, with the same colour coding. The grey histogram shows the fraction of sources for $L > L_{\rm int}$.

## 2 The CALYPSO Large Program

The CALYPSO Large Program[1] is a survey of Class 0 protostars located in the most active star-forming regions in the northern hemisphere with the IRAM Plateau de Bure Interferometer. The aim of the survey is: (i) to identify continuum and line emission from protostellar disks [7, 8]; (ii) to investigate the hot corino chemistry [9]; and (iii) to investigate the occurence and properties of protostellar jets by targeting CO (3 – 2), SiO (5 – 4), and SO ($5_6 - 4_5$) lines [10]. The first goal of the survey of protostellar jets performed by [10] is to answer a crucial question, i.e. whether jets, and, in general, mass ejection phenomena are commonly observed in protostars. The second goal is to derive the jet properties (velocity, width, mass loss rates), and the molecular abundances, which are crucial for understanding what region of the disk-protostar system jets are launched from. Finally, a third goal is to obtain a large database for follow-up observations at an extremely high-spatial resolution. The obtained results on protostellar jets are summarized in the following sections.

## 3 Jets ubiquity at the protostellar stage

The detection rate of outflows and jets emission in the molecular tracers as a function of the protostellar internal luminosity, $L_{\rm int}$, is shown in Fig. 1. Outflows in CO (2 – 1) are detected in 100% of the 21 Class 0 protostars in the CALYPSO sample, indicating that ejection phenomena are ubiquitous at the protostellar stage. Collimated high-velocity jets traced by SiO (5 – 4) are detected in 67% of the sources and 79% of these also show jet emission in SO ($5_6 - 4_5$). The detection rate of jets increases with $L_{\rm int}$, which is a probe of the mass accretion rate onto the protostar. This confirms the expected correlation between the mass accretion and the mass ejection rate (which, in turn, is proportional to the brightness of the emission lines). Hence, the non-detection of jet emission associated with the less-accreting sources could be an observational bias and deeper observations could demonstrate that jets are ubiquitous in our sample.

[1] http://irfu.cea.fr/Projects/Calypso
https://www.iram-institute.org/EN/content-page-317-7-158-240-317-0.html

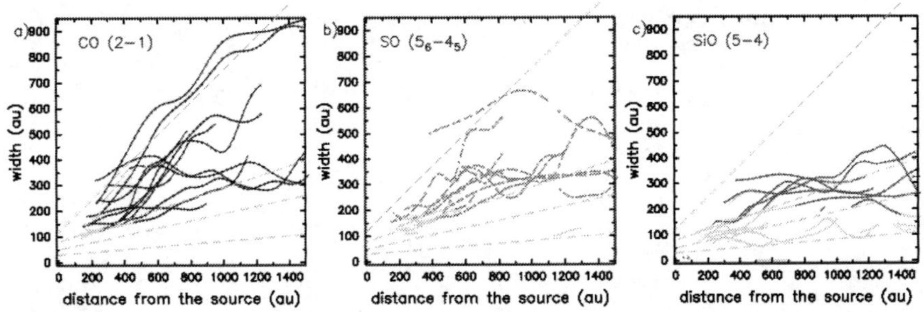


**Figure 2.** Flow widths ($2R_{\rm jet}$) estimated from the CO (a), SO (b), and SiO (c) emission from ~ 200 up to 1500 au distance from protostars. Cyan lines indicate widths that are smaller that the transverse beam and consequently considered as inclusive upper limits. The green dashed lines correspond to full opening angles of $\alpha = 3°, 8°, 12°$, and 35° and initial widths of 30, 50, 80, and 120 au, and outline the collimation properties of the flow in different tracers.

## 4 Protostellar jets: spatio-kinematical properties

The observed protostellar flows have an onion-like structure: SiO (5 − 4) emission is more collimated than SO ($5_6 - 4_5$) emission, which in turn is narrower than CO (2 − 1), with median opening angles of 10°, 15°, and 25°, respectively (Fig. 2). High-velocity CO emission is as collimated as SiO indicating that low-velocity CO probes entrained material, while high-velocity CO traces the collimated jet. At scales larger than 300 au, most of the high-velocity SiO jets are broader (~ 4° − 12°) than Class I and Class II atomic jets (~ 3°). This could be due to projection effects as well as to contamination by the bow-shock wings.

The median radial velocity of the SiO protostellar jets is 30 km $\rm s^{-1}$, that is, about two times smaller than that of atomic jets driven by Class II sources [11]. Assuming that the velocity of the jet scales with the Keplerian velocity at the launching point, the increase of the jet velocity with age is consistent with a jet launched from the same region of the disk around a central object of increasing mass. At least 33% of the SiO bipolar jets show a velocity asymmetry between the two lobes by a factor of 1.3 − 2.1, in agreement with what was found for atomic jets from T Tauri stars [12]. The similarity in knot spacings and velocity asymmetries suggests that the jet launching mechanism in protostars of $10^4$ yr might be similar to that in Class II sources ($10^6$ yr). Finally, we find that 50% of the SiO jets are precessing or wiggling.

## 5 The correlation between accretion and ejection

The mass loss rates of the protostellar jets range from $7 \times 10^{-8}$ $\rm M_\odot\,yr^{-1}$ up to values of $\sim 3 \times 10^{-6}$ $\rm M_\odot\,yr^{-1}$ for internal luminosities of the driving protostars of $\sim 1 - 50$ $\rm L_\odot$ (see Fig. 3). These values are larger by up to five orders of magnitude than those measured for atomic jets driven by Class II sources (from ~ $10^{-11}$ to a few $10^{-8}$ $\rm M_\odot\,yr^{-1}$, [e.g., 13, 14]). Moreover, we find that $\dot{M}_{\rm jet} \sim 0.1 - 0.5 \dot{M}_{\rm acc}$ for most of the jets, with the exception of the "CO-poor" and monopolar jets for which $\dot{M}_{\rm jet} \sim 0.01 - 0.1 \dot{M}_{\rm acc}$. These $\dot{M}_{\rm jet}/\dot{M}_{\rm acc}$ ratios are similar to those found for atomic Class II jets ($\dot{M}_{\rm jet} \sim 0.01 - 0.3 \dot{M}_{\rm acc}$, [13, 14]) and indicate that the correlation between ejection and accretion holds over the whole star-formation process, from protostellar objects of $10^4$ years to pre-main sequence stars of 1 Myr. The total jet power ($L_{\rm jet}$) is 10%-50% of the source internal luminosity for ~ 60% of the jets in the sample (Fig. 3-right). This indicates that the gravitational energy released by accretion onto the protostar could be efficiently extracted and converted into mechanical energy in the jet.

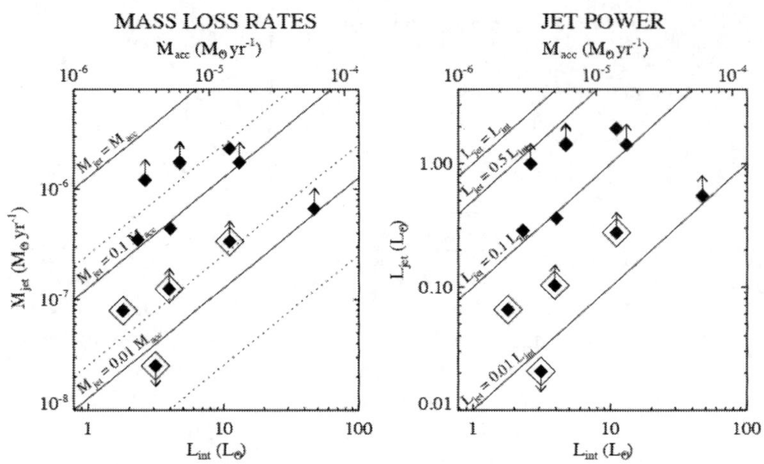


**Figure 3.** Protostellar jet mass-loss rates ($\dot{M}_{\rm jet}$ in $M_\odot\,yr^{-1}$, *left*), and mechanical luminosities ($L_{\rm jet}$ in $L_\odot$, *right*) versus the source internal luminosity ($L_{\rm int}$ in $L_\odot$). Lower and upper limits are indicated by upward and downward arrows. The jet knots where no CO is detected (the "CO-poor" jets) are indicated by larger empty diamonds. The black lines in the left panel indicate $\dot{M}_{\rm jet} = [0.01, 0.1, 1] \times \dot{M}_{\rm acc}$, with $\dot{M}_{\rm acc}$ estimated from $L_{\rm int}$, assuming a protostellar mass, $M_*$, of 0.05 $M_\odot$ or 0.25 $M_\odot$ (solid and dotted lines, respectively). The $\dot{M}_{\rm acc}$ values labeled on the upper x-axis correspond to $M_* = 0.05\,M_\odot$. The black solid lines in the right panel indicate $L_{\rm jet} = [0.01, 0.1, 0.5, 1] \times L_{\rm int}$.

## References

[1] A. Frank, et al., *Jets and Outflows from Star to Cloud: Observations Confront Theory*, in *Protostars and Planets VI*, edited by H. Beuther, R.S. Klessen, C.P. Dullemond, T. Henning (2014), p. 451, 1402.3553

[2] H. Shang, et al. *Jets and Bipolar Outflows from Young Stars: Theory and Observational Tests*, in *Protostars and Planets V*, edited by B. Reipurth, D. Jewitt, K. Keil (2007), p. 261

[3] R.E. Pudritz, et al. *Disk Winds, Jets, and Outflows: Theoretical and Computational Foundations*, in *Protostars and Planets V*, edited by B. Reipurth, D. Jewitt, K. Keil (2007), p. 277, astro-ph/0603592

[4] C. Codella, et al., A&A **568**, L5 (2014), 1407.6229

[5] C.F. Lee, et al., Nature Astronomy **1**, 0152 (2017), 1706.06343

[6] J.J. Tobin, et al., ApJ **867**, 43 (2018), 1809.06434

[7] A.J. Maury, et al., A&A **621**, A76 (2019), 1810.11221

[8] S. Maret, et al., A&A **635**, A15 (2020), 2001.06355

[9] A. Belloche, et al., A&A **635**, A198 (2020), 2002.00592

[10] L. Podio, et al., A&A **648**, A45 (2021), 2012.15379

[11] B. Nisini, et al., A&A **609**, A87 (2018), 1710.05587

[12] G.A. Hirth, et al., ApJL **427**, L99 (1994)

[13] P. Hartigan, et al., ApJ **452**, 736 (1995)

[14] L. Podio, et al., A&A **545**, A44 (2012), 1207.3435

# The emergence of molecular complexity in star forming regions as seen with ASAI

*Bertrand* Lefloch[1,*], *Charlotte* Vastel[2,3], *Eleonora* Bianchi[1], and *Rafael* Bachiller[4]

[1]Univ. Grenoble Alpes, CNRS, IPAG, 38000 Grenoble, France
[2]Université de Toulouse, UPS-OMP, IRAP, Toulouse, France
[3]CNRS, IRAP, 9 Av. Colonel Roche, BP 44346, 31028 Toulouse Cedex 4, France
[4]IGN, Observatorio Astrónomico Nacional, Apartado 1143, 28800 Alcalá de Henares, Spain

**Abstract.** The Large Program "Astrochemical Surveys At IRAM" (ASAI) investigates the emergence of molecular complexity along the different stages of the solar-type star formation process, by carrying out unbiased line surveys of a sample of ten template sources in the range 80-272 GHz with the IRAM 30m telescope. We present here an overview of the main results of the Large Program ASAI.

## 1 The motivation for ASAI

The chemical evolution of the matter during the long process that brought it from prestellar cores (PSCs), to protostars and their associated shocks, protoplanetary disks, and ultimately to the bodies of the Solar System is one of the key questions in modern Astrophysics and it was the main motivation for the Large Program "Astrochemical Surveys At IRAM" (ASAI; [15]) at the IRAM 30m telescope. More precisely, ASAI targeted two specific goals : a) to obtain an evolutionary view on chemistry, in particular Complex Organic Molecules (COMs) [18]: molecular abundances, formation and destruction pathways, b) to understand the influence of the environmental conditions, in particular the star formation feedback processes. Using the EMIR receivers, ASAI carried out unbiased line surveys of the molecular line emission of a sample of ten template sources illustrating the different chemical stages of a Sun-like star during its formation process, from the cold prestellar phase (PSC TMC1 and L1544), to solar-type protostars (First HydroStatic Core B1b, Class 0 IRAS4A and L1157mm, Class 0/I L1527, Class I SVS13A), outflows (L1157-B1, L1448-R2), and last to protoplanetary disks (Class II AB Aur). The observations covered the range 80–272 GHz for all sources but PSCs (70-110 GHz for L1544; 130-170 GHz for TMC1). They were carried out over 6 semesters between 2012 and 2015 (360 hours).

## 2 Overall results

The 3mm window was first investigated in full detail because of its high SNR. The number of the detected species (main isotopologues) is ≈ 45, with low variations between the sources of the sample : 50 in L1544, 51 in B1b, 47 L1527, 48 in IRAS4A, 35 in SVS13A. This

*e-mail: bertrand.lefloch@univ-grenoble-alpes.fr

number appears rather independent of the source evolutionary stage and luminosity, meaning that chemical complexity is present at all stages of protostellar evolution. Similar numbers of molecular species are actually found in the massive SFRs Orion (43) and SgrB2 (56). The main difference with the more luminous and massive sources resides in the spectral line density, which is lower by one order of magnitude in solar-type sources ($\sim$ 5-12 $GHz^{-1}$). No other COMs than those previously known were detected with ASAI.

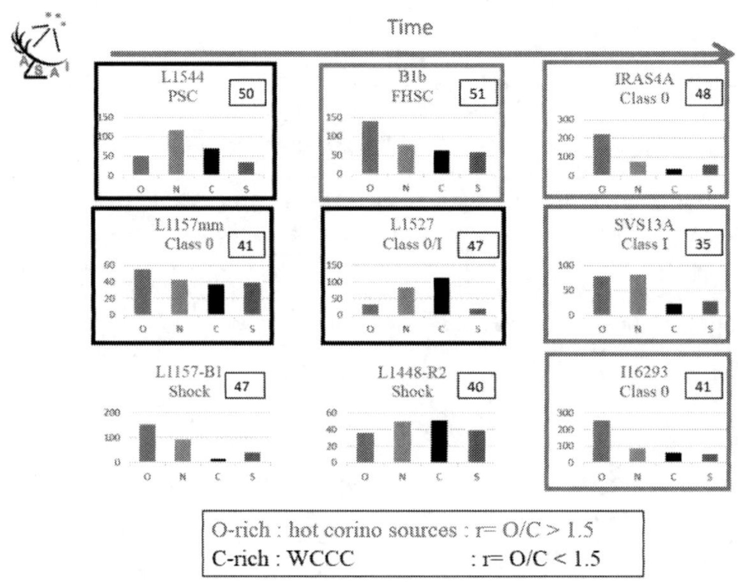


**Figure 1.** Distribution of the composition in chemical families based on the number of molecular lines in the 3mm band: Oxygen, Nitrogen, Carbon, Sulfur. Similar results are obtained when considering the number of molecular species. The ratio r of O- to C-bearing species allows to determine if protostars are hot corino (r > 1.5) or WCCC sources (r < 1.5). The evolutionary status and the number of species (main isotopologues) are given for each source. Adapted from [15].

Very significant differences are observed between the spectra of the sample sources, reflecting the evolutionary stage and the physical conditions This chemical differentiation affects not only deuteration or S-bearing species but also COMs. To determine how much the molecular content differs between sources, an inventory of the molecular species by "elemental families" was obtained: C, O, N, S, Si, P It allowed to classify the source sample into two main chemical families, depending on the ratio of O-bearing to C-bearing species: first, hot corinos (prototype : IRAS16293 – 2422) have a rich content in oxygen with respect to carbon; second, sources with a rich content in hydrocarbon with respect to oxygen; it is the signature of the Warm Carbon Chain Chemistry (WCCC) sources (prototype : L1527). The ratio r of O-bearing to hydrocarbon species allows to discriminate between hot corinos (r= 2-4) and WCCC sources (r= 1-1.4) in the sample. The question remains whether another chemical class (N- or /S-) could arise depending on the environmental/initial conditions of star formation. A study similar to ASAI on a broader sample of sources and SFRs is needed to address this question.

## 3 Cold but not poor in organics

L1544 is one of the densest and most dynamically evolved starless cores known. Its general physical-chemical structure is shown in the scheme of Fig. 2. On this Figure, the two vertical black lines show the radius below which CO and $NH_2D$ start to freeze-out onto the grain mantles, disappearing from the gas-phase.

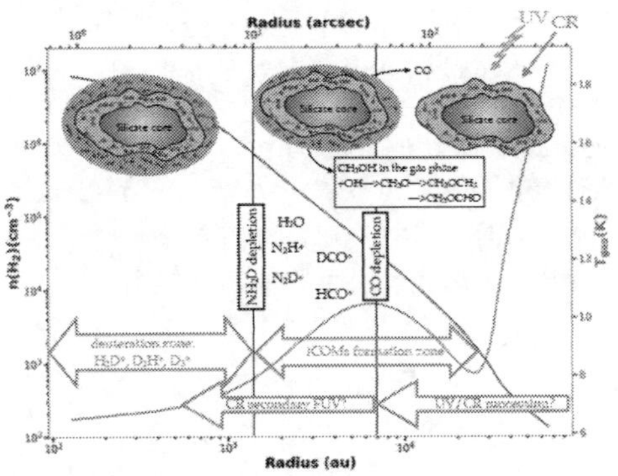


**Figure 2.** Sketch of the physical-chemical structure of the prototypical prestellar core L1544 (from the review by [6] with the permission of the authors). The dust grains are illustrated, as silicate cores, covered with the grain surface chemical composition (yellow) coated with the icy grain mantle (in blue; thicker in the cold dense interiors of the core). The temperature bump at ~ 7000 au corresponds to the so-called photo-desorption layer, a chemical zone between the FUV starlight pervading interstellar space and the FUV produced by cosmic-rays.

The L1544 survey ranges from 70 to 110 GHz and led to the detection of a very rich chemistry. For instance, more than twenty sulphur bearing species have been detected with more than 50 transitions [7, 27] such as CS, CCS, $C_3S$, SO, $SO_2$, $H_2CS$, OCS, HSCN, NS, $HCS^+$, $NS^+$, $H_2S$ and $CH_3SH$. From the chemical modelling they conclude that these species present a strong radial variation for the abundances, which mostly are emitted in the external layer where non thermal desorption of other species has previously been observed. They concluded that a strong sulphur depletion is necessary to fully reproduce the observations. Within ASAI many nitrogen bearing species have also been detected such as the hyperfine structure of the cyanomethyl radical $CH_2CN$ [25], $HNC_3$, HCCNC and $HC_3N$ [12, 28], $HC_3NH^+$ and $HCNH^+$ [20], CNCN, $NCCNH^+$, $C_3N$, $CH_3CN$, $C_2H_3CN$, and $H_2CN$ [29]. Molecular ions have also been studied in ASAI such as $HOCO^+$ [26], leading to an indirect estimate of $CO_2$ in the gas phase in the $[2\times10^{-7}-2\times10^{-6}]$ range. And COMs have also been detected [13, 24] such as methanol ($CH_3OH$), acetaldehyde ($CH_3CHO$) and dimethyl ether ($CH_3OCH_3$) with integrated intensities of just a few mK km $s^{-1}$. From these ASAI detections, the above authors built a detailed chemical network to reproduce the molecular abundances and found that most emission occur in the external envelope of the prestellar core [24, 30], while the simple deuterated species trace the internal structure where CO is depleted [e.g. 22]. The non-LTE analysis of the methanol lines [24] concluded that the bulk of the emission originates in the external (~ 8000 au from the center), relatively dense (~ $2 \times 10^4$ $cm^{-3}$) and cold (~ 10 K) layers of L1544, what is defined as molecular and COMs zone in Fig. 2, through non-thermal processes. Over the years, a total of about 140 molecules have been detected towards L1544, one fourth being COMs.

The molecular complexity found in L1544 was surprising at first, since the low temperatures and high densities did not favour neither an active surface chemistry nor the presence of COMs in the gas phase, usually highly bounded to the dust grain surfaces. COMs were indeed mainly found in warm environments, such as hot cores and hot corinos (see Sect. 4). Many detailed astrochemical models have been developed over the past few years to predict the COMs abundances in a layer at about 8000 au found in our ASAI observations. We refer to the Protostars and Planets VII chapter from [6] for a review on the models that aim to reproduce the molecular complexity in the cold regions of the ISM.

## 4 Solar-type protostars

**A Class I hot corino:** The ASAI survey revealed a rich hot corino chemistry in the Class I protostar SVS13A, thanks to the detection of more than 100 emission lines from interstellar COMs such as acetaldehyde ($CH_3CHO$), methyl formate ($HCOOCH_3$), dimethyl ether ($CH_3OCH_3$), ethanol ($C_2H_5OH$) and, formamide ($NH_2CHO$), in addition to other less complex organics like methanol ($CH_3OH$), methyl cyanide ($CH_3CN$), cyanoacethylene ($HC_3N$), formaldehyde ($H_2CO$) and, thioformaldehyde ($H_2CS$) [2–5]. The COMs abundance ratios are in good agreement with those measured in younger Class 0 sources such as IRAS 4A, suggesting that Class I hot corinos are as chemically rich as younger Class 0 ones [3]. These results provided a first evidence of chemical inheritance between the Class 0 and the Class I phase. Moreover, the comparison with the measurements in the 67P comet, obtained thanks to the ROSETTA mission, showed that some COMs (ethanol, methyl formate and acetaldehyde) are in agreement within a factor of 10 (see Fig. 3). This work suggested that these COMs in comets could be at least partially inherited from the protostellar stage [3].

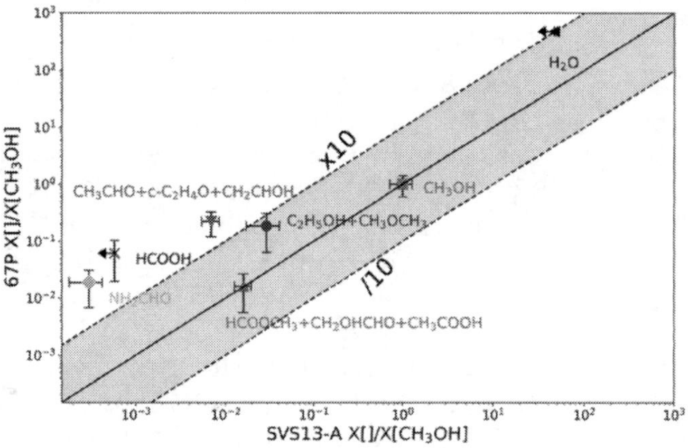


**Figure 3.** Abundance ratios, normalized to methanol, of COMs detected in SVS13A and in the 67P comet [adapted from 3]. Abundances of isomers are not distinguished since in the 67P the abundances are measured using a mass spectrometer.

**Deuterium fractionation:** We have measured the deuterium fractionation of several molecular species in SVS13A and compared it to those obtained in prestellar cores, Class 0 protostars, protoplanetary disks and comets (see Fig. 4). The comparison suggested again inheritance between the evolutionary stages, with the exception of methanol deuteration which seems to show a significant decrease in the Class I phase. Moreover, the different degree of deuteration measured between the different species gave us information about their different formation pathways. For example, we could speculate that $CH_3CN$ formation takes place probably during the prestellar stage and in gas-phase [5]. The main result of this work was to complement the deuteration measurement in the Class I phase building a bridge between the observation of younger prestellar cores and Class 0 protostars and those in Solar System objects.

**Formamide across the evolutionary stages:** A relevant transversal project was focused on formamide ($NH_2CHO$), one key prebiotic molecular species. In particular, formamide was detected in four out of the 10 ASAI sources (IRAS 4A, SVS13A, Cep E, and OMC-2), and in the Class 0 IRAS 16293–2422. This systematic study increased the number of known low and intermediate-mass formamide-emitting protostars to five and it uncovered an almost linear correlation between HNCO and $NH_2CHO$ abundances, holding for several orders of magnitude. These results contributed to shed light on the formamide formation pathways. Indeed, while it was first suggested that HNCO and $NH_2CHO$ were chemically linked, further

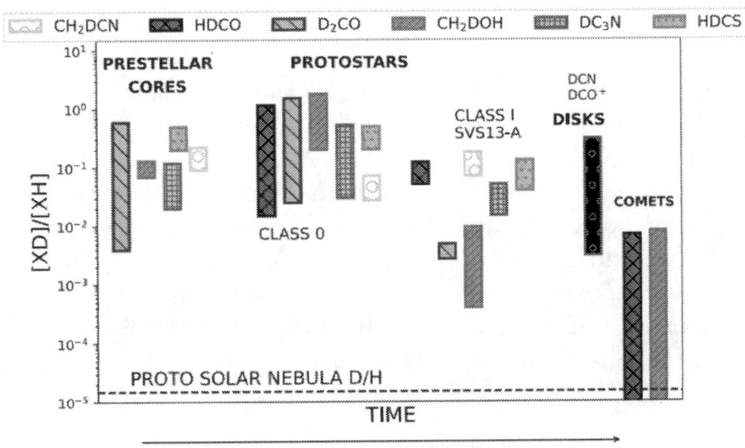


**Figure 4.** [XD]/[XH] ratio measured in organic matter in different astronomical sources: prestellar cores, Class 0 protostars, the Class I SVS13A, and protoplanetary disks [adapted from 4]. The measurements towards SVS13A fill the gap between Class 0 protostars and protoplanetary disks.

experimental and theoretical studies have shown that the correlation is probably physical rather than chemical because the two species are formed through similar physical processes across different environments [17, and references therein.].

## 5 Shocks

The prototypical outflow shock region L1157-B1 provides a remarkable example of the chemical richness and differentiation induced by shocks. The excellent sensitivity of the survey led to identify a large number (10–100) of lines from many COMs, some detected for the first time in a shock [14]) : $CH_3OH$, $H_2CCO$, $C_2H_5OH$, **$CH_3CHO$**, $HCOOCH_3$, **$HCOCH_2OH$**, **$CH_3OCH_3$**, t-HCOOH, $CH_3CN$, **$C_2H_3CN$**, **$HC_5N$**, **$NH_2CHO$**, **$CH_3SH$** (species in boldface are first detections in a shock). The chemical richness discovered in L1157-B1 compares very well to that of protostellar envelopes at the scale of a few 1000 au probed by the beam of the IRAM 30m. COM abundances measured in L1157-B1 are usually a factor 2-10 higher than those measured in hot corinos. Abundances of O-bearing COMs are typically $\sim 10^{-8}$, i.e. 2%-5% of $CH_3OH$.

As mentioned in Sect. 4, linear correlations are sometimes observed between molecular species abundances, which has been proposed to trace a common chemical origin between pair members. The study of L1157-B1 has revealed such correlations for various pairs of COMs and organic species [14, 19]. The case of glycolaldehyde $HCOCH_2OH$ and ethanol $C_2H_5OH$, measured in low-mass protostars and shocks, led to explore a new formation pathway of the former species, poorly constrained until then. Based on the ASAI results, [23] proposed a scheme in which glycolaldehyde forms in the gas phase from ethanol, once the latter is released in the gas phase either through sputtering (shock) or sublimation (hot core) of grain mantles. Interestingly, this new scheme successfully accounts for the observed correlation $HCOCH_2OH/C_2H_5OH$ towards hot corinos and L1157-B1 [14].

Molecular species with a peptide link -NH-C(=O)- $NH_2CHO$ and HNCO were detected in L1157-B1 with abundances as high as those measured in the most abundant Galactic sources [19]. The importance of these species in the synthesis of amino acids has stimulated a lot of observational and theoretical work on their formation route. The definite answer however came from NOEMA observations which unambiguously showed that the $NH_2CHO$ emission originates from the post-shock gas [9] and not from the sputtering region in the shock where the grain mantle material is released into the gas phase.

## 6 Conclusion

ASAI was a very successful project which involved about 30 scientists from 7 laboratories in France, Spain, Italy, UK and Japan. It is worth noticing that it also stimulated and contributed to theoretical and experimental works in Astrochemistry. The ASAI Legacy has provided a vast observational data set which was the basis for the successful Large Program SOLIS at the IRAM-NOEMA interferometer. Some of the results presented here show that the synergy between the two Large Programs is extremely powerful for the characterization of Solar-type protostars as it allows at the same time to resolve the molecular emission, thanks to the high-angular resolution of SOLIS and to accurately derive the physical gas parameters thanks to the detection of a large number of molecular transitions provided by ASAI.

## References

[1] Bachiller, R., et al., A&A, **372**, 899-912 (2001)
[2] Bianchi, E., et al. MNRAS, **467**, 3011-3023 (2017)
[3] Bianchi, E., et al., MNRAS, **483**, 1850-1861 (2019)
[4] Bianchi, E., et al. ACS Earth and Space Chem., **3**, 2659-2674 (2019)
[5] Bianchi, E., et al. A&A, arXiv:2202.09285 (2022)
[6] Ceccarelli, C., et al. PPVII, Sp. Sc. Series of Univ. of Arizona Press (2022)
[7] Cernicharo, J., et al. ApJ, **853**, L22-L26 (2018)
[8] Codella, C., et al. MNRAS, **462**, L75-L79 (2016)
[9] Codella, C., et al., A&A, **605**, L3-L9 (2017)
[10] Garrod, R.T., et al. ApJ, **682**, 283-302, (2008)
[11] Herbst, E., van Dishoeck, E.F., 2009, ARA&A, 47, 247
[12] Hily-Blant, P. et al., MNRAS, 480, 1174-1186 (2018)
[13] Jiménez-Serra, I. et al., ApJL 830, 6 (2016)
[14] Lefloch, B., et al., MNRAS, **469**, L73-L77 (2017)
[15] Lefloch, B., et al. MNRAS, **477**, 4792-4809 (2018)
[16] López-Sepulcre, A., et al., MNRAS, **449**, 2438–2458 (2015)
[17] López-Sepulcre, A., et al., ACS Earth and Space Chem., **3**, 2122-2137 (2019)
[18] Herbst, E., van Dishoeck, E., ARA&A, **47**, 427-480 (2009)
[19] Mendoza, E., et al., MNRAS, **445**, 151-161 (2014)
[20] Quénard, D., et al. MNRAS 470, 3194-3205 (2017)
[21] Roberts, H., et al., ApJL 591, 41-44 (2003)
[22] Sipilä, O. & Caselli, P., A&A 615, 15 (2018)
[23] Skouteris, D., et al. ApJ, **854**, 135-144 (2018)
[24] Vastel, C., et al., ApJL 795, 2 (2014)
[25] Vastel, C., et al., A&A 582, 3 (2015)
[26] Vastel, C., et al., A&A 591, 2 (2016)
[27] Vastel, C., et al., MNRAS 478, 5514-5532 (2018)
[28] Vastel, C., et al., MNRAS 474, 76-80 (2018)
[29] Vastel, C., et al., A&A 625, 91 (2019)
[30] Vasyunin, A. I. et al., ApJ 842, 33 (2017)

# Chemical Richness of protoplanetary Disks and related physical Properties

*Anne* Dutrey[1,*]

[1]Laboratoire d'Astrophysique de Bordeaux, Allée Geoffroy Saint Hilaire, 33600 Pessac, France

**Abstract.** There are now several observational proofs that protoplanetary disks orbiting around TTauri stars are planet forming sites. Studying planet formations in disks requests both high sensitivity and high angular resolution (at Taurus distance, 0.1" means 15 au or 3 times the distance of Jupiter to the Sun). Moreover, $H_2$, the main gas component remains difficult to observe, its mid-IR transitions only trace warm gas near the disk surface. Our knowledge on gas disk relies on trace molecules (CO, CN, CS, HCN, $HCO^+$...) observed by powerful large interferometers such as NOEMA and ALMA. I present here some recent results from ALMA and NOEMA showing that we start to quantitatively unveil the physical and chemical properties of planet forming disks.

## 1 Introduction

In the early 2000's, the input of many ground-based and spatial facilities provided the first constraints on radial and vertical structures of gas and dust disks orbiting low mass TTauri stars. It is now established that dust disks contain grains which have already grown up to sizes of the order of at least 1 mm, these large grains being settled onto the mid-plane.The upper layer of the disk still harbors small grains, dynamically coupled to the gas, which absorb the stellar and interstellar radiations. They create a warm layer at intermediate altitude, while the interior of the shielded disk is colder with temperatures as low as 10 K in the dense mid-plane at radii > 50-100 au. With the exception of the upper disk surface, the gas temperature is essentially controlled by the dust temperature. At the disk upper surface, the incident UV flux from the central object photo-dissociates molecules making a (dense) PDR. On the contrary, the outer mid-plane corresponds to a region where most molecules stick onto dust grains and are severely depleted in the gas phase by combined effects of low temperatures, high densities and high extinction limiting photo-desorption. Typically, the CO snowline (~17-25 K, e.g. [1]) in a TTauri disk is expected at a radius of ~ 20-40 au. Closer to the star, the mid-plane is warm enough, molecules are in the gas phase as above the cold mid-plane where the temperature remains high enough (T > 17-25 K) and thermal (and photo) desorption creates a molecular rich layer from which most mm/submm molecular emission originates. Many recent observations support such a layered molecular disk structure. One of the most famous example being given by the ALMA observations of CO in HD163296 by [2]. They directly image the CO layers above and below the mid-plane, confirming at the same time, the existence of the molecular layer and the vertical temperature gradient. Such a

*e-mail: anne.dutrey@u-bordeaux.fr

peculiar structure makes the retrieval of disk properties from observations not straighforward, often requiring the support of a detailed modeling of dust and gas, including chemical models and all their complexity. Moreover, observations need to have sufficient spatial (better than 40-50 au) and spectral (0.15-0.05 km/s) resolutions to properly constrain the altitude and the kinematics of the observed molecular layer.

## 2 Molecular Complexity in Disks

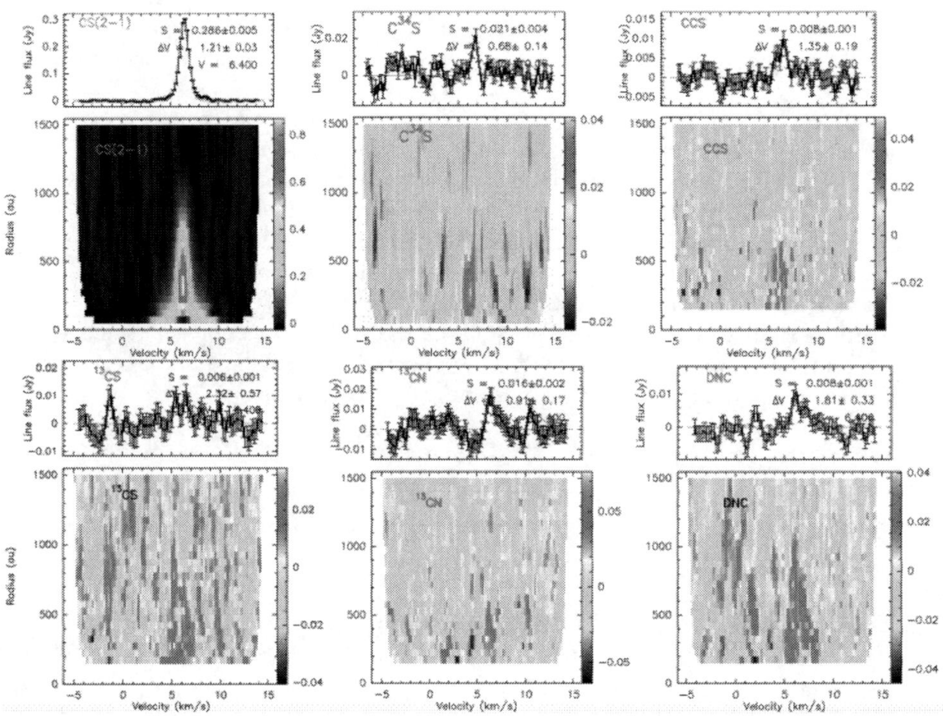


**Figure 1.** Adapted from Phuong et al., 2021 [10]. Velocity-Radius plots of several molecule detected at 3mm with NOEMA and PolyFix in the GG Tau A disk (first detection of CCS in a disk). The data have been locally corrected from the value of the Keplerian speed (the mass of the central stars being known, the Keplerian speed can be properly calculated everywhere in the disk), de-projected from the disk inclination and each radius has been then azimuthally averaged.

A compilation of molecular detection in disks by [9] reveal that there are about 45 molecules (including isotopologues) detected in low-mass (TTauri) and intermediate mass (HAe) disks. Most detections arise from the warm molecular layer. A few species, such as $H_2O$, are detected in warm inner disks. Only two species with 6 atoms ($CH_3OH$ and $CH_3CN$) have been detected, most of the species have two or three atoms.

ALMA images at moderate angular resolution (typically 0.3") of abundant molecules such as CS, $C_2H$, HCN or CN reveal the presence of rings (e.g. [14]). These rings likely originate from a mix of local variations in density, column density and temperature together with changes in the UV penetration and grains properties. Detailed chemical analyses of the observations also tend to favor in TTauri disks a C/O ratio around $\sim$ 1, larger than in the interstellar medium, at least in a significant area in the disks in which it has been measured ([14, 15]).

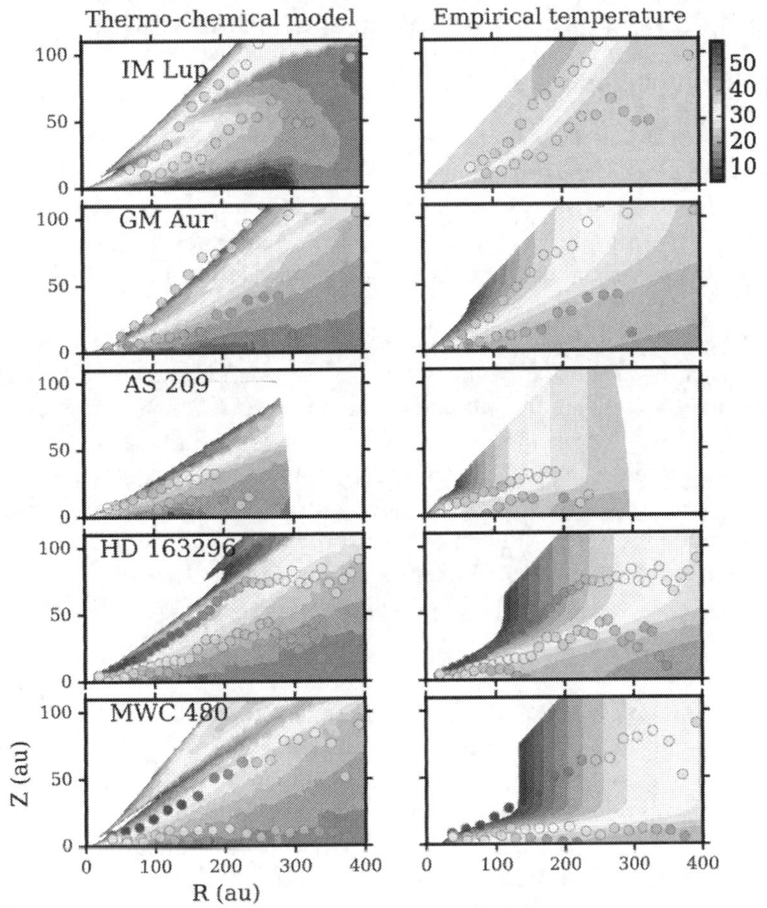


**Figure 2.** Adapted from Zhang et al., 2021 [6]: comparison between the ALMA (MAPS) observations of CO lines and a thermo-chemical modeling (left) and a simple parametric model for the disk structure (right). Surface brightness temperature measurements for CO, $^{13}CO$, and $C^{18}O$ 2-1 lines are given by circles (highest circles correspond to CO 2-1).

## 2.1 An example of a molecular Survey: the GG Tau dense Ring

Using NOEMA and thanks to the flexibility of PolyFix, Phuong et al. 2021, [10] have recently reported the first deep survey done in a proto-planetary disk in the 3mm range (from 2.6 to 4.2mm) that contains many fundamental transitions of simple molecules. They observed the Keplerian outer disk around the triple star GG Tau A located in Taurus ([3]). Thanks to its high mass (0.15 $M_{\odot}$) and large size (outer radius 800 au), it is a good laboratory to study the "cold chemistry" occuring in TTauri disks.

Among the thirty-eight molecules which were observed, only seventeen were detected, including CCS seen for the first time in a TTauri disk (there is also a marginal detection of OCS). Fig.1 presents Sulfur and Nitrogen bearing species (DNC, $^{13}CN$, CS, $C^{34}N$, $^{13}CS$ and CCS). With the exception of CS, most them where detected at low SNR (of the order of 6-8). Very interestingly and despite a low angular resolution, most sulfur-bearing species seem to originate from the dense ring (180-260 au from the stars) while deuterated species are observed in the outer, colder, less dense disk (radius beyond 260 au and up to 800 au) as expected from chemical models.

## 3 High angular and spectral Resolution Imaging of bright Lines

### 3.1 Gas structure from resolved CO maps

Mapping several lines from CO isotopologues appears as a powerful method to the retrieve the gas disk structure, in particular its temperature. In TTauri disks, the first rotational levels of CO are thermalized and optically thick, typically sampling surfaces at 3 scale-heights. Optically thinner lines from $^{13}CO$ and $C^{18}O$ 2-1 reach an opacity around 1 closer to the disk plane. This opacity difference was used for the first derivation of the vertical kinetic temperature profile ([4] and [5]).

Thanks to the ALMA resolving power and sensitivity, imaging CO lines in a few TTauri and Herbig Ae disks in the context of the MAPS large program, Law et al., 2021 [7] were recently able to directly compare the altitude of the observed CO layers with a simple parametric model for the gas temperature ([4]), as it is shown right side of Fig.2. Zhang et al., 2021 [6] used a thermo-chemical model to derive the whole disk structures by comparing with these observations (Fig.2). They conclude that the observational method and the thermo-chemical modeling agree within a temperature range of 15-40 K that corresponds to the typical temperature of the molecular layer of TTauri disks.

### 3.2 Gas Mass

Going one step further, Schwarz et al., 2021 [8] studied the disk orbiting the TTauri GM Aur by using all the CO lines observed by ALMA on this object (1-0, 2-1, 3-2 and 6-5 lines of several CO isotopologues) and the HD 1-0 line observed by the Herschel Satellite. They found that the disk is more massive than originally thought with a gas mass of about 0.2 $M_\odot$. In this specific case, the mass derivation was obtained using a thermo-chemical code to model the HD and CO lines. This impressive study, using in total eleven CO lines, shows that deriving accurate disk masses remains a complex task and suggests that at least a fraction of disks might be massive enough to be gravitationally unstable.

### 3.3 Deriving the Turbulence

Dartois et al., 2003 and Pietu et al., 2007 [4, 5] have shown that the local turbulence $\delta v_{turb}$ in a disk can be derived from the local line-width provided the thermal component is accurately known. Using the PdBI, Guilloteau et al., 2011 [16] used the "heavy" molecule CS to minimize the impact of the thermal component and found subsonic turbulence in the DM Tau disk. With the advent of ALMA, several studies show that disks have a low turbulence. Teague et al., 2016 [17] demonstrated that absolute calibration errors limit the method accuracy to Mach numbers above about 0.1. Using several molecular tracers such as CO, $DCO^+$ and CS, Flaherty et al., 2017, 2020 [18, 19] found low turbulence in the disks surrounding HD163296, MWC480, and V4046 Sgr, DM Tau remaining the only firm detection (Mach ~ 0.3).

## 4 Edge-on Disks

Contrary to an inclined disk, an edge-on disk directly probes the vertical structure. In the Position-Velocity (PV) diagram of a Keplerian disk, there is a direct link between position and velocity. All radial lines correspond to constant disk radii and averaging along them directly gives the average molecular brightness at each specific radius. Cuts at various altitudes provides a direct, model-independent, tomography of the 2-D gas structure (Fig.3).

### 4.1 The Flying Saucer

Fig.3 shows the reconstruction of the CO and CS brightness distributions $T_B(r,z)$ for the Flying Saucer (a TTauri disk located in $\rho$ Oph and inclined by $87^o$ along the line of sight) obtained by Dutrey et al., 2017 [11] using ALMA data at 0.5″ (70 au at the source distance). Thermalized, optically thick lines as CO J=2-1 give directly access to the kinetic temperature of the gas, provided the angular resolution is high enough to reduce the beam dilution effect. The observed kinetic temperature from CO 2-1 by [11] is in good agreement with that reported by [6, 7] in similar TTauri disks. In the Flying Saucer [11] also observed a clear depression around the mid-plane, where CO is expected to freeze onto grains. However, the limited angular resolution precludes any accurate determination of the temperature in this area.

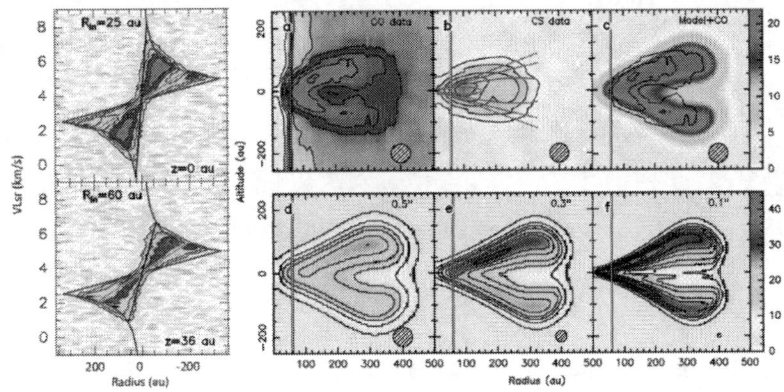

**Figure 3.** Adapted from Dutrey et al., 2021 [11], ALMA observations (CO 2-1 and CS 5-4) of the Flying Saucer edge-on disk. Left: two PV diagrams for CO at altitude z=0 and 36 au. Right - Top panel CO, CS observations and the best model for CO with observations in black contours - Bottom panel: impact of the angular resolution, the north-south dissymetry is due to the $3^o$ departure from edge-on.

### 4.2 Chemical Stratification and Models

The molecular stratification is directly visible in an edge-on disk. Fig.3 shows that the density tracer CS arises from a lower layer (about one scale-height), while CO, which characterizes the overall extent of the molecular gas, extends up to four scale heights. Such distributions can be unambiguously compared to model predictions. The models in Fig.4 are a first attempt of this, using two different chemical approaches. On the left, Gavino et al., 2021 [12] did a simple model using a single grain size (0.1 $\mu$m). On the right, the disk physical model is similar (same $H_2$ distribution and mass) but includes grain growth, dust settling and the proper dependence of dust temperature with grain sizes. By comparing with Fig.3, the differences on the CS and CO distributions are obvious. The simple model does not reproduce the observations, contrary to the second one. Edge-on disks can also be useful tools to directly calibrate thermo-chemical models.

The above examples illustrate the increasing accuracy of mm/submm studies of disks. These studies will allow in the forthcoming years a real understanding of planet forming disks, provided the integration time of the observations will be high enough to sample the molecular complexity and/or produce images of high quality.

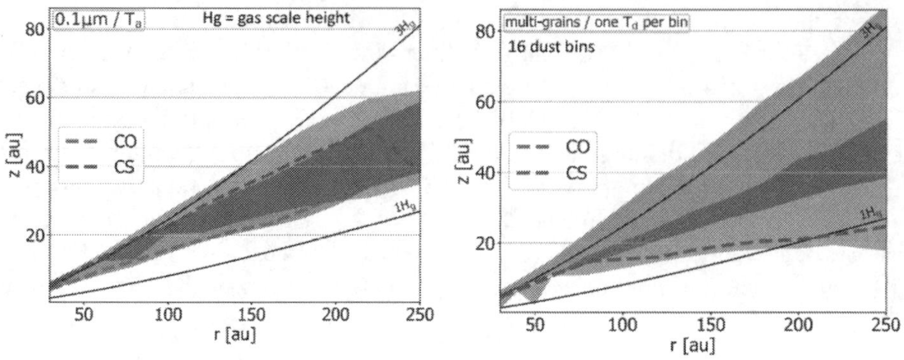


**Figure 4.** Derived from Gavino et al., 2021 [12]: two chemical models for the edge-on disk of the Flying Saucer. The CO (red) and CS (green) distributions are shown. Dust temperatures are derived from a Monte-Carlo radiative transfer code. Left: a chemical model with a single grain size (0.1 $\mu$m). Right: model for the same disk ($H_2$ distribution and mass) with dust growth, settling and a grain temperature depending on grain sizes.

**Acknowledgments:** Anne Dutrey thanks Ke Zhang, Thi Phuong Nguyen and Sacha Gavino for providing figures for this article.

## References

[1] Qi, Ch. et al., Science **341**, 630-632 (2013)
[2] de Gregorio-Monsalvo, I., A&A **557**, 133-140 (2013)
[3] Dutrey, A., et al., A&A **286**, 149-159 (1994)
[4] Dartois, E., A&A **399**, 773-787 (2003)
[5] Pietu, V., A&A **467**, 163-178 (2007)
[6] Zhang, K., et al., ApJS, **257**, 1-29 (2021)
[7] Law, C., et al., ApJS, **257**, 1-24 (2021)
[8] Schwarz, K., et al 2021, ApJS, **257**, 1-24 (2021)
[9] McGuire, B., ApJS, **259**, 30-81 (2022)
[10] Phuong,N.T., et al., A&A **653**, L5-10 (2021)
[11] Dutrey, A., et al., A&A **607**, L130-135 (2017)
[12] Gavino, S., et al., A&A **654**, 65-102 (2021)
[13] Ruaud, M., et al., MNRAS **459**, 3756-3767 (2016)
[14] Bergin, E. et al., ApJ **831**, 101-119 (2016)
[15] Semenov, D., et al., A&A **617**, 28-43 (2018)
[16] Guilloteau, S., et al., A&A **548**, 70-79 (2012)
[17] Teague, R.,et al., A&A **592**, 49-63 (2016)
[18] Flaherty, K., et al., A&A **843**, 150-170 (2017)
[19] Flaherty, K., et al., A&A **895**, 109-127 (2020)

EPJ Web of Conferences **265**, 00040 (2022) https://doi.org/10.1051/epjconf/202226500040
*Multi-line Diagnostics of the Interstellar Medium*

# AB Aur: A Rosetta stone for planet formation theories

*Pablo* Rivière-Marichalar[1],*

[1]Observatorio Astronómico Nacional (OAN,IGN), Calle Alfonso XII, 3. 28014 Madrid, Spain

**Abstract.** AB Aur is a Herbig Ae star that hosts a well-known protoplanetary disk that depicts a plethora of features that can be attributed to the presence of a forming planet, such as a large inner cavity, a dust trap, and spiral arms. We present NOEMA observations of molecular species towards this transition disk around AB Aur, including CO, $^{13}CO$, $C^{18}O$, HCN, $HCO^+$, SO, $H_2CO$, and $H_2S$. These observations reveal complex chemistry with strong chemical differentiation and the presence of a cavity-crossing filament.

## 1 Introduction

Planet formation takes place in circumstellar disks made of gas and dust surrounding young stellar objects. Therefore, characterizing the physical conditions and the chemical composition of protoplanetary disks is key for understanding planet formation. AB Aur is a widely studied Herbig A0 star that hosts a transitional disk. At a distance of ~160 pc [1], the system is well suited to study the distribution of gas and dust in circumstellar environments. The system depicts a plethora of features that can be attributed to planet formation, such as a large inner cavity [2–5], spiral arms [4, 6–8], and a dust trap [2, 5]. Low-resolution spectral observations of AB Aur are available from ~1 $\mu$m up to ~ 400 $\mu$m. The detection of several O I, C II, C, OH, and rovibrational lines of CO and $H_2$ demonstrates the existence of warm gas. The origin of these emission lines is, however, uncertain due to the moderate angular resolution of the observations. Our team is performing a large observational effort to study the dust and gas content in this interesting object [5, 8, 10].

## 2 Accretion within the dust cavity?

We observed AB Aur with NOEMA in 2017 [10] to map the emission from $HCO^+$ 3-2 and HCN 3-2. Velocity integrated intensity maps are shown in Fig. 1. There are prominent differences between the maps of the two species. While HCN is only detected in an annular ring spatially coincident with the dust ring, $HCO^+$ is detected in the outer ring and also features intense emission toward the central position. Furthermore, a bright $HCO^+$ emission bridge connects the outer disk with the inner structure.

We show in the top-central panel of Fig. 1 the first-moment map of $HCO^+$ J=3-2. The outer parts of the disk are close to Keplerian rotation, but twisted isophotes are observed towards the central regions (r<0.6"). It has been proposed that twisted isophotes can be reproduced using either two misaligned Keplerian disks [7] or an outflow [12]. A simple

*e-mail: p.riviere@oan.es

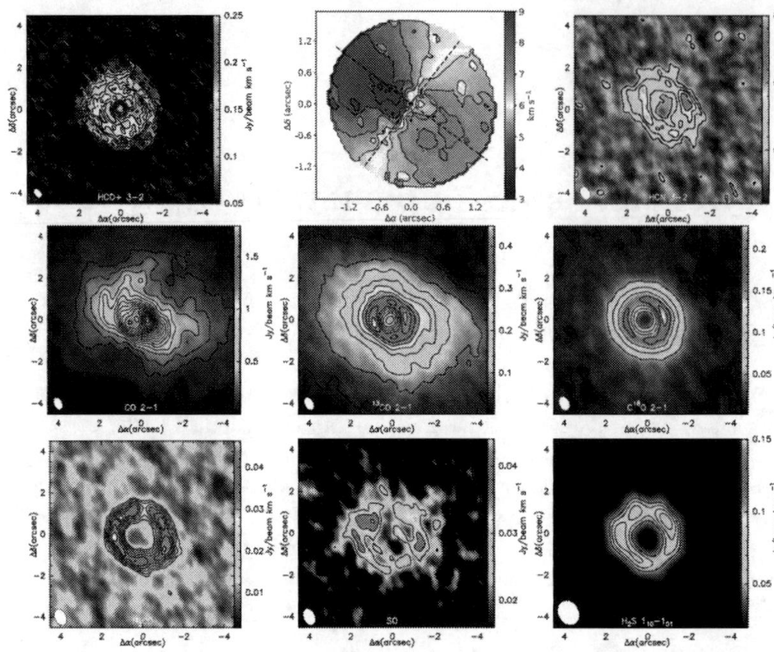


**Figure 1.** Top left: $HCO^+$ integrated intensity map. Top center: $HCO^+$ first moment map. The dashed lines indicate the direction of the major and minor axes. The blue circle contains a region with large anomalies with respect to a Keplerian rotation velocity field. Top right: HCN integrated intensity map. Middle and bottom panels: from left to right and top to bottom the panels show the velocity integrated intensity maps of $^{12}CO$ 2-1, $^{13}CO$ 2-1, $C^{18}O$ 2-1, $H_2CO$ $3_{03}$-$2_{02}$, SO $5_6$-$4_5$, and $H_2S$ $1_{10}$-$1_{01}$. Th synthetized beam is shown in the bottom left corner of each map.

model consisting of two misaligned disks provides a good fit across the disk except in the position of the bridge region, where prominent residuals are observed (see the blue circle in Fig. 1, top central panel). Visual inspection of spectra at this position demonstrates the presence of material moving at forbidden velocities, most likely infalling material moving at free-fall velocities. By including a free-fall velocity component in our model we get a much better fit to the first-moment map. At the position of the bridge, the free-fall model has a velocity of 5.5 km $s^{-1}$, compared to the velocity of 5.3 km $s^{-1}$ of the blue-shifted component observed towards the bridge. Assuming that this component is due to accretion, we derived a mass accretion rate $3\times10^{-8}$ to $3\times10^{-7}$ $M_\odot$ $yr^{-1}$ comparable to previous measurements [13].

## 3 Radial segregation of chemical species

In a later paper [8] we presented observations of $^{12}CO$, $^{13}CO$, $C^{18}O$, HCN, $HCO^+$, SO, and $H_2CO$ with very different radial and azimuthal distributions. The species showed strong radial segregation, with differences as large as 100 au in the positions of their peaks. The only species that peaked toward the center was $HCO^+$. All the species showed emissions that were more extended than that of the dust, and the most extended species were $^{12}CO$ and $^{13}CO$, with $5\sigma$ emission detected at distances larger than 4" due to pollution by envelope emission. It would be reasonable to postulate that this radially layered structure is related to the snow lines of the different species, which are determined by their binding energies. If thermal desorption is the main desorption mechanism in the mid-plane, the radius of the snow line would be expected to anti-correlate with the binding energy. However, this was not true for the observed species, suggesting that thermal desorption is not the main driver of the chemical evolution of the disk. Our data reveal that different species are probing different regions of the disk and the envelope. Species such as $^{12}CO$ and $^{13}CO$ trace the disk surface and the remnant envelope. Other species, such as $C^{18}O$ and HCN, are tracing the dusty disk, with a radial profile that almost overlaps with that of the continuum. $H_2CO$ and SO trace the outer regions of the disk. The differences observed in the radial distribution of the different species surveyed were also present in their azimuthal profiles. $C^{18}O$ peaks close to the position of the dust trap, while SO peaks at $\sim$180° from the position of the dust trap, and other species such as HCN, $H_2CO$, and $HCO^+$ are mostly flat along the ring. We used the $H_2CO$ lines observed to derive a mean gas temperature of 39 K, in good agreement with the temperature of 37 K that we obtained using SO lines. This mean temperature of 39 K is larger than the typical value for disks around T Tauri stars. We derived a gas-to-dust map of the disk by combining observations of $^{13}CO$, $C^{18}O$, and continuum emission at 1 mm. The mean gas-to-dust ratio of the disk was 40, well below the canonical value of 100 assumed for the ISM.

We computed a set of six 1+1D Nautilus simulations [14] to compare with the radial profiles of the observed species [8]. Nautilus is a time-dependant, three-phase astrochemical model in which gas, grain surface, and grain mantle phases are considered. We tested two values for three parameters that were relevant to the chemistry of the system: the gas-to-dust ratio (40 and 100), the C/O abundance ratio (0.7 and 1.0), and the sulfur abundance ($1.5\times10^{-5}$, which is the cosmic abundance [17], and a depleted value of $8\times10^{-8}$). The high abundance of observed $H_2CO$ favored models with a gas-to-dust ratio of 40, in agreement with the value derived from our observations. HCN is very sensitive to the C/O ratio, and its radial profile was better adjusted assuming C/O=1. The SO radial profile, however, was better reproduced by a model with strong sulfur depletion ([S/H] = $8 \times 10^{-8}$) and C/O=0.7. This model ([S/H] = $8 \times 10^{-8}$, C/O=0.7) did indeed reproduce all the observed column densities exception made of HCN.

## 4 Resolved observations of $H_2S$

We observed the $H_2S$ $1_{10}$-$1_{01}$ transition with NOEMA (see the velocity integrated intensity map in Fig. 1 bottom right panel) with an angular resolution of 1.19" × 0.97" [15]. This is the second time that resolved observations of $H_2S$ towards a protoplanetary disk were performed after the observations toward GG Tau [16]. We detected emission in a ring extending from 0.67" ($\sim$109 au) to 1.69" ($\sim$ 275 au) with strong azimuthal asymmetries. The position of the peak in azimuth was coincident with that of the continuum at 1 mm and $C^{18}O$. Both the radial and azimuthal distributions of $H_2S$ are very similar to that of $H_2CO$. Our recent study of $H_2S$ emission toward young stars in Taurus identified a tentative correlation between $H_2S$ $1_{10}$-$1_{01}$ and $H_2CO$ $2_{11}$-$1_{10}$ velocity integrated line intensities. Our maps point to very similar spatial origins for both emission lines, providing further observational support for such correlation. The two species, $H_2S$ and $H_2CO$, have formation routes on the surface of grains, a fact that could explain the correlation. Assuming LTE we derived a mean $H_2S$ abundance with respect to H nuclei of $(1.9\pm0.4)\times10^{13}$ cm$^{-2}$

To study the sulfur budget in the disk we produced a Nautilus model. Given the moderate angular resolution of our data, we focused on the vertical profile at a radius r=200 au, representative of the $H_2S$ emission ring. The model is based on the one that we used previously (see Sect. 3) to study the chemical segregation of species in the disk [8]. We assumed a gas-to-dust ratio of 40, no sulfur depletion ([S/H] = $1.5 \times 10^{-5}$), and C/O=0.7. The details of the model can be found in the original papers [8, 15]. Our model results in an $H_2S$ column density of $3.5\times10^{13}$ cm$^{-2}$, in good agreement, given the uncertainties, with the value derived assuming LTE. Furthermore, assuming LTE we derived an SO mean abundance of $3.2\times10^{-9}$ $cm^{-2}$, again in good agreement with an abundance of $9.8\times10^{-9}$ predicted by our Nautilus model. Our model showed that 99.99% of $H_2S$ is locked in the surface of dust grains. Furthermore, 90.5% of the total sulfur content is locked in the surface of grains, and 98% of the sulfur content is contained in only six species: $H_2S$, $SO_2$, $H_2S_3$, and $CS_2$ in the surface of grains and gaseous S and $S^+$. Gaseous species (S and $S^+$) become relevant only at around 15 au over the mid-plane, where the visual extinction is only 1.5 mag. Regarding the overall composition of the grain surfaces, $H_2S$ is the third most abundant species in the model after $H_2O$ and CO, which highlights the importance of these species for the chemistry of protoplanetary disks.

## 5 Summary

The AB Aur protoplanetary disk presents a series of features that makes it a good candidate to look for young and embedded forming planets, including a cavity in the inner disk, a dust trap, and spiral arms. We are driving a large observational effort using the NOEMA observatory to characterize the gaseous content of this system. Observations of $HCO^+$ revealed the presence of a filamentary structure that is crossing the cavity, connecting the inner and outer disk. The material in this bridge appears to be moving at free-fall velocities and it is likely tracing material being accreted onto the central source. Our NOEMA observations point to strong radial segregation of chemical species in the disk. The positions of the peaks in the radial profiles of the surveyed species suggest that thermal desorption is not driving the chemical evolution of the disk. We derived a mean gas-to-dust ratio of 40, significantly smaller than the canonical value of 100 assumed for the ISM, in agreement with recent studies of the gas content of protoplanetary disks. We further used observations of $H_2CO$ and SO to estimate a mean temperature of 39 K for the gas in the disk. Finally, we used observations of SO and $H_2S$ to study the sulfur budget in the disk. According to our NAUTILUS model, $H_2S$ is the main sulfur carrier on the surface of grains. Future observations with NOEMA will help us to grow our database to study the protoplanetary disk around AB Aur.

## References

[1] Gaia collaboration, Vizier Online Data Catalog, I345 (2018)
[2] Piétu, V., Guilloteau, S., Durtey, A., A&A **443**, 945-954 (2005)
[3] Tang, Y. -W., Guilloteau, S., Piétu, V., et al., A&A **547**, A84 (2012)
[4] Tang, Y. -W., Guilloteau, S., Durtey, A., et al., ApJ **84**, 32 (2017)
[5] Fuente, A., Barutearu, C., Neri, R., et al., A&A **846**, L3 (2017)
[6] Fukagawa, M., Hayashi, M., Tamura, M., et al., ApJL **605**, L53-L56 (2004)
[7] Hashimoto, J., Tamura, M., Muto, T., et al., ApJ **729**, L17 (2011)
[8] Rivière-Marichalar, P., Fuente, A., Le Gal, R., et al., A&A **642**, A32 (2020)
[9] Fuente, A., Cernicharo, J., Agúndez, M., et al., A&A **524**, A19 (2010)
[10] Rivière-Marichalar, P., Fuente., A., Baruteau, C., et al., ApJL **879**, L14 (2019)
[11] Rodríguez, L. F., Zapata, L., A., Dzib, S. A., et al., ApJ **793**, L21 2014
[12] Rosenfeld, K., A., Chiang, E., Andrew, S. M., ApJ **782**, 62 (2014)
[13] Salyk, C., Herczeg, G. J., Brown, J. M., et al., ApJ **769**, 21 (2013)
[14] Wakelam, V., Ruaud, M., Hersant, F., et al., A&A **594**, A35 (2016)
[15] Rivière-Marichalar, P., Fuente, A., Esplugues, G., A&A, submitted.
[16] Phuong, N. T., Chapillon, E., Majumdar, L., et al., A&A **616**, L5 (2018)
[17] Daflon, S., Cunha, K., de la Reza, R., et al., AJ **138**, 1577-1583 (2009)

# The QUIJOTE[1] line survey of TMC-1

*José* Cernicharo[1,*], *Marcelino* Agúndez[1], *Carlos* Cabezas[1], *Nuria* Marcelino[2,3], *Belén* Tercero[2,3], *Juan Ramón* Pardo[1], *Raúl* Fuentetaja[1], and *Pablo* de Vicente[3]

[1]Instituto de Física Fundamental, Dpt. Molecular Astrophysics, CSIC, C/Serrano 121, 28006 Madrid, Spain

[2]Observatorio Astronómico Nacional, IGN, C/Alfonso XII 3, 28014 Madrid, Spain

[3]Observatorio de Yebes, IGN, Cerro de la Palera s/n, 19141 Yebes, Guadalajara, Spain

**Abstract.** We present the recent results obtained with the QUIJOTE line survey of the cold dark core TMC-1. The observations have been carried out with the YEBES 40m radio telescope (see Figure 1) in the Q-band (31-50 GHz). A new set of receivers have been installed in the telescope within the frame of the ERC synergy Nanocosmos project that allows to cover the whole 31-50 GHz band in dual polarization [1]. The spectral resolution is 38.15 kHz. The sensitivity achieved so far varies between 0.12 and 0.25 mK, and allows to search for new molecules in a line by line (no stacking) detection procedure. These new data have permitted to detect the many relatively small protonated species of abundant molecules and eight sulfur-bearing species. The most exciting result is the detection of hydrocarbon chains and cycles with low permanent dipole moment, such as $CH_2CHCCH$, $CH_2CCHCCH$, the propargyl radical ($C_3H_3$), cyclopentadiene, indene (the first PAH detected in space), ortho-benzyne and two ethynyl derivatives of cyclopentadiene (c-$C_5H_5CCH$) ([2–7]). We have found that the gas-phase chemistry of hydrocarbons in TMC-1 has to be revisited in depth.

## 1 The QUIJOTE[1] line survey

The QUIJOTE[1] line survey of TMC-1 [3] performed with the Yebes 40m radio telescope has permitted to detect near 40 new molecular species in the last two years, most of which are hydrocarbons and cycles such as indene, benzyne, cyclopentadiene, and two isomers of ethynyl cyclopentadiene (see, e.g., [3–7] and references therein). Propargyl ($CH_2CCH$) has been found to be one of the most abundant hydrocarbon radicals in this source [2, 8]. Other hydrocarbons such as ethynylallene ($CH_2CCHCCH$), vinylacetylene ($CH_2CHCCH$), and butadiynylallene ($CH_2CCHC_4H$) have been found to have also large abundances [4, 6, 9]. Moreover, cyano derivatives of benzene and naphthalene have also been found towards TMC-1 [10, 11]. This suggests that benzene and naphthalene are also very abundant in this cold prestellar core.

Although most discoveries performed with QUIJOTE are based on data from the Yebes 40m radio telescope, several species have been identified thanks to high sensitivity data at 3mm gathered with the IRAM 30m radio telescope. Species such as $CH_3CO^+$, $HCCS^+$, $HCCCO^+$, and $HCCNCH^+$ are examples of molecules detected with both radio telescopes [12–15]

*e-mail: jose.cernicharo@csic.es

[1]**Q**-band **U**ltrasensitive **I**nspection **J**ourney to the **O**bscure **T**MC-1 **E**nvironment

**Figure 1.** The Yebes 40m radio telescope observing TMC-1 in January 2021. The Yebes observatory is placed at 950 m of altitude in the land of Castilla la Mancha (The land of Don Quijote). It is 60 km East of Madrid. The aperture efficiency of the telescope across the Q-band varies between 0.7 at 30 GHz and 0.6 at 50 GHz.

### 1.1 Observations

New receivers, built within the Nanocosmos project[2], and installed at the Yebes 40m radiotelescope, were used for the observations of TMC-1 ($\alpha_{J2000}$ = $4^h41^m41.9^s$ and $\delta_{J2000}$ = $+25°41'27.0''$). A detailed description of the system is given by [1]. Details of the QUIJOTE line survey are provided in [3, 12]. The observations were carried out during different observing runs between November 2019 and May 2022. The receiver consists of two cold high electron mobility transistor amplifiers covering the 31.0-50.3 GHz band with horizontal and vertical polarizations. Receiver temperatures in the runs achieved during 2020 vary from 22 K at 32 GHz to 42 K at 50 GHz. Some power adaptation in the down-conversion chains have reduced the receiver temperatures during 2021 to 16 K at 32 GHz and 25 K at 50 GHz. The backends are $2 \times 8 \times 2.5$ GHz fast Fourier transform spectrometers with a spectral resolution of 38.15 kHz providing the whole coverage of the Q-band in both polarisations.

The last version of the QUIJOTE data corresponds to 546 hours of observing time on the source, of which 293 and 253 hours were acquired with a frequency switching throw of 8 MHz and 10 MHz, respectively. The last results of QUIJOTE concern the detection of five isomers of cyano propene and are given by [18], and the detection of $CH_2CCHCCCH$ by [9]. The intensity scale used in this work, antenna temperature ($T_A^*$), was calibrated using two absorbers at different temperatures and the atmospheric transmission model ATM [16, 17]. The antenna temperature has an estimated uncertainty of 10 %. All data were analyzed using the GILDAS package[3].

## 2 Main results

QUIJOTE has reached now a level of sensitivity (0.12-0.25 mK per 38.15 kHz channel across the Q-band) that permits to detect new isotopologues and derivatives of abundant species. Examples of detection and complete spectral characterization of rare isotopologues through

[2]`https://nanocosmos.iff.csic.es/`
[3]`http://www.iram.fr/IRAMFR/GILDAS`

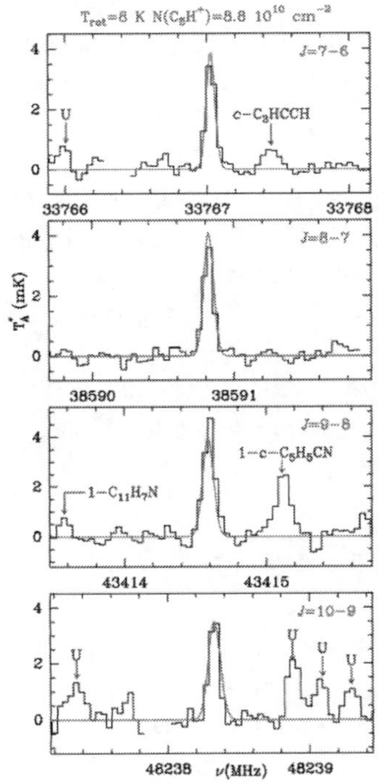


**Figure 2.** Observed lines of $C_5H^+$ with QUIJOTE towards TMC-1. The abscissa corresponds to the rest frequency assuming a local standard of rest velocity of 5.83 km $s^{-1}$. The ordinate is the antenna temperature corrected for atmospheric and telescope losses in mK. The red line shows the synthetic spectrum derived with MADEX [22] for $T_{rot}$=8 K and $N(C_5H^+)=8.8\times10^{10}$ $cm^{-2}$. Blank channels correspond to negative features produced in the folding of the frequency switching data (from [25]).

QUIJOTE data are: HDCCN [19], $CH_2DC_3N$ [20], $CH_2DC_4H$ [21]. Although QUIJOTE has not yet reached the confusion limit, special care has to be taken when assigning lines to a given molecule as blending with other features often occurs. Line identification in this work was done using the following catalogues: MADEX [22], CDMS [23], and JPL [24]. By July 2022 the MADEX catalogue has 6435 spectral entries corresponding to the ground and vibrationally excited states, together with the corresponding isotopologues, of 1736 molecules.

Contrarily to the GOTHAM survey that detects species through spectral stacking (see, e.g., [10]), all QUIJOTE detections are based on the classical and old fashion method of line by line identification of all expected transitions of a given molecule within the Q-band. With the present level of sensitivity of QUIJOTE we have detected 1591 features above the 1 mK level (5-7 $\sigma$ below 42 GHz). Of these lines 188 remain unidentified. We note, however, that the number of unknown spectral features above the $3\sigma$ level is much larger (around 1200 by July 2022).

As an example of identification of a new molecule with QUIJOTE without previous information of the frequencies of the transitions, we present the detection of the cation $C_5H^+$ [25]. Inspecting the U-lines we have found that only four of these U-lines have intensities above 3 mK. The frequencies of these four lines are, in addition, in perfect harmonic relation with $J_u$=7, 8, 9, 10. They are shown in Fig. 2. They do not show any hyperfine structure, and no other nearby lines are present with similar intensities, which discard a symmetric rotor, or a linear radical, as possible carrier. We are, hence, dealing with a linear molecule with a $^1\Sigma$ ground electronic state, or with a slightly asymmetric rotor with electronic state $^1A$. By fitting the observed frequencies to the standard Hamiltonian for a linear molecule ($\nu(J \rightarrow J-1) = 2B_0J - 4D_0J^3$) we derived $B_0$=2411.94397±0.00055 MHz and $D_0$=138±3 Hz. The standard deviation of the fit is 4.4 kHz. The possibility that these four strong lines appear by chance in harmonic relation with such a precision is negligible. Hence, we have detected a new molecular species in TMC-1. We selected all possible configurations of 4-5 atoms of H, C, N, O and S that could reproduce the observed rotational constant and conclude that $C_5H^+$ was the carrier of these lines. Although the definitive confirmation could arrive from laboratory experiments, the high level of theory ab initio calculations permit, nowadays, to assign with a high degree of confidence the observed lines to $C_5H^+$ [25]. In the same work we also reported the detection, for the first time in a cold starless core, the cation $C_3H^+$.

Since June 2020 QUIJOTE has detected near 40 new molecules in space (see [25] and references therein). The detection of cyclopentadiene and of indene [5], together with very abundant hydrocarbons such as $CH_2CCH$, $CH_2CHCCH$, $CH_2CCHCCH$, $CH_2CCHC_4H$ [2–4, 9], indicates that many reactions have to be included in the chemical models. In particular reactions between cations and radicals, and cations with large hydrocarbons are missing so far in the best models of cold astrophysical environments [26]. The large abundances of hydrocarbons and PAHs in these unexpected environments require a sensitive survey such as QUIJOTE to unveil all the aspects of the chemical complexity of cold prestellar cores [26].

## References

[1] Tercero, F., López-Pérez, J. A., Gallego, et al. 2021, A.&A., 645, A37
[2] Agúndez, M., Cabezas, C., Tercero, B., et al. 2021a, A.&A., 647, L10
[3] Cernicharo, J., Agúndez, M., Cabezas, C., et al. 2021a, A.&A., 647, L2
[4] Cernicharo, J., Cabezas, C., Agúndez, M., et al. 2021b, A.&A., 647, L3
[5] Cernicharo, J., Agúndez, M., Cabezas, C., et al. 2021c, A.&A., 649, L15
[6] Cernicharo, J., Agúndez, M., Kaiser, R., et al. 2021d, A.&A., 652, L9
[7] Cernicharo, J., Agúndez, M., Kaiser, R., et al. 2021e, A.&A., 655, L1
[8] Agúndez, M., Marcelino, N., Cabezas, C. et al. 2022, A.&A., 657, A96
[9] Fuentetaja, R., Cabezas, C., Afúndez, M. et al. 2022, A.&A., 663, L3
[10] McGuire, B. A., Burkhardt, A. M., Kalenskii, S., et al. 2018, Science, 359, 202
[11] McGuire, B. A., Loomis, R. A., Burkhardt, A. M., et al. 2021, Science, 371, 1265
[12] Cernicharo, J., Cabezas, C., Bailleux, S., et al. 2021f, A.&A., 646, L7
[13] Cabezas, C., Agúndez, M., Marcelino, N. et al., 2022a, A.&A., 657, L4
[14] Cernicharo, J., Marcelino, N., Agúndez, M., et al. 2020, A.&A., 642, L17
[15] Agúndez, M., Cabezas, C., Marcelino, N. et al. 2022b, A.&A., 659, L9
[16] Cernicharo, J. 1985, Internal IRAM report (Granada: IRAM)
[17] Pardo, J. R., Cernicharo, J., Serabyn, E. 2001, IEEE Trans. Antennas and Propagation, 49, 12
[18] Cernicharo, J., Fuentetaja, R., Cabezas, C., 2022a, A.&A., in pres, 663, L5
[19] Cabezas, C., Endo, Y., Roueff, E. et al. 2021a, A.&A., 646, L1
[20] Cabezas, C., Roueff, E., Tercero, B. et al. 2021b, A.&A., 650, L15
[21] Cabezas, Fuentetaja, R., Roueff, E. et al. 2022b, A.&A., 657, L5
[22] Cernicharo, J., 2012, *in ECLA 2011: Proc. of the European Conference on Laboratory Astrophysics, EAS Publications Series, 2012, Ed.: C. Stehl, C. Joblin, & L. d'Hendecourt (Cambridge: Cambridge Univ. Press)*, 251; `https://nanocosmos.iff.csic.es/?page_id=1619`
[23] Müller, H.S.P., Schlöder, F., Stutzki, J., Winnewisser, G. 2005, J. Mol. Struct., 742, 215
[24] Pickett, H.M., Poynter, R. L., Cohen, E. A., et al. 1998, J. Quant. Spectrosc. Radiat. Transfer, 60, 883
[25] Cernicharo, J., Agúndez, M., Cabezas, C. et al. 2022b, A.&A., 657, L16
[26] Cernicharo, J., Agúndez, M., Cabezas, C. et al. 2022c, A.&A., 663, L9

# Tracing episodic accretion with NOEMA: FU Orionis-type stars and their millimeter environment

*O.* Fehér[1,*], *Á.* Kóspál[2,3], *F.* Cruz-Saenz de Miera[2], *P.* Ábrahám[2], *M. R.* Hogerheijde[4], *Ch.* Brinch[5], and *D.* Semenov[3]

[1]Institut de Radioastronomie Millimétrique, 38400 Saint-Martin-d'Hères, 300 Rue de la Piscine, Grenoble, France

[2]Konkoly Observatory, Research Centre for Astronomy and Earth Sciences, Hungarian Academy of Sciences, 1121 Budapest, Konkoly Thege Miklós út, 15-17, Hungary

[3]Max Planck Institute for Astronomy, Königstuhl 17, 69117 Heidelberg, Germany

[4]Leiden Observatory, Leiden University, Niels Bohrweg 2, 2333 CA Leiden, The Netherlands

[5]Niels Bohr International Academy,The Niels Bohr Institute, University of Copenhagen, Blegdamsvej 17, 2100 Copenhagen Ø, Denmark

**Abstract.** The earliest phases of star formation are characterised by intense mass accretion from the circumstellar disk to the central star. One group of low-mass young stellar objects, FU Orionis-type stars (FUors) exhibit accretion rate peaks accompanied by bright eruptions in the optical and infrared regime. The occurance of these outbursts might solve the luminosity problem of protostars, play a key role in accumulating the final stellar mass, and have a significant effect on the parameters of the envelope and the disk. We are performing a systematic investigation of FUors with millimeter interferometry using NOEMA and ALMA to study the outburst events and examine whether FUors represent normal young stars in exceptional times or are unusual objects. The targeted FUors show very diverse circumstellar morphologies with envelope parameters similar to those of both Class I and Class II systems, but their disks are more massive and more compact than T Tauri disks. To shed light onto the process of disk-formation, accretion, and to what role FUors play in low-mass star-formation, we require the identification and light curve monitoring of as many of these stars as possible, together with the multi-wavelength and multi-scale mapping of their circumstellar environment.

## 1 Introduction

The process of low-mass star formation is the history of interstellar material accretion and infall, first during the gravitational collapse of cold, dense cloud cores, then the infall from the envelope to a circumstellar disk and the accretion of material from the disk to the surface of the forming star. Temporarily enhanced accretion events have been detected around young stellar objects as high-amplitude outbursts since the eighties [10, 11] with two historical groups of objects named after their archetype stars. One group is the so-called FUors (after FU Ori), pre-main sequence stars that show 5-6 magnitude brightness increases in the optical and IR regime, and spectral signitures indicating a rotating, accreting disk, while they

*e-mail: feher@iram.fr

exhibit an accretion rate of $10^{-4}$ $M_\odot$ year$^{-1}$ compared to the typical rate of $10^{-8}$ $M_\odot$ year$^{-1}$ for T-Tauri stars. The length of these FUor-type eruptions can reach years or decades, have been observed at all stages of star formation and theorized to be quasi-recurring, together with their shorter-timescale, fainter counterparts called EXor-outbursts (named after EX Ori) which show accretion rates of at most $10^{-7}$ $M_\odot$ year$^{-1}$ during eruptions. This episodic eruption phenomenon was regarded as a possible solution for the so-called protostellar luminosity problem. The problem states that protostars are observed to be less luminous than models predict they need to be to reach the main sequence in the time available for them ([6] and references therein). However, if high-accretion rate events are recurring during their early formation, it might explain this discrepancy. The analysis of IR emission and the 10 $\mu$m silicate feature towards FUors by [9] and [16] also supported an "evolutionary sequence." In this picture, younger, more deeply embedded Class I stars (showing the silicate feature in absorption) suffer FUor-eruptions triggered by e.g. gravitational, viscous-thermal, or magneto-rotational instabilities (e.g. [2, 4]) during which a significant amount of material accretes to the star from the disk. Then the disk replenishes by infall from the envelope. After repeated outbursts and enhanced accretion rate events, the eruptions become less powerful (EXor-type outbursts) as the system develops into a Class II, T-Tauri-type star with a remnant envelope and the silicate feature in emission [17].

However, recent results have complicated our understanding of these objects and these peculiar events. Only a few dozen of FUors are known today [3], but high spectral and spatial resolution observations are revealing the large diversity of circumstellar environments present towards these stars. An ALMA study of the object V346 Nor identified a Keplerian disk and a pseudo-disk around the star, and calculations of mass infall rate from envelope to disk is higher than the accretion rate from disk to star, which could possibly trigger an instability in the disk, starting an eruption event [12]. Other examinations with ALMA showed that the circumstellar disks around FUors are both more compact and more massive than the disks around EXors and regular T Tauri stars, and a significant ratio of them (but not all) are gravitationally unstable [5, 13–15]. These findings raise interesting questions about the types of mechanisms that are responsible for triggering episodic accretion and whether the difference between FUor-disks and T Tauri-disks is intrinsic or evolutionary.

## 2 Eruptive young stars through millimeter interferometry

High angular resolution and high sensitivity interferometry allows us to study the morphology and kinematics of FUor-disks and envelopes. The presence or lack of an envelope can help place individual FUors into an evolutionary timeline. Since observational evidence about the frequency of FUor outbursts is limited, it is still unclear whether the bursts really occur periodically, or stochastically in time. Rather than a periodical phenomenon, structured/clumpy infalling envelopes may occasionally load extra material into the disk faster than the accretion rate onto the star, making the disk unstable and triggering a burst. Since clumpiness in the circumstellar environment is directly observable with millimeter observations, the interferometric spatial scale mapping of the envelope is a key method in clearing out this question.

Our survey with the NOEMA (NOrthern Extended Millimeter Array) interferometer consists of several observing runs starting from 2012 with the latest datasets arriving in January 2022 and covers the line and continuum emission in the 100-200 GHz frequency range of more than 16 northern FUors reaching spatial resolutions of around 1000 au. In each case, the main targeted lines are transitions of $^{13}$CO and C$^{18}$O in order to map both the more diffuse and denser gas and assess excitation temperatures, opacities and possibly column densities around the objects. On-the-fly observations with the IRAM 30 m telescope are merged to

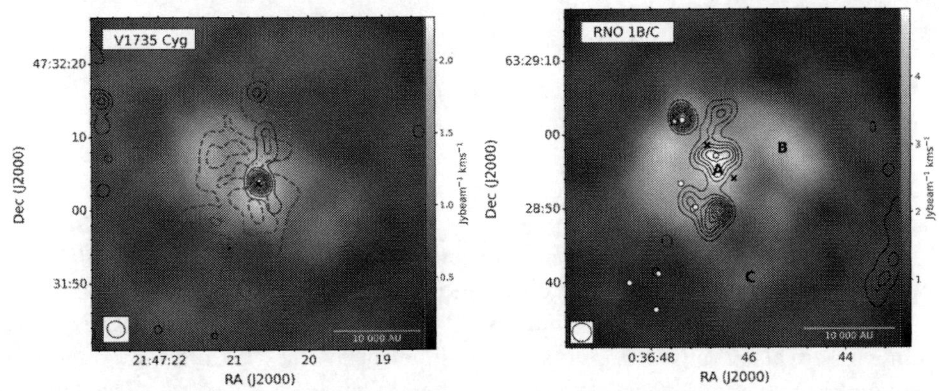


**Figure 1.** NOEMA CO and continuum maps of northern FUors (adapted from [7] and [8]). Left: $^{13}$CO integrated intensity map of V1735 Cyg. The cross marks the optical position of the FUor, the black contours are the 2.7 mm continuum emission in units of Jy beam$^{-1}$ at the 10...90% of the peak. The dashed contour is the blue-shifted (integrated between -0.2–2 km s$^{-1}$) and the dotted contour is the red-shifted lobe (integrated between 7–9.4 km s$^{-1}$) of the outflow at 50, 70 and 90% of their peak emission. Right: $^{13}$CO integrated intensity map of RNO 1B/C. The black contours are the 2.7 mm continuum emission at the 10...90% of the peak. The two crosses are the members of the binary FUor-system (RNO 1B is at the center) and the star symbol marks an IRAS source. The letters mark clumps identified on the CO map and circles mark some of the other IR/sub-mm/mm sources in the vicinity.

these datasets to correctly detect the large-scale gas emission the interferometer would filter out.

Our maps revealed extended, fragmented gas clouds around all our targets on a few thousand au spatial scales with local peaks at or close to the FUors [7]. Some FUors like V1057 Cyg or V1735 Cyg only show one central clump, others like V1515 Cyg or V2492 Cyg are associated with several other clumps in their neighbourhood, some occasionally corresponding to other IR or radio sources if the FUor is not isolated e.g. V710 Cyg, V2493 Cyg. The central clumps which we identify as envelopes are roughly circular with radii of 1500–5000 au and with masses of 0.02–0.5 $M_\odot$. Their interior is often warmer than their surroundings with average temperatures of 20–50 K. $^{13}$CO(1-0) velocity channel maps towards V1057 Cyg revealed a velocity gradient in the envelope, indicating rotation. A large scale velocity gradient was also measured in the environment of V2492 Cyg, caused by the shock front appearing as a bright H$\alpha$ rim close to the star. Even outflows could be identified on some of the maps: the integrated intensities of the $^{13}$CO(1-0) line wings around V1735 Cyg appear in two lobes on opposite sides of the FUor, one red-shifted, one blue-shifted (Fig. 1, left). The continuum emission of our targets is just as varied as the gas environment: some envelopes appear as strong, circular sources (e.g. V1057 Cyg, V2492 Cyg), others show many continuum features. A good example is RNO 1B/C or V710 Cas (Fig. 1, right). Clump A is located between the two members of the binary FUor-system and appears both in CO and the continuum as a strong, resolved source. Clump B is seen in CO but not in the continuum, however, the bright continuum source to the south of clump A does not appear in CO, and the other bright continuum source to the north-east appears only as a faint local CO maximum. A few of the FUors are weakly or not detected (e.g. V1515 Cyg, V733 Cep), yielding only an upper limit for the envelope mass. Maps of V582 Aur revealed that the star is located at the edge of an extended cloud structure with two well-identified velocity components, and the unresolved continuum source around the FUor is possibly a disk with a mass of 0.04 $M_\odot$ [1].

## 3 Future directions

Thanks to the large bandwidth and high spectral resolution of the PolyFiX correlator of NOEMA, our more recent measurements of FUor objects like RNO 127, HH 354 IRS or Gaia 17bpi cover not only the usual CO transitions with high spectral resolutions, but many other molecular species as well, providing a detailed look into the distribution of different gas tracers around the stars on interferometric scales. Synergy with ALMA continuum and CO data makes it possible to map the circumstellar vicinity from the envelope through the envelope-disk interface to the accreting disk. Due to the capabilities of the IRAM 30m back-ends, the single-dish maps of about 40 molecular transitions of e.g. CN, SO, $HC_3N$, $N_2H^+$ in the 3 mm band are also available for most of our targets, providing an interesting opportunity to build a multi-scale, multi-wavelength view of these diverse environments. The data allows us to examine the circumstellar morphology tracing structures of different density, physical size, and chemical history, and by utilizing chemical models, we aim to draw conclusions about the progenitors and the evolutionary path of the examined FUors. The discovery of new eruptions and the monitoring of the light curves and spectral features of already known outbursting sources will lead us to have more insight on the triggering and evolution of these intriguing events.

## References

[1] Ábrahám et al.,ApJ, **853**, 28 (2018)
[2] Armitage et al., MNRAS, **324**, 705 (2001)
[3] Audard et al., Protostars and Planets VI, 387 (2014)
[4] Bell & Lin, ApJ, **427**, 987 (1994)
[5] Cieza et al., MNRAS, **474**, 4347 (2018)
[6] Dunham et al., Protostars and Planets VI, 195 (2014)
[7] Fehér et al., A&A, **607**, A39 (2017)
[8] Fehér et al., Proceedings of the IAU, 345, 87-90 (2020)
[9] Green et al., ApJ, **648**, 1099 (2006)
[10] Hartmann et al., ARA&A, **34**, 207 (1996)
[11] Herbig, ApJ, **217**, 693 (1977)
[12] Kóspál et al., ApJ, **843**, 45 (2017)
[13] Kóspál et al., ApJS, **256**, 30 (2021)
[14] Li et al., ApJ, **840**, 72 (2017)
[15] Liu et al., A&A, **612**, A54 (2017)
[16] Quanz et al., ApJ, **668**, 359 (2007)
[17] Vorobyov & Basu, ApJ, **805**, 115 (2015)

# Searching For the t=0 of Planetary System Formation

*Edwin* Bergin[1,*], *Merel* van 't Hoff[1], and *Jes* Jørgensen[2]

[1]Department of Astronomy, University of Michigan, Ann Arbor, USA
[2]Niels Bohr Institute, University of Copenhagen, Copenhagen K., Denmark

**Abstract.** The composition of bodies in the solar system points to strong gradients in the volatile content within solid bodies hinting at the presence of gas-ice transitions across sublimation fronts in the young formative stages when the gas-rich disk was present. Terrestrial worlds are constructed out of the disk solids which are primarily silicate and water, but might also contain a significant fraction of organic material. These refractory organics are the source of carbon to Earth-like worlds, but have the potential to be destroyed if temperatures exceed 300-500 K (depending on pressure). These temperatures are most readily prevalent during the early stages of planetary system formation where the seeds of terrestrial worlds are potentially assembled. Here we present an ongoing observational search for refractory carbon grain destruction. We also discuss the implications on the overall gas phase chemistry within sublimation zones and on the ultimate composition of planetary bodies forming from available materials.

## 1 Hot Gas and Terrestrial Planet Formation

One of the most important tasks for astronomy during the coming decade(s) will be to characterize potentially habitable planets. A central facet of habitability is whether the elements needed to foster life: C, H, O, N, S, P, are present or absent. For the Earth it is believed that water and carbon (with other elements) were provided by impacts of asteroid-like materials (i.e. C I chondrites and other meteorite classes), perhaps with some cometary contribution [1]. However, when viewed from the interstellar context there is a more fundamental problem with the disposition of carbon in terrestrial worlds and even primitive (unaltered) meteorites in that, for example, the Earth is four orders of magnitude depleted in carbon when compared to the carbon content of interstellar grains. Fig. 1 demonstrates this graphically, comparing the C/Si ratios within in the solar system to the Sun and the interstellar medium (ISM).

In the ISM about 50% of the carbon is present in refractory form [3]. These ISM carbon grains are seen in the refractory organic material within both comet Halley and 67P. Heading towards the inner solar system, carbonaceous chondrites are a class of primitive meteorites. The C I chondrites are the most pristine material delivered from the asteroid belt as they are largely thermally unaltered and generally reflect solar abundances [4]. One glaring exception to this is that the C/Si ratio in C I chondrites is lower than that carried by ISM grains by a factor of 10! Even in this context, relative to chondrites, we stress that the Earth is even more significantly depleted in carbon.

*e-mail: ebergin@umich.edu

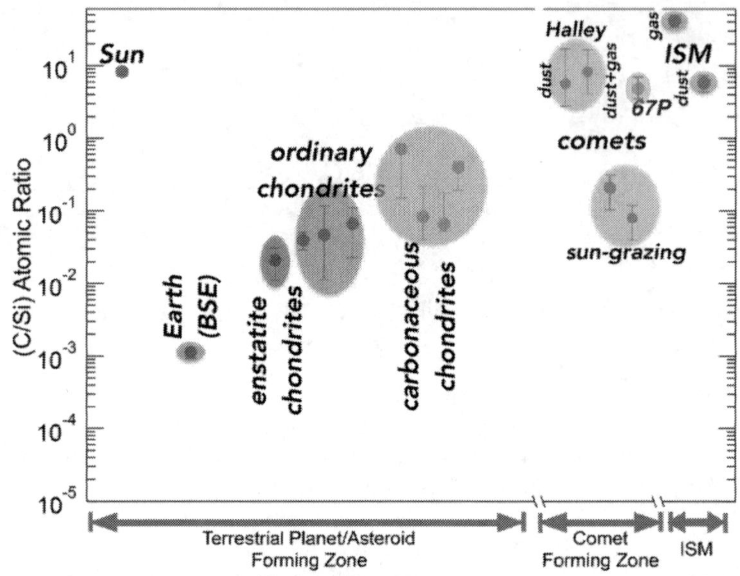


**Figure 1.** Atomic ratios of bulk carbon relative to silicon in various solar system bodies and the ISM. The shaded ovals represent the range/errors in the measured data for each class. For chondrites the errors represent the range of determined values, while for other bodies the errors are $1\sigma$ measurement uncertainties. Adapted from [2].

A central facet in the *cosmochemical and geophysical* understanding of the composition of solids in the inner solar system is the process of equilibrium condensation [5]. In this model, the primary theory for the origin of Earth's composition, all elements are initially assumed to be in chemical equilibrium within hot (>2000 K) gas. As the gas cools various minerals form in a sequence dictated by their condensation temperature and precipitate [6]. At minimum, two pieces of evidence support this picture. First, according to chemical equilibrium, the first solids to condense will be minerals containing calcium and aluminum [7] and calcium-aluminum rich inclusions (CAI) in meteorites represent the oldest solids in these bodies, thereby setting t=0 for the evolution of the Solar System [8]. Second, when normalized to Magnesium (Mg) and solar composition there is a trend in relative abundances for lithophile elements (rock loving elements found in the mantle and not in the metal dominated core) with their half-mass condensation temperature, the temperature at which 50% of the mass in an element is condensed from the gas into solids. This is shown graphically via the blue-shaded band in Fig. 2.

However, refractory organics are not a product of equilibrium condensation [4, 9]. The only solution to this conundrum is that the carbon carried by ISM grains in the inner few au is returned to the gas. When the star is born these materials are released into the gas within a volatility trend linked to the sublimation temperature of the primary elemental carbon carriers from volatile ices to more refractory carriers. This sequence is provided as the grey thick line in Fig. 2. In order to account for the bulk Earth's carbon upper limit [9], ISM refractory carbon must be destroyed [e.g. 10] prior to the formation of the Earth's initial materials.

The unifying theme for both condensation and sublimation is that the largest effects *must* be found in phases where temperatures are elevated above ~300–500 K in the inner disk, with even higher temperatures required for heavier elements. This project seeks to merge astro-

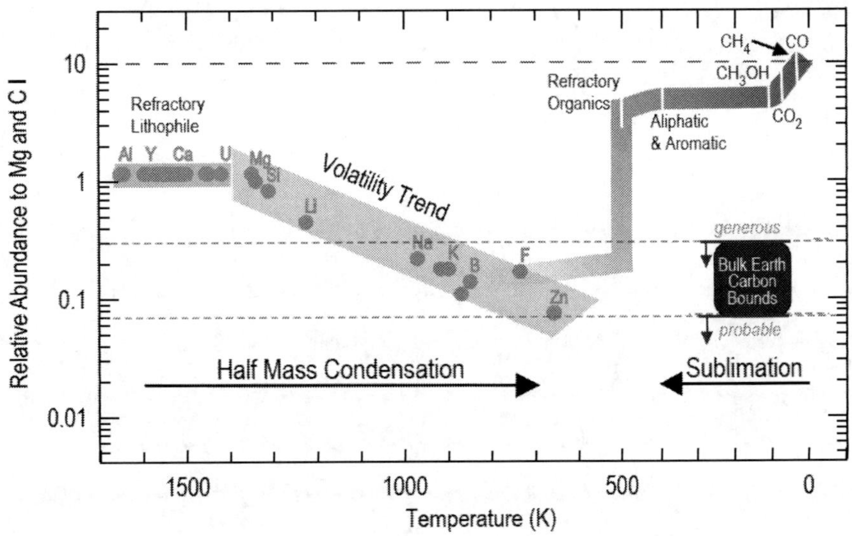


**Figure 2.** *Elemental relative abundances on Earth compared to measurements in CI chondrites (representing solar abundances) in both cases measured relative to Mg. The* **volatility trend associated with condensation (blue-shaded band)** *describes the relative abundances of lithophile ("rock loving") elements in the "bulk silicate Earth" (its mantle and crust) as a function of their half-mass condensation temperatures. On the low-temperature side,* **the sublimation sequence of carbon (gray thick line)** *traces the falling relative abundance of condensed carbon in the solar nebula as the disk warms up with refractory and icy carbon carriers released into the gas. Adapted from [9].*

physics and geophysics to find this gas and set new and unique constraints on the formation of solids in the inner parts of the Solar System and *exo-planetary systems*.

## 2 The t=0 Project

Sublimation readily operates in the midplane and can destroy carbonaceous mm/cm-sized pebbles that constitute the first steps toward planetesimals at temperatures between 300-500 K (pressure dependent) [9, 11]. This location is called the soot-line [12] analogous to the water ice-line which lies at greater distances from the star. Class II disks are generally dominated by stellar irradiation and the soot-line resides interior to the meteorite formation zone near a few au [13–15]. A soot line that lies inside of the meteorite formation zone would produce bodies with substantial amounts of carbon, comparable to that of 67P or Halley (C/Si ~ 5). This is well above the carbon content of even the most primitive meteorites (~0.2; Fig. 1). Thus, the pre-cursors to meteoritic material (and the Earth) must have formed primarily during earlier protostellar stage [9]. In this case, the soot-line lies at greater distances due to the higher amount of energy released from the active accretion. This gas might be more detectable and is the focus of our work.

We have begun a systematic search for carbon grain destruction in the hopes of constraining its presence or absence. Our first goal is to locate high temperature (>300 K) gas in close proximity to protostars (presumably in a young disk) via spatially unresolved NOEMA observations. Once this gas is isolated we aim to search for whether it is chemically distinct in comparison to gas above the water sublimation temperature (>100 K). For example, van 't Hoff[11] proposed N released from C-rich grains could be a signature as it would produce abundant nitriles in hot gas, but see also [16]. A final, longer-term, step would be to resolve the emission.

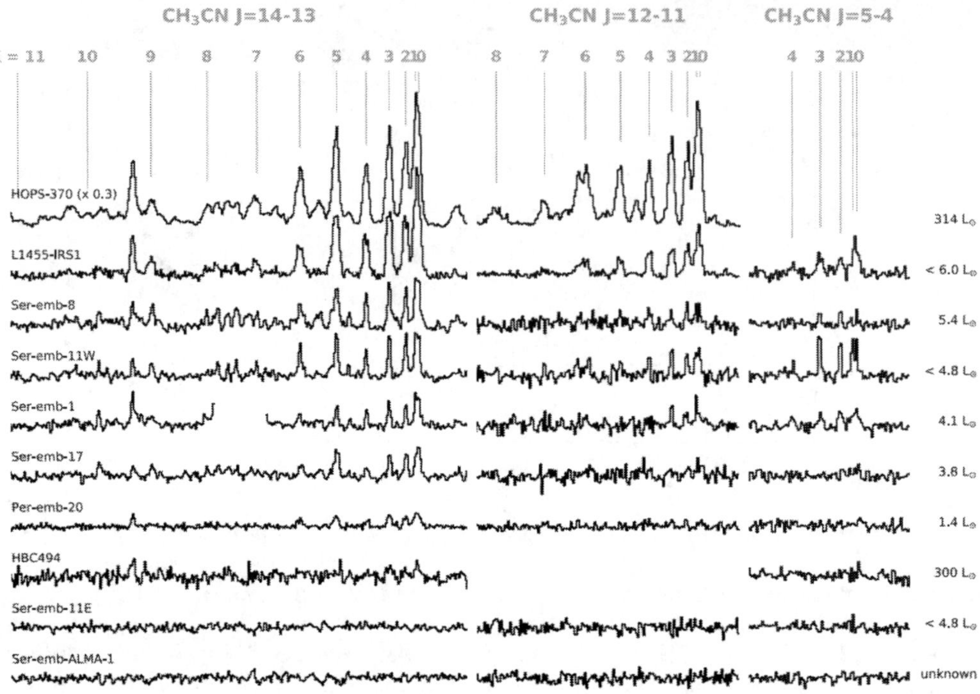


**Figure 3.** *Results from the NOEMA pilot suvey of $CH_3CN$ towards young protostars. All spectra are shown at a uniform velocity resolution of 2 km/s. For a given J transition the K components have the same flux scaling, but the flux scale is different between J transitions. The 5-4 and 12-11 data are obtained with ~1.0″ resolution, while the 14-13 transitions are observed with ~0.4″ resolution. Data will be published by van 't Hoff et al. (in prep.).*

### 2.1 NOEMA Pilot Survey

Over the past 3 years, our team has been compiling an extensive dataset of high resolution (~0.4″ at 1mm, ~1.0″ at 2mm, ~1.0″ at 3mm) NOEMA observations of 10 protostars. Eight of these have luminosities commensurate with low mass (solar mass or below) star formation, one is a FU Ori type object (HBC494) and one is likely an intermediate-mass star (HOPS-370). These sources have been pre-selected to be rich in organic emission based on detection of $CH_3OH$ [17, 18], or their outbursting nature. Our 1mm data are taken at 0.4″ (~hundred au at source distances) and do not spatially resolve the soot line. So we will use the central pixel to isolate the central regions and then detect the hot gas via emission lines that arise from high energy states ($E_u$ > 300 K). This does not guarantee the presence of hot gas, as these lines would still be emissive at 100–200 K; but, multiple lines can be used to derive the gas temperature as described below. Because of concerns about high dust optical depth in the innermost regions of protostellar envelopes or in the young disk we also obtained observations at 3mm with NOEMA (image cubes) and single point observations at 4mm with the GBT.

To find, and characterize hot gas we use $CH_3CN$, a symmetric top [19] that is a well known probe of temperature in hot gas and has been used to explore temperature more recently in disk sources [20]. In Fig. 3 we show an example of why $CH_3CN$ is a fantastic thermal probe. Symmetric tops allow for observation of multiple emission lines arising from different energy states (labelled with $J_K$) over a small window. Further, lines with different K for a given J state are only connected via collisions and have relative populations that trace the gas kinetic temperature [see, e.g., 21]. Multiple lines of $CH_3CN$ are detected in nearly every system. Amazingly, for many systems, we detect emission from the $14_9$–$13_9$ transition with an upper state energy level of 670 K. This line requires hot gas (> 200 K) for excitation.

In sum, hot gas appears to be present in these systems. We are in the process of compiling and understanding the rotation diagrams for each source and all transitions. It is complicated by the fact that the emission appears optically thick for lower-J lines. Intriguingly, the column density required for lower-J lines (at 3mm) is higher than inferred for the transitions detected at 1mm. This may hint that dust optical depth is obscuring a complete sampling of the $CH_3CN$ column at shorter wavelengths [22], but thermal gradients might contribute as well. This work will be published by van't Hoff et al. (in prep.).

## 3 Implications

Models of planet formation suggest two modes of planetary construction: one dominated by accretion of planetesimals (~10–100 km-sized bodies) and another proceeding through accretion of mm to cm-sized pebbles [23, 24]. Depending on many factors, in particular the presence or absence of a cold gas giant in the system (i.e., Jupiter-analog) and on the inwards drift of volatile-rich pebbles[24] a range of outcomes, including terrestrial worlds with more bulk carbon content, is possible. Whether these worlds are habitable will be an interesting question for future exploration; but the presence/absence of refractory carbon material is central.

In our initial analyses for several systems, temperatures appear to be well above that expected for carbon grain sublimation. Formally speaking, if all carbon grains sublimate in the inner disk, it is just a factor of two extra carbon. This factor of two is, of course, bulk and the speciation of carbon can lead to order of magnitude enhancements in specific molecules [16]; understanding this issue is a key goal of this effort. We stress that a null result (i.e., no evidence for carbon grain destruction) in a sense has the most far-reaching implications as it implies that carbon-rich pebbles are available for planet formation.

## References

[1] B. Marty, K. Altwegg, H. Balsiger, A. Bar-Nun, D.V. Bekaert, J.J. Berthelier, A. Bieler, C. Briois, U. Calmonte, M. Combi et al., Science **356**, 1069 (2017)

[2] E.A. Bergin, G.A. Blake, F. Ciesla, M.M. Hirschmann, J. Li, Proceedings of the National Academy of Science **112**, 8965 (2015), 1507.04756

[3] A. Mishra, A. Li, **809**, 120 (2015), 1507.06599

[4] K. Lodders, *Solar System Abundances of the Elements*, in *Principles and Perspectives in Cosmochemistry*, edited by A. Goswami, B.E. Reddy (2010), p. 379

[5] W.F. McDonough, S.S. Sun, Chemical geology **120**, 223 (1995)

[6] L. Grossman, **36**, 597 (1972)

[7] A.M. Davis, *Volatile Evolution and Loss* (2006), p. 295

[8] Y. Amelin, A.N. Krot, I.D. Hutcheon, A.A. Ulyanov, Science **297**, 1678 (2002)

[9] J. Li, E.A. Bergin, G.A. Blake, F.J. Ciesla, M.M. Hirschmann, Science Advances **7**, 3632 (2021), 2104.02702

[10] H.P. Gail, M. Trieloff, **606**, A16 (2017), 1707.07611

[11] M.L.R. van 't Hoff, E.A. Bergin, J.K. Jørgensen, G.A. Blake, **897**, L38 (2020)

[12] M.E. Kress, A.G.G.M. Tielens, M. Frenklach, Advances in Space Research **46**, 44 (2010)

[13] P. D'Alessio, N. Calvet, D.S. Woolum, *Thermal Structure of Protoplanetary Disks*, in *Chondrites and the Protoplanetary Disk*, edited by A.N. Krot, E.R.D. Scott, B. Reipurth (2005), Vol. 341 of *Astronomical Society of the Pacific Conference Series*, p. 353

[14] S.M. Andrews, J. Huang, L.M. Pérez, A. Isella, C.P. Dullemond, N.T. Kurtovic, V.V. Guzmán, J.M. Carpenter, D.J. Wilner, S. Zhang et al., **869**, L41 (2018), 1812.04040

[15] C.P. Dullemond, T. Birnstiel, J. Huang, N.T. Kurtovic, S.M. Andrews, V.V. Guzmán, L.M. Pérez, A. Isella, Z. Zhu, M. Benisty et al., **869**, L46 (2018), 1812.04044

[16] C.E. Wei, H. Nomura, J.E. Lee, W.H. Ip, C. Walsh, T.J. Millar, **870**, 129 (2019), 1811.10194

[17] J.B. Bergner, R. Martín-Doménech, K.I. Öberg, J.K. Jørgensen, E. Artur de la Villarmois, C. Brinch, ACS Earth and Space Chemistry **3**, 1564 (2019), 1907.07791

[18] T.H. Hsieh, N.M. Murillo, A. Belloche, N. Hirano, C. Walsh, E.F. van Dishoeck, J.K. Jørgensen, S.P. Lai, **884**, 149 (2019), 1909.02706

[19] R.B. Loren, L.G. Mundy, **286**, 232 (1984)

[20] R.A. Loomis, L.I. Cleeves, K.I. Öberg, V.V. Guzman, S.M. Andrews, **809**, L25 (2015), 1508.07004

[21] E.A. Bergin, P.F. Goldsmith, R.L. Snell, H. Ungerechts, **431**, 674 (1994)

[22] M. De Simone, C. Ceccarelli, C. Codella, B.E. Svoboda, C. Chandler, M. Bouvier, S. Yamamoto, N. Sakai, P. Caselli, C. Favre et al., **896**, L3 (2020), 2006.04484

[23] A. Johansen, M. Lambrechts, Annual Review of Earth and Planetary Sciences **45**, 359 (2017)

[24] A. Izidoro, S.N. Raymond, *Formation of Terrestrial Planets* (2018), p. 142

# Water Silhouettes against the Cosmic Microwave Background from the Most Distant Starburst Galaxies

*Dominik A.* Riechers[1,*], *Axel* Weiss[2], *Fabian* Walter[3,4], *Christopher L.* Carilli[4], *Pierre* Cox[5], *Roberto* Decarli[6], and *Roberto* Neri[7]

[1]I. Physikalisches Institut, Universität zu Köln, Zülpicher Strasse 77, 50937 Köln, Germany
[2]Max-Planck-Institut für Radioastronomie, Auf dem Hügel 69, 53121 Bonn, Germany
[3]Max-Planck-Institut für Astronomie, Königstuhl 17, 69117 Heidelberg, Germany
[4]National Radio Astronomy Observatory, Pete V. Domenici Array Science Center, P.O. Box O, Socorro, NM 87801, USA
[5]Sorbonne Université, UPMC Université Paris 6 and CNRS, UMR 7095, Institut d'Astrophysique de Paris, 98bis boulevard Arago, 75014 Paris, France
[6]INAF - Osservatorio di Astrofisica e Scienza dello Spazio, via Gobetti 93/3, 40129 Bologna, Italy
[7]Institut de RadioAstronomie Millimétrique, 300 Rue de la Piscine, Domaine Universitaire, 38406 Saint Martin d'Hères, France

**Abstract.** Astrophysical objects can imprint distortions on the observed Cosmic Microwave Background (CMB) that give access to information for cosmology research that cannot be obtained otherwise. ΛCDM cosmology implies a linear scaling of the CMB temperature ($T_{\rm CMB}$) with redshift $z$, but departures of this linear scaling behavior are allowed in more complex, but currently poorly observationally constrained cosmological models, such as those that include an evolution of physical constants, decaying dark energy, or axion-photon-like coupling processes. We here introduce a new method to directly measure $T_{\rm CMB}$ out to $z > 6$ based on $H_2O$ absorption against the CMB, and describe our findings based on an initial detection towards the massive dusty starburst galaxy HFLS3 at $z$=6.34. This far exceeds the redshift range where direct $T_{\rm CMB}$ measurements across cosmic time have been previously possible, providing a crucial test of standard cosmology.

## 1 Introduction

The Cosmic Microwave Background (CMB) is the perhaps most important observable in cosmological studies, and it provides one of the key pieces of evidence for the Hot Big Bang model. Since the CMB acts as a background to all astrophysical objects we observe in the universe, different classes of objects can imprint different kinds of distortions onto the CMB, as the CMB photons travel on their path to Earth. The most common distortions at the wavelengths where the CMB is strongest are due to thermal emission from dust from our own Milky Way all the way to the first galaxies that form within the first billion years after the Big Bang, as well as non-thermal radio synchrotron emission in particular from active galactic nuclei.[7] Such signals locally enhance the observed radiation intensity, and need to be

*e-mail: riechers@ph1.uni-koeln.de

accounted for when studying the CMB. On the other hand, there also are processes that lead to a local reduction in the radiation intensity, leading to an absorption signal at certain wavelengths. The most well-known of these processes is the Sunyaev-Zel'dovich (SZ) effect observed toward massive galaxy clusters, which is dominated by a thermal component (and smaller kinetic component due to bulk cluster motion and relativistic corrections).[8, 12] The SZ effect occurs due to the fact that the intra-cluster medium in massive galaxy clusters consists of a hot plasma, which contains significant numbers of high-energy electrons. A fraction of CMB photons crossing the galaxy cluster potential thus are subject to Thomson scattering off the fast electrons, which increases their energy. Thus, at the position of the galaxy cluster, we observe a deficit in photons at wavelengths longward of 1.4 mm (i.e., an absorption signal), and a surplus at shorter wavelengths (i.e., an emission signal). Since the strength of the effect for a galaxy cluster of a given mass depends on the energy of the CMB photons (but not its distance, or redshift), and hence, the CMB temperature ($T_{\rm CMB}$), observations of the SZ effect can be used as a cosmological probe, in particular to constrain $T_{\rm CMB}$ as a function of redshift $z$. To date, this has been done successfully out to a redshift of $z{\sim}1$, but too few sufficiently massive galaxy clusters are known at higher redshift to obtain reliable measurements to date.[2, 3]

In the next sections, we present a new cosmological tool to constrain $T_{\rm CMB}$ out to $z > 6$, i.e., extending the range to >13 billion years of cosmic history (for more details, see [10]). However, a key question to ask is what makes such measurements scientifically interesting. In standard ΛCDM cosmologies, $T_{\rm CMB}$ simply linearly increases as $(1{+}z)$, i.e., $T_{\rm CMB}(z){=}T_{\rm CMB}(z{=}0)\,(1{+}z)$. However, this scaling behavior actually depends on a number of assumptions on the underlying physical laws and processes, some of which are not tested thoroughly on cosmological time and distance scales. For example, the dark energy density in standard models is assumed to be constant with time, but given our limited knowledge about the nature of dark energy, it remains entirely possible that it, for example, could decay over time.[5] This would introduce a scaling of $\Lambda$ with redshift, which could then also impact the $T_{\rm CMB}$ evolution with $z$. Other possibilities to alter the standard $(1{+}z)$ scaling behavior include an evolution of physical constants with time,[13] or axion-photon-like coupling processes.[4] As such, investigations of the $T_{\rm CMB}(z)$ scaling behavior implicitly constrain potential deviations from the standard cosmology. For simplicity, such deviations will be expressed in the following as deviations from the power law index of 1 in the $(1{+}z)^1$ term. The results in this work were first published by [10].

## 2 Water absorption against the CMB: Discovery

In the summer of 2020, we have used the IRAM NOEMA interferometer to spectrally scan the 3 mm window, targeting the $z{=}6.34$ massive dusty starburst galaxy HFLS3. In 2013, this galaxy was reported to be the most distant such object at the time based on the detection of CO $J{=}6{\to}5$ and $7{\to}6$, [CI] $2{\to}1$, and para-$H_2O$ $2_{11}{\to}2_{02}$ emission, redshifted to 3 mm, in a 77 hr-long spectral scan with the CARMA interferometer.[9] Given the heterogeneous depth and limited sensitivity of the scan by today's standards, and the substantial upgrade to NOEMA in the meantime, we decided to re-visit this system to more comprehensively characterize the molecular complexity in this early massive galaxy, observed only 880 million years after the Big Bang. Figure 1 shows a comparison of the previous to the new data, demonstrating the dramatic improvement in data quality, and the richness of the broad-band spectrum.[10] For the remainder of this article, we focus on a single line, redshifted to 75.9 GHz, or 3.95 mm. This line corresponds to the ortho-$H_2O$ $1_{10}{\to}1_{01}$ transition at rest-frame 538 $\mu$m, and is seen in absorption (Fig. 2). The absorption is about twice as strong as the galaxy's dust continuum emission, which suggests that it absorbs into the CMB at the redshift of HFLS3.

## 3 Water absorption against the CMB: Mechanism

To understand the nature of this line absorption signal into the CMB, we need to take into account the properties of the CMB and starburst, as well as the structure of the ortho-$H_2O$ rotational line ladder. First, the expected $T_{CMB}$ at the redshift of HFLS3 is 20.0 K. Since the CMB acts as a background to the galaxy, a measurement of zero flux in the line would correspond to a line brightness temperature of 20 K, in equilibrium with the CMB. Second, the CMB can act as a heating source to the $H_2O$-bearing material, and excite the $H_2O$ rotational ladder. The energy difference of the $1_{10}{\rightarrow}1_{01}$ transition is only $\Delta E/k_B$=26.7 K, where $k_B$ is the Boltzmann constant. Given the Boltzmann distribution of CMB photons, this means that a 20 K CMB can substantially excite the $H_2O$ rotational ladder. RADEX[14] calculations suggest that, in such a configuration, 77.2% of the molecules will be in the ground $1_{01}$ state, while 20.3% will be in the first excited $1_{10}$ state. This level of excitation will bring the transition line in thermal equilibrium with the CMB. As a result, no net flux would be observed in any of the $H_2O$ transitions.

However, the situation changes when accounting for the infrared radiation field contributed by the dust emission in the starburst HFLS3 itself. The dust in this galaxy has a temperature of $T_{dust}$=63 K, which acts as a background to the cold $H_2O$ associated with HFLS3 (hence, causing absorption into the dust continuum), while at the same time carrying enough energy to also excite the $H_2O$ rotational ladder. RADEX calculations suggest that, with the addition of the starburst radiation field, 68.0% of the molecules will be in the ground $1_{01}$ state, while only 14.6% will be in the first excited $1_{10}$ state. This excitation level corresponds to an excitation temperature of only $T_{ex}$=17.4 K, i.e., 2.6 K below the CMB temperature – explaining the absorption into the CMB. To understand this effect, it is necessary to consider the next higher transition of ortho-$H_2O$, i.e., $2_{21}{\rightarrow}1_{10}$. This transition occurs at 108 $\mu$m, i.e., substantially closer to the peak of the dust spectral energy distribution (SED) of HFLS3 of 73 $\mu$m than the $1_{10}{\rightarrow}1_{01}$ transition. This is important, because the starburst SED does not have a black-body shape, but a greybody shape, since the dust optical depth at shorter wavelengths on the Rayleigh-Jeans tail is significantly higher than at longer wavelengths. As a result, more 108 $\mu$m photons are available from the starburst to excite the $2_{21}{\rightarrow}1_{10}$ transition than 538 $\mu$m photons to excite the $1_{10}{\rightarrow}1_{01}$ transition. Under different circumstances, this would cause a bottleneck in the excitation of the $2_{21}{\rightarrow}1_{10}$ transition due to availability of molecules in the lower state. However, due to the availability of "additional" molecules excited by the CMB, the observed deficiency of molecules in the $1_{10}$ state compared to thermal excitation can occur.

## 4 How to measure $T_{CMB}(z)$ from water absorption

To explain the mechanism in the previous section, we have assumed that $T_{CMB}(z)$ is as predicted by standard cosmology. We now explain how the measurement can be used to test this assumption. Considering $T_{CMB}(z{=}6.34)$ at the distance of HFLS3 as the unknown, we need to know the dust SED shape of HFLS3, which is constrained well from observations, to determine the availability of 538 and 108 $\mu$m photons. We also need to know the column density of $H_2O$ molecules to predict the depth of the absorption. Lastly, to obtain a line flux, we need to know the size of the absorbing region (i.e., the observed size of the dust continuum of HFLS3), and the filling factor of the absorbing clouds in the corresponding surface area. RADEX models with these inputs suggest that the data is consistent with $T_{CMB}(z{=}6.34)$=16.4–30.2 K ($1\sigma$ range), i.e., compatible with the 20.0 K expected in standard cosmology (Fig. 3). The models also suggest a minimal filling factor of almost 60%, which implies that the cold $H_2O$ screen in front of the dust continuum is not highly clumpy, but rather smooth. Given the implied gas masses, they also suggest that it is unlikely that the gas densities are high enough to significantly collisionally excite $H_2O$.[10]

In any case, the models also indicate that any significant contribution due to collisional excitation would weaken the effect, and thus, place more stringent constraints on the radiative excitation calculations considered here. Furthermore, they suggest that, for galaxies with dust SED shapes similar to HFLS3, the effect should only become observable at $z$~4.5, but that where observable, the line flux should be nearly independent of redshift out to at least $z$~12.

## 5 Conclusions

Our initial discovery of $H_2O$ absorption against the CMB toward the massive dusty starburst galaxy HFLS3 is consistent with a CMB temperature at $z$=6.34 as expected in standard cosmology, and thus, with a non-evolving dark energy density with time.[10] Our models suggest that it should be possible to observe this effect rather commonly toward massive dusty starburst galaxies at $z$>4.5, where the exact redshift cutoff depends on how similar the dust SED shapes of this population are to each other. Preliminary findings from an ongoing NOEMA follow-up campaign of this intial result appear to confirm these model predictions, which will be reported in future work.

## References

[1] Avgoustidis, A. et al., Phys. Rev. D **93**, 043521 (2016)
[2] de Martino, I. et al., Astrophys. J. **808**, 128 (2015)
[3] Hurier, G. et al., Astron. Astrophys. **561**, A143 (2014)
[4] Jaeckel, J. & Ringwald, A., Annu. Rev. Nucl. Part. Sci. **60**, 405-437 (2010)
[5] Jetzer, P. et al., Gen. Relativ. Gravitation **43**, 1083-1093 (2011)
[6] Klimenko, V. V., et al., Astron. Lett. **46**, 715-725 (2020)
[7] Planck Collaboration, Astron. Astrophys. **641**, A1 (2020)
[8] Rephaeli, Y., Annu. Rev. Astron. Astrophys. **33**, 541-580 (1995)
[9] Riechers, D. A., et al., Nature **496**, 329-333 (2013)
[10] Riechers, D. A., et al., Nature **602**, 58-62 (2022)
[11] Saro, A. et al., Mon. Not. R. Astron. Soc. **440**, 2610-2615 (2014)
[12] Sunyaev, R. A., & Zel'dovich, Annu. Rev. Astron. Astrophys. **18**, 537-560 (1980)
[13] Uzan, J.-P., Living Rev. Relativ. **14**, 2 (2011)
[14] van der Tak, F. F. S., et al., Astron. Astrophys. **468**, 627-635 (2007)

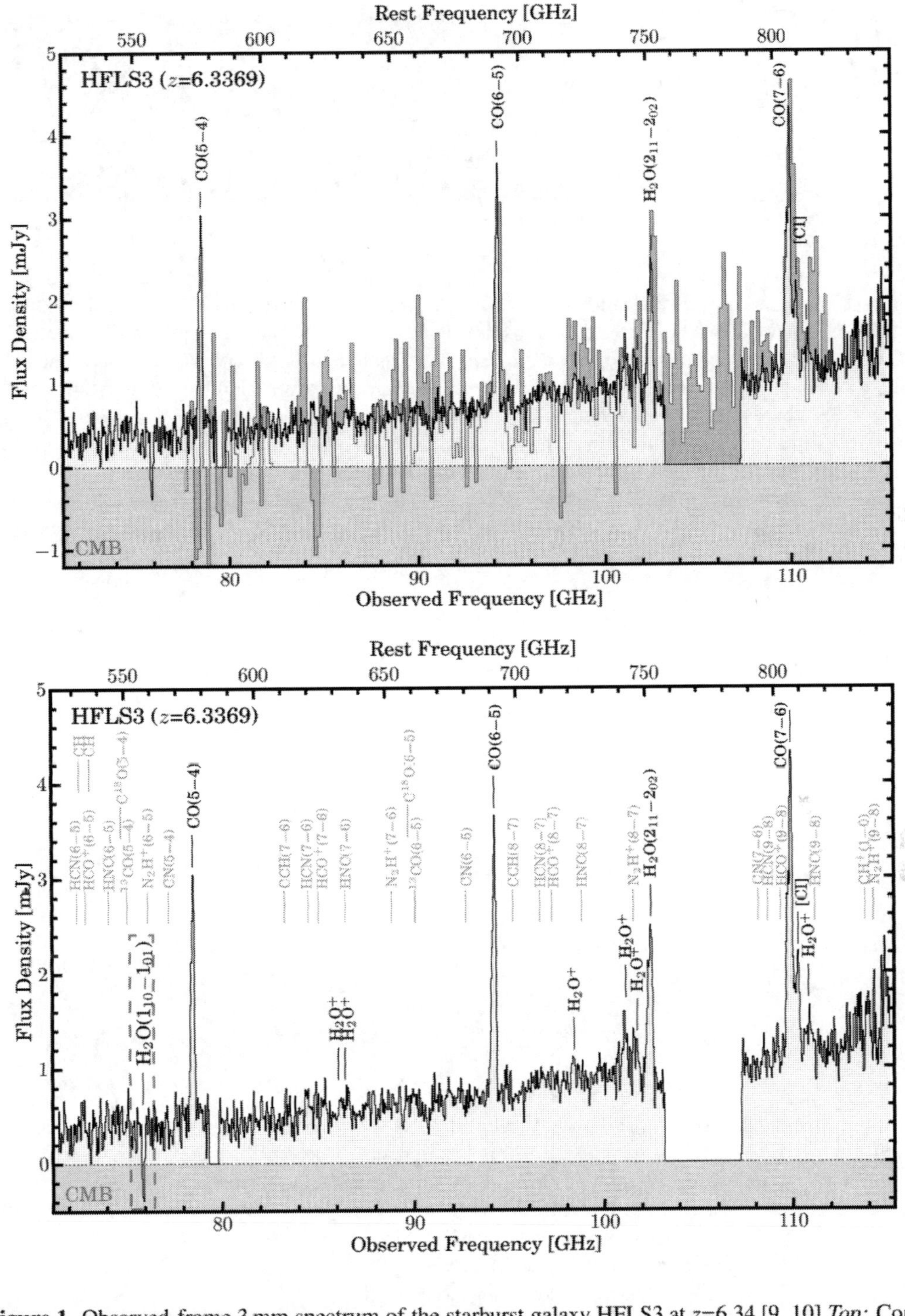


**Figure 1.** Observed-frame 3 mm spectrum of the starburst galaxy HFLS3 at $z$=6.34.[9, 10] *Top:* Comparison of the original CARMA "discovery" spectrum (gray histogram) and the new, more sensitive NOEMA spectrum (black/yellow histogram), demonstrating the major improvements possible with the greatly improved bandwidth and sensitivity of the upgraded array. The brightest lines are labeled. *Bottom:* NOEMA spectrum only, with additional lines labeled. The focus of this manuscript is the redshifted $H_2O$ $1_{10}\rightarrow 1_{01}$ transition highlighted by the dashed red box, which absorbs through the galaxy's dust continuum into the CMB (see Fig. 2 for additional details).

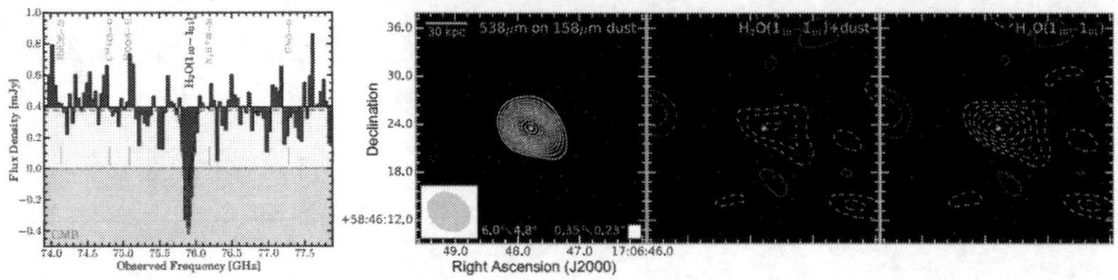


**Figure 2.** Zoom-in to the redshifted $H_2O$ $1_{10}{\rightarrow}1_{01}$ transition toward HFLS3 (*left*), and moment maps (*right*).[10] *Left:* Line spectrum at 40 MHz (158 km s$^{-1}$) resolution (blue), on top of the dust continuum of HFLS3 (yellow) and the CMB (gray; same labeling as Fig. 1). The black curve shows a fit to the line emission, and the dashed red curve shows the best-fit radiative transfer model. *Right:* Dust continuum at the wavelength of the line (538 $\mu$m rest frame; green contours), absorption before subtraction of the galaxy's dust continuum, and after continuum subtraction (blue dashed contours; left to right). All maps are overlaid on an intensity scale map of the rest-frame 158 $\mu$m dust continuum observed with NOEMA at high spatial resolultion.[9] The beam sizes are indicated at the bottom of the first panel. Continuum contours are shown in steps of $1\sigma$, starting at $\pm 3\sigma$. Line contours are shown in steps of $1\sigma$, starting at $\pm 2\sigma$. Negative intensity contours are dashed, and in the case of the $H_2O$ line, indicate absorption strength.

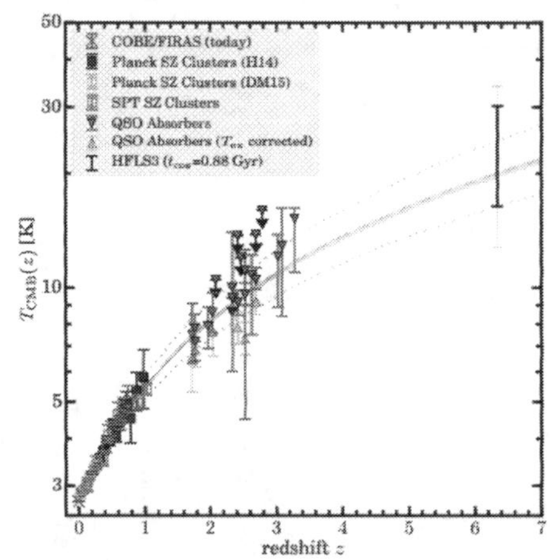


**Figure 3.** Measurements of $T_{\rm CMB}(z)$.[10] Symbols at $0 < z \leq 1$ are based on the thermal SZ effect.[1–3, 11] Symbols at $1.8 < z < 3.3$ are indirect constraints based on UV absorption line measurements toward background quasars, where different symbols at the same redshift correspond to different predictions for the same systems based on different assumptions for excitation due to collisional excitation of the tracer.[6] The symbol at $z$=6.34 corresponds to HFLS3, with $1\sigma$ (black) and $2\sigma$ (gray) uncertainty ranges shown. The orange dashed curve and shaded region show the best fit and uncertainty to all data, which remain consistent with standard cosmology (i.e., a (1+$z$) scaling behavior). The dotted lines indicate a $\pm$10% deviation from a power-law index of 1 for reference.

# $CH^+$(1-0) in a z~2.8 galaxy group: Probe of multi-phasic turbulent gas reservoirs

*Alba* Vidal-García[1,*], *Edith* Falgarone[1,**], *Fabrizio* Arrigoni Battaia[2], *Benjamin* Godard[1], *Rob J.* Ivison[3], *Martin A.* Zwaan[3], *Cinthya* Herrera[4], *David* Frayer[5], *Paola* Andreani[3], *Quong* Li[6], *Raphaël* Gavazzi[7], *Edwin* Bergin[8], *Fabian* Walter[9], and *Alain* Omont[7]

[1]LPENS, Université PSL, CNRS, Sorbonne Université, Université de Paris, Paris, France
[2]Max Planck Institute fur Astronomie, Garching, Germany
[3]European Southern Observatory, Garching, Germany
[4]Institut de radioastronomie millimétrique, Saint Martin d'Hères, France
[5]Green Bank Observatory, Green Bank, WV, USA
[6]Kavli Institute for Astronomy and Astrophysics, Peking University, Beijing, People's Republic of China
[7]Sorbonne Université, CNRS, Institut d'Astrophysique de Paris, Paris, France
[8]University of Michigan, Ann Arbor, MI, USA
[9]Max Planck Institute fur Astronomie, Heidelberg, Germany

**Abstract.** Starburst galaxies at redshifts z~2 to 4 are among the most intensely star-forming galaxies in the universe. The way they accrete their gas to form stars at such high rates is still a controversial issue. We have detected the $CH^+$(1-0) line in emission and/or in absorption in all the gravitationally lensed starburst galaxies observed so far with ALMA in this redshift range. The unique spectroscopic and chemical properties of $CH^+$ allow its rotational transition to highlight the sites of dissipation of mechanical energy. Whilst the absorption lines reveal highly turbulent reservoirs of low-density molecular gas extending far out of the galaxies, the broad emission lines with widths up to a few thousands of km/s, arise in myriad molecular shocks powered by the feedback of star formation and possibly active galactic nuclei. The $CH^+$(1-0) lines therefore probe the sites of prodigious energy releases, mainly stored in turbulent reservoirs before being radiated away. These turbulent reservoirs act as extended buffers of mass and energy over timescales of a few tens to hundreds of Myr. Their mass supply involves multi-phasic gas inflows from galaxy mergers and/or cold stream accretion, as supported by Keck/KCWI Ly$\alpha$ observations of one of these starburst galaxies.

## 1 Introduction

In numerical simulations, galaxies grow by accreting cold streams [1]. The momentum exchange between the streams and the galaxies at the bottom of the dark matter potential well is so violent that a large turbulent region is created in the circumgalactic medium (CGM) around the galaxies. While the feedback from Active Galactic Nuclei (AGN) and star formation is

*e-mail: alba.vidal@ens.fr
**e-mail: edith.falgarone@ens.fr

observed through powerful outflows [2], cold stream accretion is still elusive. Such large turbulent reservoirs of diffuse gas around galaxies are therefore signposts of cold stream accretion. Accretion of diffuse matter onto galaxies is traced by redshifted absorption lines. The $CH^+$ molecule has been a valuable tracer of the CGM motions around high-redshift lensed starburst galaxies [3].

$CH^+$ is a most fragile but precious molecular tracer. It has a high endothermic formation ($E_{form}$ ~0.5eV) and is highly reactive, so it is observed where it forms. Once formed, its lifetime is so short (~1 year), that a warm chemistry activated by bursts of turbulent dissipation is needed to overcome its destruction rate. Unlike most molecules, $CH^+$ is not photodissociated, but destroyed by collisions. Finally, it has a high dipole moment, so absorption lines of its $J$=1-0 transition trace diffuse gas and emission lines trace high density gas seen in shocks and photodissociation regions [4, 5]. All these properties make $CH^+$ a unique tracer of dissipation of mechanical energy in turbulence [3, 6].

## 2 SMM J02399-0136 galaxy group

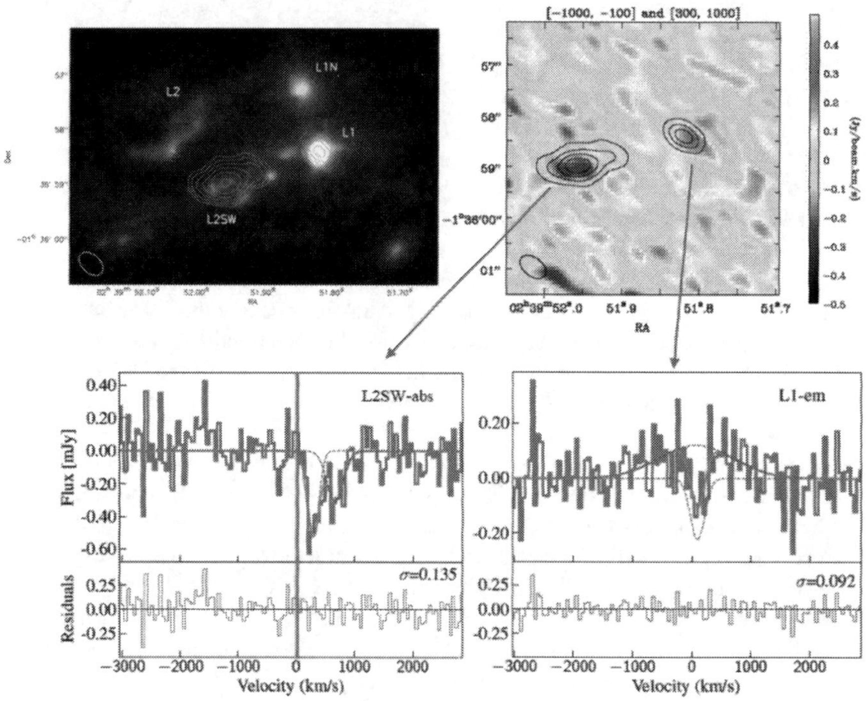


Figure 1: *Upper row: Left:* ALMA rest-frame 360-$\mu$m continuum image of SMM J02399−0136 over a false-colour composite from Frontier Fields *HST* frames. *Right:* Map of the $CH^+$ line integrated area (i.e. moment-0) over the velocity range [-1000,-100] and [300,1000] km s$^{-1}$. *Lower row:* Two $CH^+$(1-0) continuum-subtracted ALMA spectra, the residuals and the r.m.s noise level are also shown. The dashed lines show individual Gaussian fits to the emission and absorption components. The spectra are obtained over the L2SW continuum peak (*left*), and over a solid angle of ~ 1 arcsec$^2$ south-west of L1 (*right*).

We have observed and detected the $J$=1-0 transition of $CH^+$ in 18 starburst galaxies at z≃2-3. One of the targets is SMM J02399−0136, a galaxy group lensed by the Abell 370 cluster [7]. The group comprises a starburst galaxy, L2SW, and an AGN, L1, shown in the continuum observations of the top panels of Fig. 1 and two reflection nebulae unseen in the sub-milimeter. The average redshift of the two galaxies, $z_{ref} = 2.8041 \pm 0.0004$, inferred from the ALMA CO(7-6) image at a resolution (0.48"×0.46"), is used as a reference for the velocity scale [6]. The spectra towards L2SW and L1 show broad lines of widths of ~600 and ~300 $km\,s^{-1}$ respectively for $CH^+$. The absorption line against L2SW is redshifted by ~ 600 $km\,s^{-1}$, and therefore traces inflowing gas towards the galaxy. From the linewidth, we estimate the radius of the turbulent CGM to be ~20 kpc (see [6] for the details of the calculation), which corresponds to the integral scale of the CGM turbulence assuming that the width of the line is only due to turbulence. The line opacity provides the column density of $CH^+$ $N(CH^+) \sim 6 \times 10^{14}\,cm^{-2}$ and from these observables we derive the mass of the turbulent CGM to be $\sim 4 \times 10^{10}\,M_\odot$.

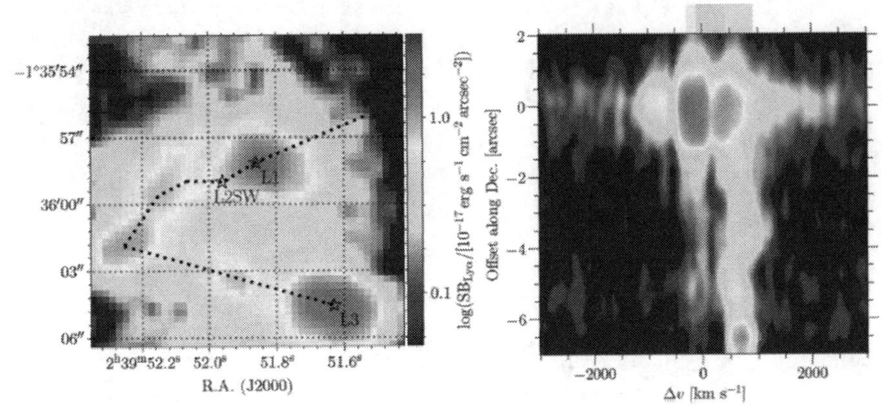


Figure 2: Position-velocity map of the Ly$\alpha$ nebula on the *right* taken along the track shown in the *left* panel (black dots). The blue and green boxes at the top show the full velocity coverage of the $CH^+$ absorption lines in front of L2SW and L1, respectively.

The Ly$\alpha$ observations obtained with Keck/KCWI of this galaxy group are presented in [8]. The brightest part is concentrated around the starburst and AGN but the faintest parts extend to regions up to over 80 kpc. Fig. 2 shows a position-velocity map of the Ly$\alpha$ emission across the nebula. The cut shows that the brightest parts are also those with the largest widths, with FWZI up to ~ 6000 $km\,s^{-1}$. In contrast, in the most extended parts, lines are much narrower (FWHM~400 $km\,s^{-1}$) and appear redshifted by ~ 600 $km\,s^{-1}$, similarly to the $CH^+$ absorption lines which are indicated by the shaded boxes.

## 3 What we learn from the comparison of Ly$\alpha$ and $CH^+$ observations

In the left panel of Fig. 3, the agreement of the size of the turbulent reservoir of diffuse molecular gas seen in $CH^+$(1-0) absorption against the L2SW and L1 and that of the Ly$\alpha$ nebula not only validates the assumptions made to calculate the CGM radius but also suggests that the CGM is at least biphasic, with a cool molecular phase of low density that we detect in $CH^+$(1-0) absorption and a warmer phase emitting in Ly$\alpha$. Note that the lensing shear direction makes the kpc/arcsec correspondence different along the RA and Dec axis.

The Ly$\alpha$ spectra in the direction of L1, L2, L2SW and L3 are shown in the right panel of Fig. 3. The line profiles are asymmetric and different at each positions. The blue and green

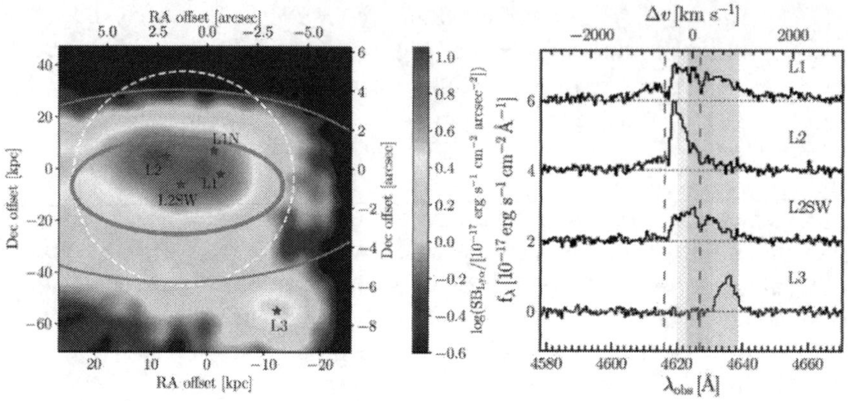


Figure 3: *Left:* Continuum-subtracted Ly$\alpha$ emission of the gas surrounding SMM J02399−0136. The two magenta ellipses show the size of the turbulent reservoir traced by $CH^+$(1-0) absorption, derived for two durations of the starburst phase, 50Myr (thick) and 100Myr (thick). *Right:* Ly$\alpha$ spectra observed in the direction of L1, L2, L2SW and L3. The blue and green shaded regions mark the velocities of the $CH^+$(1-0) absorption lines towards L2SW and L1.

shaded boxes show that the velocity range of the $CH^+$(1-0) absorption lines covers the red wing of the Ly$\alpha$ emission at these positions and the velocity range of the extended emission. The agreement of the velocity of the $CH^+$ absorbing gas and the extended Ly$\alpha$ emission means that both phases are dynamically coupled and inflowing towards the galaxies. This redshifted gas scatters photons back from the observer and contributes to the asymmetry of the Ly$\alpha$ line profiles. From these comparisons and a large set of multi-transition CO ancillary data [6], we conclude that the 20 kpc-scale CGM in SMM J02399−0136 is multi-phasic and inflowing towards the galaxies.

Several kpc-scale shocks are detected tentatively in $CH^+$(1-0) emission in the environment of the starburst galaxy and AGN [6]. Their specific location in space and velocity with respect to the high-velocity Ly$\alpha$ emission suggests that they lie at the interface of the inflowing CGM and the high-velocity Ly$\alpha$ emission, and signpost the feeding of CGM turbulence by AGN- and stellar-driven outflows. The mass and energy budgets of the CGM require net mass accretion at a rate commensurate with the star formation rate. From this similarity, we infer that the merger-driven burst of star formation in this galaxy group is ultimately fuelled by large-scale gas accretion.

## References

[1] Madau P. and Dickinson, M. ARA&A **52**, 415-486 (2014)
[2] Veilleux, S. et al. Astronomy and Astrophysics Review, **28**, 2-170 (2020)
[3] Falgarone, E. et al. Nature, **548**, 430-433 (2017)
[4] Godard, B. et al. A&A, **622**, A100 (2019)
[5] Lehmann, A. et al. A&A, **658**, A165 (2022)
[6] Vidal-García, A. et al. MNRAS, **506**, 2551-2573 (2021)
[7] Richard, J. et al. MNRAS, **444**, 268-289 (2014)
[8] Li, Q. et al. ApJ, **875**, 130 (2019)

# The Interstellar and Circumgalactic Media at low and high redshift as traced by Atomic Carbon and Carbon Monoxide

*Paola* Andreani[1,*], *Lazaros* Souvaitzis[2], *Padelis* Papadopoulos[3], *Thomas* Bisbas[3,4], *Carlos* De Breuck[1], *Bjorn* Emonts[5], *Zhi-Yu* Zhang[6], *Yusuke* Miyamoto[6], and *Allison* Mann[7]

[1]European Southern Observatory, Garching, Germany
[2]Max-Planck for Astrophysics, Garching, Germany
[3]Department of Physics, Aristotle University of Thessaloniki
[4]Physikalisches Institut, Universität zu Köln, Köln, Germany
[5]National Radio Astronomy Observatory, Charlotteville, VA, USA
[6]School of Astronomy and Space Science, Nanjing University, Nanjing, China
[7]Fukui University of Technology, Fukui, Japan
[8]Department of Physics & Astronomy, University of British Columbia, Canada

**Abstract.**
A different chemistry of the interstellar medium (ISM) is expected in the circumgalactic medium (CGM) gas where high-energetic particles (i.e. cosmic rays) seem to be produced in-situ by the hot X-ray gas, as it is observed in the Perseus cluster. This very different astrochemistry, where extreme gas-dust thermal decoupling is expected, and where CO can be destroyed over large mass-scales, is the subject of the investigation briefly reported here. We introduce an on-going project aiming at studying the properties of the CGM of two clusters at low and high redshift using their molecular gas tracers and thermal emission from dust.

## 1 Circumgalactic Medium in Galaxy clusters: Why?

The circumgalactic medium (CGM) is the site where galaxies exchange their gas with the surrounding environment via inflows from the CGM to the galaxy and/or outflows powered by AGN and star formation which may deposit chemically enriched material in the CGM. The CGM therefore represents a site where physical conditions might be different from the less extreme interstellar medium of galaxy discs. It is expected that in galaxy clusters (GCs) with intense SF and/or AGN activities the CGM (and intracluster medium, ICM) contain large amounts of lower-density, galaxy-expelled, molecular gas and dust, which would then be subjected to a high-energy particle-dominated chemistry [see, 4–7, 13, 21, 22].

Strong $H_2$ winds are expected from galaxy cluster members with SFR ~ 150-250 $M_{\odot} yr^{-1}$, injecting gas and dust into the CGM/ICM [7]. In these media, dust is very efficiently cooled by thermal emission while no longer heated by the far-UV photons (from the distant SF inside the BCG) nor by ambient CRs.

*e-mail: pandrean@eso.org

## 2 Gas tracers

In this scenario, the traditional way to estimate the amount of molecular gas, $M(H_2)$, and its physical properties may not be valid and need further investigations.
For decades molecular gas has been estimated inside galaxies from the Carbon Monoxide (CO) emission luminosity (from the $X_{CO}$ factor) and from the dust emission. But, how will the standard $H_2$-tracing methods be impacted in extreme environments? How does star-formation happen away from the central galaxy and how is the initial mass function (IMF) affected?

In fact, high-energetic particles, like cosmic rays, which are copious in sites of intense star formation, have the power to destroy the CO molecules, enhancing neutral and ionised Atomic Carbon (CI, $C^+$). The high-energetic particles may become the fundamental regulators and power sources in these environments. The thermal state of UV-shielded dense gas is altered, dust might be too cold and undetected because less hit by FUV photons which are too far away. This might have an effect on the Jeans mass, the gas fragmentation properties and the characteristic mass of young stars, affecting the condition of star formation and very likely the stellar initial mass function (IMF) [12, 16, 19].

Indeed in the past decade chemical-physical models associated to simulations show how CO loses its molecular hydrogen tracing capability in low metallicity environments and in violent media (due to starbursts/AGN activities) and CI and $C^+$ remain the dominant form of carbon [2, 8, 9, 17]. Therefore a thorough investigation of the interstellar/circumgalactic and intergalactic media must span a wide range of available tracers, to probe density, temperature, ionisation state, etc.

With the advent of improved receivers (at IRAM, APEX) and larger facilities (ALMA), the access to broader frequency ranges and higher sensitivity allows to observe the two CI forbidden lines (492, 809 GHz) and the ionised Carbon, $C^+$ (1.898 THz). These lines have been proven to be a fair tracer of the total molecular gas under a large range of physical conditions. Furthermore, the two CI system is much simpler than the CO, it allows to derive more easily the gas physical parameters (excitation temperature, mass, density,...)[1, 15], tracing the CO-invisible $H_2$ gas where energetic particles destroy CO (leaving behind Atomic Carbon) and because its high-frequency emission is less affected by the CMB[23].

## 3 Observational data

In light of the above, in order to investigate the star formation conditions in extreme environments we have undertaken a study of the gas tracers towards two proto-type galaxy clusters, hosting a powerful AGN in their centre and showing a very turbulent CGM environment. These are excellent candidates to study $H_2$ gas and dust during an episode of elevated SF and AGN feedback.

### 3.1 MACS1931.8-26

MACS1931.8-26 is a cluster at z=0.35 whose core contains a bright central galaxy, a cool core and an AGN. A reservoir of $T_d$=10 K dust is detected in its CGM regions at ~30 kpc from its centre, concomitant with warm gas with high CO(3-2)/(1-0) ($R_{31}$=0.93) ratios with a total cold gas mass (as marked by the CO(1-0) emission) of $\sim 2 \times 10^9$ $M_\odot$ [7].

This result indicates a large ratio between the kinetic and the dust temperature, $T_{kin}/T_d$ and thus a power source other than SF-generated FUV radiation. In this cluster, as in the Phoenix cluster [21], there is either a superdusty CGM/ICM with gas-to-dust ratio of ~10-25 (and discrepancies in the relative distributions of their $H_2$ gas and dust), or we are missing

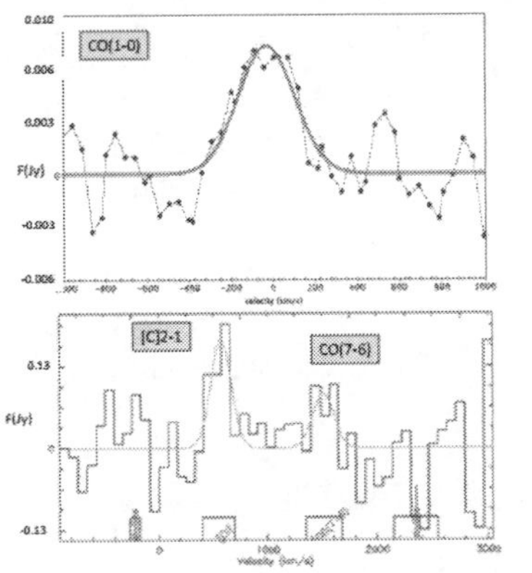


**Figure 1.** Preliminary analysis of the CO(1-0) emission observed at the NRO45m dish and of the APEX/SEPIA660 data of the [CI]2-1 and CO(7-6) lines (Project ID 0106.A-0997(A)) (Andreani et al, in preparation).

$H_2$ gas that would restore the gas-to-dust ratio back to ~100-150. Missing $H_2$ gas could be due to the interferometric observations filtering out the extended diffuse emission and/or to CO-dark gas.

To investigate this issue further, we recently obtained single dish observations with APEX/SEPIA660 of the [CI]2-1 and CO(7-6) spectra and with NRO 45m of the CO(1-0) emission. Preliminary spectra are shown in Figure 1. A quick look at the CO spectrum shows that the single dish CO(1-0) line luminosity is larger than derived from the ALMA 12m data [7], likely indicating that extended emission is missed in the interferometric observations. In addition the APEX data towards [CI]2-1 and CO(7-6) (see, Figure 1) show an unexpectedly bright [CI] compared to the CO(7-6), indicative of cosmic ray heated gas. Additional observations to complement the available data have been requested. The availability of several CO lines and the two Atomic Carbon lines will allow to make a full analysis of the thermal state of the gas, breaking the degeneracy inherent to the CO analysis only.

### 3.2 MRC1138-262

MRC1138-262 is a radio galaxy with a powerful radio-loud AGN hosted by one of the most massive galaxies in formation in the heart of the Spiderweb protocluster. It is one of the best studied high-z objects [11, 14, 20] with a stellar mass of $M_\star \sim 2 \times 10^{12} M_\odot$. It has a complex morphology of merging satellite galaxies surrounded by a giant (~250 kpc) Ly-$\alpha$ halo, tracing a rich proto-intercluster medium in the cluster core. Past ALMA observations of [CI]1-0, and CO(4-3) and APEX observations of $C^+$ show emission across ~50 kpc, while low-surface-brightness CO(1-0) emission in the CGM detected with ATCA reveals the presence of a massive ($\sim 10^{11}$ $M_\odot$) extended $H_2$ gas reservoir [3, 4] which fuels in-situ star formation in the CGM/ICM. All transitions span a velocity distribution far wider than that corresponding to the velocities of the satellite galaxies that lie within the overall molecular distribution.

The detection of extended emission outside the central galaxy was confirmed and enhanced by combining data from the ALMA Compact Array (ACA) with the 12m observations (see Fig 2). Despite the low signal-to-noise ratio the combined data show indeed that the emission is extending far away (~50kpc) from the central galaxy. Further investigation to derive the properties of this extended emission in a quantitative way are underway. The combination of the CO spectral line distribution, the ratio of the Carbon lines and the information on the dust temperature offers a unique set of data to break the degeneracy to derive the gas physical quantities [see for instance 10, 18].

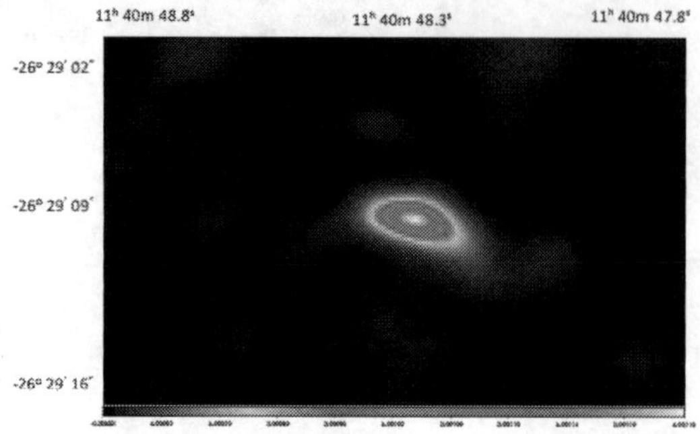


**Figure 2.** Preliminary analysis of the combination of the ALMA ACA and 12m arrays of the CO(4-3) lines (Souvaitzis et al, in preparation).

## 4 Conclusions

We presented a preliminary analysis of the molecular gas towards two clusters of galaxies, which might be considered a good template to investigate physical processes in extreme environments. We argue that the combined use of various gas tracers, together with the combination of observations from diverse instruments accessing different spatial scales, shows its power to infer the physical status of the gas in star forming regions far away from the ISM of galaxy discs. The chemical-physical status of this gas may largely affect the star formation process and the resulting stellar initial mass function.

## References

[1] Bisbas T. et al, ApJ **803**, 37 (2015)
[2] Clark P. et al, MNRAS **486**, 4622 (2019)
[3] De Breuck C. et al, A&A **658**, L2 (2022)
[4] Emonts B. et al, MNRAS **477**, L60 (2018)
[5] Ferland et al, MNRAS **386**, L72 (2008)
[6] Ferland et al, MNRAS **392**, 1475 (2009)
[7] Fogarty K. et al, ApJ **879**, 103 (2019)
[8] Glover S. and Clark P., MNRAS **456**, 3596 (2016)
[9] Gaches B. et al, A&A **658**, A151 (2022)
[10] Harrington K. et al, ApJ **908**, 95 (2021)
[11] Hatch N. et al, MNRAS **395**, 114 (2009)
[12] Hocuk S. & Spaans M., A&A **536**, A41 (2011)
[13] Lim et al, ApJ **850**, 31 (2017)
[14] Miley G. et al, ApJL, **650**, 29 (2006)
[15] Papadopoulos P.P., et al, MNRAS **351**, 147 (2004)
[16] Papadopoulos P.P., et al, MNRAS **414**, 1705 (2011)
[17] Papadopoulos P.P., et al, MNRAS **478**, 1716 (2018)
[18] Papadopoulos P.P., et al, MNRAS **510**, 725 (2022)
[19] Papadopoulos P.P. & Thi W.F., arXiv:1207.2048v3 (2012) 2013ASSP...34...41P
[20] Pentericci L. et al, A&A **326**, 580 (1997)
[21] Russel H. et al, MNRAS **472**, 4024 (2017)
[22] Salome P. et al, A&A **484**, 317 (2008)
[23] Zhang Z-Y et al, R. Soc. Open Sci. **3**, 160025 (2016)

# Molecular gas dynamics around nuclei of galaxies

*Françoise* Combes[1,*]

[1]Observatoire de Paris, LERMA, Collège de France, CNRS, PSL University, Sorbonne University, Paris, France

**Abstract.** Recent molecular line observations with ALMA in several nearby Seyferts have revealed the existence of molecular tori, and the nature of gas flows at 10-20 pc scale. At 100 pc scale, or kpc-scale, previous NOEMA work on gravitational torques had shown that only about one third of Seyfert galaxies experienced molecular inflow and central fueling, while in most cases the gas was stalled in rings. At higher resolution, i.e. 10-20 pc scale, it is possible now to see in some cases AGN fueling due to nuclear trailing spirals, influenced by the black hole potential. This brings smoking gun evidence for nuclear fueling. In our sample galaxies, the angular resolution of up to 60 mas allows us to reach the black hole (BH) sphere of influence and the BH mass can be derived more directly than with the M-sigma relation.

## 1 Introduction

Our vision of the close environment of the black hole and its accretion disk has evolved significantly in the recent years. While the geometrical orientation paradigm helped to understand the difference between type 1 and 2 AGN, through the presence of a dusty torus at parsec scale (e.g. Urry & Padovani 1995), it is possible that the required obscuration is only due to outflowing dust in a hollow polar cone (Asmus et al. 2016). Around the accretion disk, and the ionised Broad Line Region (BLR), dust cannot survive due to high temperatures, but at the sublimation radius, radiation pressure pushes the dust in the polar cone. Gas and dust are inflowing through a molecular disk, where sometimes water masers are seen, and also $H_2$ and CO emission lines. The disk is likely to be flaring, i.e. its thickness is increasing strongly with radius (Hönig, 2019).

In the following, I will describe:

1- how the AGN feeding occurs, how the gas angular momentum is transferred through gravity torques from dynamical features, nuclear bars and spirals, and how the gas accumulates in a molecular torus

2- how the AGN feedback develops, aided by supernovae feedback. The two main modes, radiative or kinetic mode, through winds and radio jets, may occur simultaneously, and result in significant molecular outflows at intermediate scales.

## 2 Fueling

One of the first galaxy mapped with ALMA at high angular resolution is NGC 1566 (Combes et al. 2014). This is a barred spiral galaxy, with an inner Lindblad resonance ring inside the

*e-mail: francoise.combes@obspm.fr

bar. The field of view (FOV) of ALMA at the CO(3-2) line at 345.796 GHz is 18", ending just at the border of the ring, which has then a low signal-to-noise ratio.. Inside the resonant ring, there exists a nuclear disk, revealed by ALMA. This disk shows a nuclear spiral, which is trailing, i.e. the same winding sense as the spiral structure outside of the bar, as seen in figure 1.

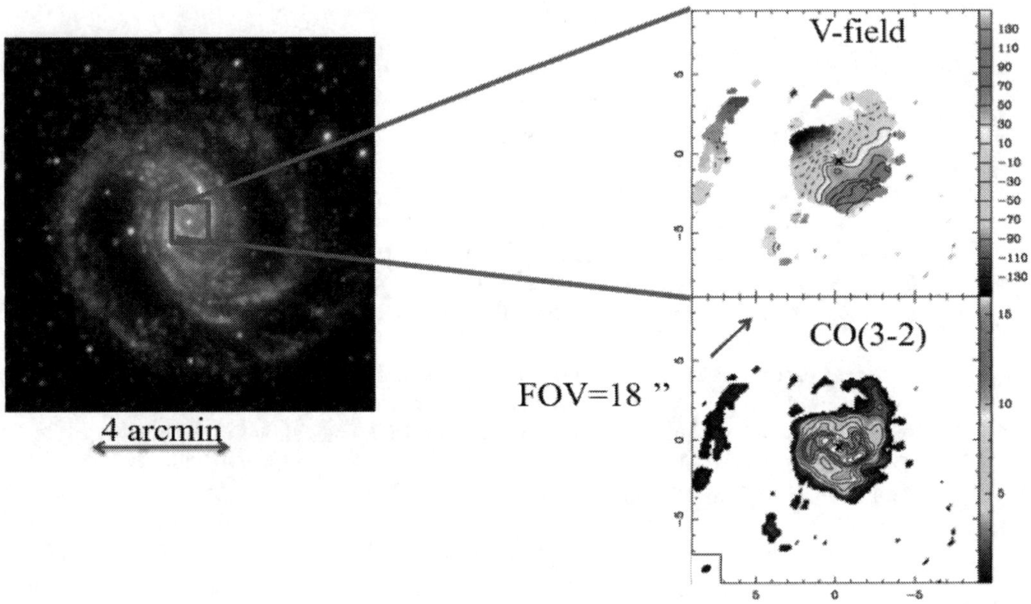


**Figure 1.** NGC 1566 observed in CO(3-2) with ALMA. The left image shows the large-scale barred spiral structure, and the inner Lindblad resonance (ILR) ring inside the bar, corresponding to the FOV of 18". The two first moments at right reveals a trailing nuclear spiral, inside a nuclear disk, with relatively regular velocity field (adapted from Combes et al. 2014).

### 2.1 Gravity torques

The main issue in the fueling process is to get rid of the high angular momentum of the gas. While viscous torques are inefficient in galactic disks down to sub-pc scales, gravity torques from tidal forces in the outer parts, and bars in the inner parts, are able to exchange angular momentum in between resonances (e.g. Buta & Combes 1996). Bars are ending just before their corotation (CR) radius, where the bar pattern speed equals the angular speed of the tars and gas. Then the torques exerted on spiral arms just outside the bar are positive, and drive the gas to the OLR (Outer Lindblad Resonance). Inside the CR radius, torques are negative down to the ILR, but then change sign inside ILR (Inner Lindblad resonance), and may give rise to a leading spiral. This ensures that gas is accumulating at the ILR ring.

However, the precessing rate $\Omega-\kappa/2$ of the elliptical orbits, which is usually an increasing function of radius inside ILR, may change behaviour close to the nucleus, due to the gravitational influence of the black hole (BH). The rotational velocity $\Omega$, the epîcyclic frequency $\kappa$, and the precessing rate $\Omega - \kappa/2$ are then dominated by the keplerian potential, and decrease with radius in the central 10-20 pc. This reverses the winding sense of the gas spiral, and trailing nuclear spirals are expected in the sphere of influence (SoI) of the BH cf Figure 2.

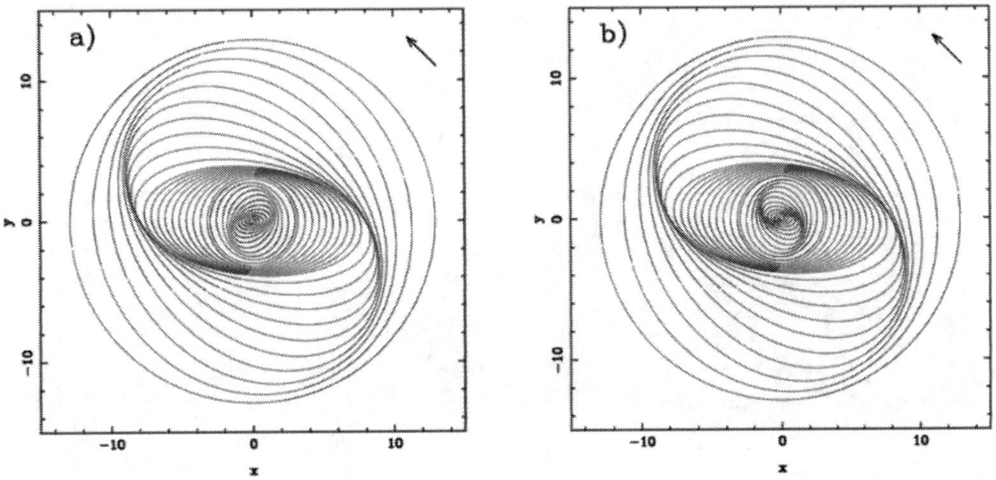


**Figure 2. Left:** Schematic representation of the gas streamlines in a barred galaxy, where ILRs exist. The bar is horizontal. When there is no central mass concentration, and the precessing rate $\Omega - \kappa/2$ decreases towards the center, the gas is crowding in a leading nuclear spiral, and the torque exerted is positive. **Right:** When there is a supermassive black hole dominating inside the ILR, the precessing rate $\Omega - \kappa/2$ increases again towards the center, and the dissipating gas follows now a trailing spiral. The torque is negative, and the gas may fuel the nucleus. The arrow indicates the sense of rotation. The colors indicate the sign of the torques, blue: negative and red: positive.

This also reverses the sign of torques, which are now negative, and drive the gas to the center, fueling the BH.

### 2.2 Trailing nuclear spirals, molecular tori

The trailing nuclear spiral detected in NGC 1566 explains the feeding of the AGN. These nuclear spirals are frequently observed, when the angular resolution is sufficient. It is also detected in NGC 613 with 0.09"×0.06" resolution (5 pc, Combes et al. 2019, Audibert et al. 2019) but not with 0.5" resolution (Miyamoto et al. 2017). It is detected in NGC 1808 (Audibert et al. 2021), and in dense gas tracers HCO+, HCN and CS, as seen in figure 3.

Inside these nuclear spirals, a molecular disk is detected, with apparent decoupled morphology and kinematics. We call these nuclear disks molecular tori, they are detected in most of our observed sample of Seyferts (Combes et al. 2019). They might be related to the dusty tori of the classical AGN unification scheme. Their orientation appears randomly distributed with respect to that of the large-scale galaxy disk. The sizes of molecular tori are typically 10-20 pc, and are contained in the SoI. Their molecular mass is of the order of a few $10^7$ $M_\odot$.

When the resolution is enough inside the BH sphere of influence, it is possible to determine the BH mass, with a proper modelisation of the tilted molecular torus (e.g. Poitevineau et al. 2022). Figure 4 shows examples of galaxies from the GATOS survey (Garcia-Burillo et al. 2021), where this is possible, albeit the tilted orientation of the nuclear disk.

## 3 Feedback

Molecular outflows are frequently observed in nearby Seyferts, and can be quite small (<1kpc), with low velocity ~ 100 km/s (NGC 1433, Combes et al. 2013). They are often

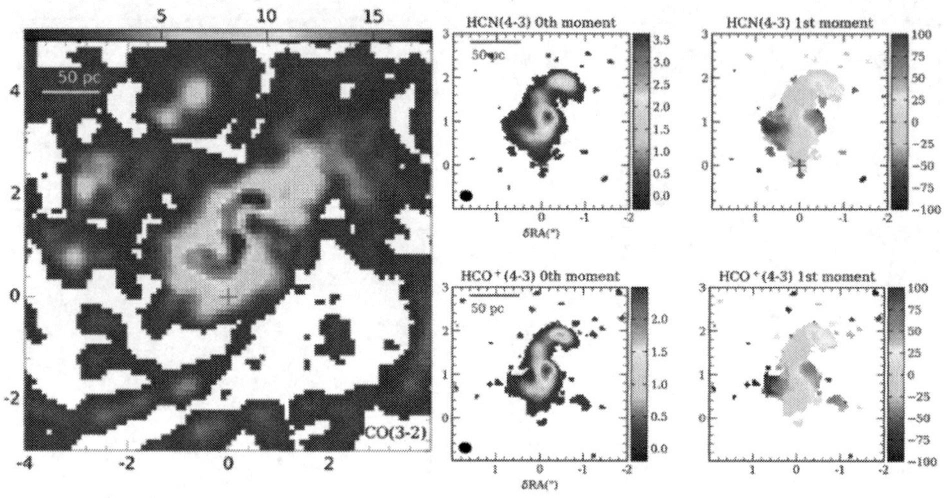


**Figure 3.** Nuclear spiral detected in NGC 1808 in CO(3-2) (left) and HCN(4-3), HCO+(4-3) (right), with a beam of 0.08" = 4 pc. The velocity field inside the nuclear spiral has a different major axis orientation from the large-scale one (adapted from Audibert et al. 2021).

associated to small radio jets (NGC 613, Audibert el al. 2019), although sometimes very collimated molecular outflows can occur without a detectable radio jet (NGC 1377, Aalto et al. 2016, 2019).

Several mechanisms have been invoked to produce these molecular outflows, some are purely due to star formation feedback, but then their power is not related to the AGN luminosity (Cicone et al. 2014). If the AGN is more luminous than 0.01 Eddington luminosity, radiation pressure on the ionised gas triggers a fast outflow in the nucleus, which then entrains the molecular gas. In the case of low-luminosity AGN (lower than 0.01 Eddington), only radio jets can entrain the molecular gas (kinetic mode). At the transition luminosities, both can be present, in particular radio jets and winds driven by radiation pressure on dust (Sadowski et al. 2013).

With high angular resolution, it is possible to disentangle AGN from supernovae feedback: for instance, the galactic wind in NGC 1808, must be a starburst driven wind, since there is no molecular outflow towards the nucleus (Audibert et al. 2021).

Radio jets appear quite coupled to galaxy disk, to produce molecular outflows, since they are not in general perpendicular to the disk. Frequently they can sweep out gas from the inner disk, as in NGC 1068 where the outflow rate reaches 63 $M_\odot$/yr (Garcia-Burillo et al. 2014, 2016). This prototypical Seyfert-2 reveals an edge-on molecular torus, perpendicular to the radio jet, and a molecular outflow, plus a hollow polar cone seen beautifully in polarised near-infrared light (Gratadour et al. 2015). The column density in the molecular torus and extended dust cone is sufficient to account for the BLR obscuration.

The most collimated molecular outflow has been found in the early-type galaxy NGC 1377 (Aalto et al. 2016, 2019). The molecular jet is almost in the plane of the sky, precessing by about 10°, and therefore it alternates in radius blue and red-shifts. While a radio source is detected in the center at cm wavelengths, there is no radio jet within the detection limit, which might yield insight in the jet life-time. The relativistic jet could disappear in less than 3000 yr, while the molecular jet will remain for more than 3 Myr.

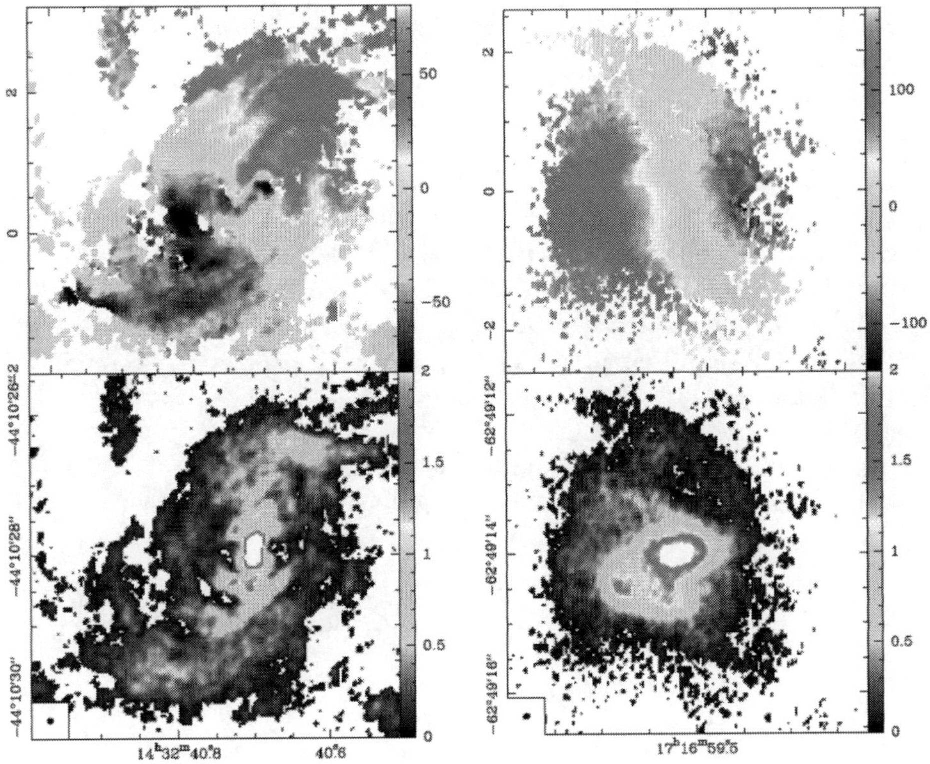

**Figure 4.** Two galaxies, NGC 5643 (left) and NGC 6300 (right) observed in CO(3-2) with ALMA, from the GATOS sample (Garcia-Burillo et al. 2021). The two first moments show the more or less tilted kinematical major axis in the nucleus (adapted from Poitevineau et al. 2022).

## 4 Summary

High angular resolution has brought a lot of informations about the AGN fueling. Non-axisymmetric features (bars, spirals, interactions) produce gravity torques, allowing exchange of angular momentum. The gas may be quicky driven towards the center, through a trailing nuclear spiral, and accumulate in a molecular torus, with decoupled morphology and kinematics.

High resolution can also help disentangle the origin of the feedback, AGN or supernovae. Molecular outflows are frequent and may be driven by radio jets or winds (or both). In case of AGN feedback, the molecular outflows are misaligned with the rest of the large-scale disk. The outflows are more likely energy-driven with AGN, while momentum-driven with starbursts. In the latter case, the feedback is weaker, with the ejected gas rapidly infalling back after a fountain episode, only postponing star formation without quenching it.

## References

Aalto, S., Costagliola, F., Muller, S. et al., 2016, A&A 590, A73
Aalto, S., Muller, S., Koenig, S. et al., 2019, A&A 627, A147
Asmus, D., Hoenig, S. F., Gandhi, P.: 2016, ApJ 822, 109
Audibert, A., Combes, F., Garcia-Burillo, S. et al.: 2019, A&A 632, A33

Audibert, A., Combes, F., Garcia-Burillo, S. et al.: 2021, A&A 656, A60
Buta, R., Combes, F.: 1996, FCPh. 17, 95
Hönig, S.F., 2019, ApJ 884, 171
Cicone, C., Maiolino, R., Sturm, E. et al.: 2014, A&A 562, A21
Combes, F., Garcia-Burillo, S., Casasola, V. et al.: 2013, A&A 558, A124
Combes, F., Garcia-Burillo, S., Casasola, V. et al.: 2014, A&A 565, A97
Combes, F., Garcia-Burillo, S., Audibert, A. et al.: 2019, A&A 623, A79
Garcia-Burillo, S, Combes, F., Usero, A. et al.: 2014, A&A 567, A125
Garcia-Burillo, S, Combes, F., Ramos Almeida, C. et al.: 2016, ApJ 823, L12
Garcia-Burillo, S, Alonso-Herrero, A., Ramos Almeida, C. et al.: 2021, A&A 652, A98
Gratadour, D., Rouan, D., Grosset, L. et al.: 2015, A&A 581, L8
Miyamoto, Y., Nakai, N., Seta, M. et al.: 2017, PASJ 69, 83
Poitevineau, R., Combes, F., Garcia-Burillo, S. et al.: 2022, in prep
Sądowski, A., Narayan, R., Penna, R., Zhu, Y. : 2013, MNRAS 436, 3856
Urry, C.M., Padovani, P.: 1995, PASP 107, 803

# Revealing which Combinations of Molecular Lines are Sensitive to the Gas Physical Parameters of Molecular Clouds

## Astrophysics Meet Data Science towards the Orion B Cloud

*Jérôme* Pety[1,2,*], *Maryvonne* Gerin[2], *Emeric* Bron[2], *Pierre* Gratier[3], *Jan H.* Orkisz[4], *Pierre* Palud[5,2], *Antoine* Roueff[6], *Lucas* Einig[1,7], *Miriam G.* Santa-Maria[7], *Victor* de Souza Magalhaes[1], *Sébastien* Bardeau[1], *Jocelyn* Chanussot[7], *Pierre* Chainais[5], *Javier R.* Goicoechea[7], *Viviana V.* Guzman[7], *Annie* Hughes[7], *Jouni* Kainulainen[4], *David* Languignon[2], *François* Levrier[7], *Darek* Lis[7], *Harvey S.* Liszt[7], *Jacques* Le Bourlot[2], *Franck* Le Petit[2], *Karin* Oberg[7], *Nicolas* Peretto[7], *Evelyne* Roueff[2], *Albrecht* Sievers[1], *Pierre-Antoine* Thouvenin[5], and *Pascal* Tremblin[7]

[1]IRAM, 300 rue de la Piscine, 38406 Saint Martin d'Hères, France
[2]LERMA, Observatoire de Paris, PSL Research University, CNRS, Sorbonne Universités, France
[3]Laboratoire d'Astrophysique de Bordeaux, Univ. Bordeaux, CNRS, France
[4]Chalmers University of Technology, Department of Space, Earth and Environment, Sweden
[5]Univ. Lille, CNRS, Centrale Lille, UMR 9189 - CRIStAL, 59651 Villeneuve d'Ascq, France
[6]Aix Marseille Univ, CNRS, Centrale Marseille, Institut Fresnel,Marseille, France
[7]See https://www.iram.fr/~pety/ORION-B/team.html for the other affiliations

**Abstract.** Atoms and molecules have long been thought to be versatile tracers of the cold neutral gas in the universe, from high-redshift galaxies to star forming regions and proto-planetary disks, because their internal degrees of freedom bear the signature of the physical conditions where these species reside. However, the promise that molecular emission has a strong diagnostic power of the underlying physical and chemical state is still hampered by the difficulty to combine sophisticated chemical codes with gas dynamics. It is therefore important 1) to acquire self-consistent data sets that can be used as templates for this theoretical work, and 2) to reveal the diagnostic capabilities of molecular lines accurately. The advent of sensitive wideband spectrometers in the (sub)-millimeter domain (e.g., IRAM-30m/EMIR, NOEMA, ...) during the 2010s has allowed us to image a significant fraction of a Giant Molecular Cloud with enough sensitivity to detect tens of molecular lines in the 70 – 116 GHz frequency range. Machine learning techniques applied to these data start to deliver the next generation of molecular line diagnostics of mass, density, temperature, and radiation field.

## 1 The ORION-B program

The ORION-B project (Outstanding Radio-Imaging of OrioN-B, PI: Jérôme Pety & Maryvonne Gerin) is a Large Program of the IRAM 30 meter telescope that has imaged 13 × 18 pc of the Orion B massive star-forming cloud, located at 400 pc, over 40 GHz of

*e-mail: pety@iram.fr

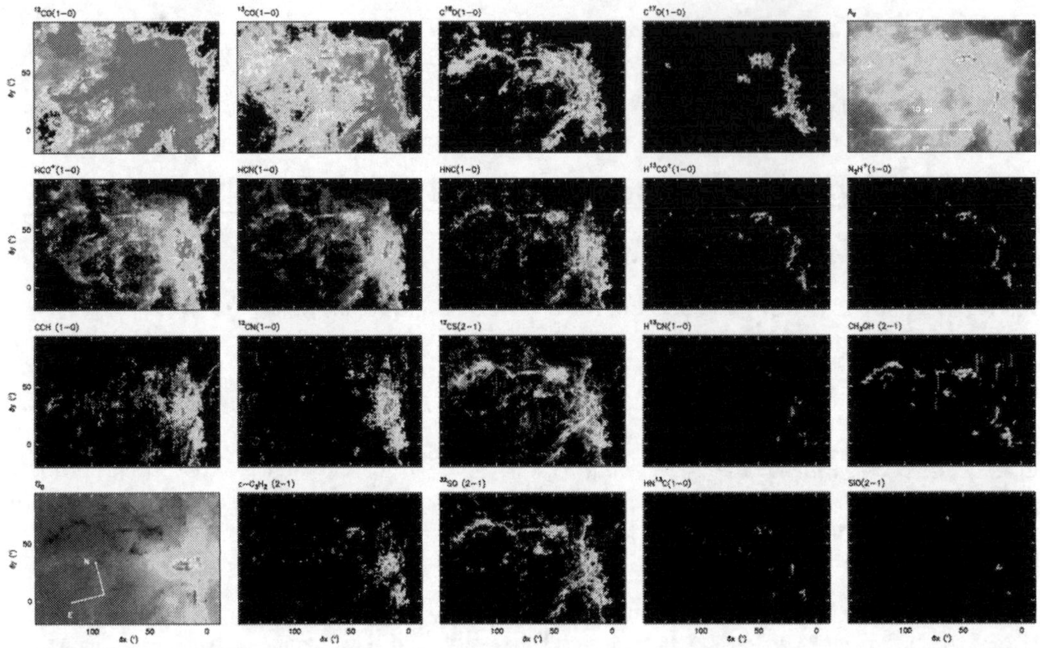

**Figure 1.** Spatial distribution of the line peak intensity of selected lines in the 3 mm band, plus the dust temperature **(bottom left panel)** and visual extinction **(top right panel)**. Continuum data comes from the publicly available analysis of the Herschel Gould Belt Survey data (PI: P. André) [1]. Only reliably detected pixels are shown. The color-scales are logarithmic to reveal the distribution of faint signal.

bandwidth at a typical resolution of 50 mpc and 0.6 km $s^{-1}$. An integration time of 850 hours yields about 200 000 images of 800 000 pixels at a typical sensitivity of 0.1 K.

Figure 1 shows the spatial distribution of the line peak brightness for a subset of 18 different molecular transitions. One of the ORION-B main goals is to study how it is possible to constrain the physical conditions (e.g., temperature, column and volume density, thermal pressure, ...) based on such observations. In a pilot study towards the first observed square degree [2], we showed how tracers of different optical depths like the CO isotopologues allow one to fully trace the molecular medium, from the diffuse envelope to the dense cores, while various chemical tracers can be used to reveal different environments. A large fraction of the flux of previously thought "dense gas tracers", as the HCN (1 – 0) line, is coming from translucent gas. Unambiguous detection of the gas whose volume density is larger than $10^4$ $cm^{-3}$ requires to detect the $N_2H^+(1-0)$ line.

A clustering algorithm was then applied to the intensities of selected molecular lines, and revealed spatially continuous regions with similar molecular emission properties, corresponding to different regimes of volume density or far-UV illumination [4]. We showed that the (1 – 0) lines of the three main isotopologues of CO are sufficient to cluster the molecular gas in three distinct density regimes ($n_H \sim 100$, 500, and $> 1000\,cm^{-3}$). Adding the $HCO^+$ and CN (1 – 0) lines reveals clusters 1) of higher volume density ($n_H \sim 7 \times 10^3$ and $4 \times 10^4\,cm^{-3}$) because the effective critical densities of these lines are between one and two order of magnitude higher than those of the CO isotopologues, and 2) clusters that match the GMC interfaces with the NGC 2023, NGC 2024, and IC 342 H II regions.

A principal component analysis of the twelve brightest lines in the 3 mm wavelength range showed that the first component is well correlated with the column density [5]. The second and third components are correlated at various levels with the volume density of dense cores and the dust temperature, respectively. These results suggest that the wide field

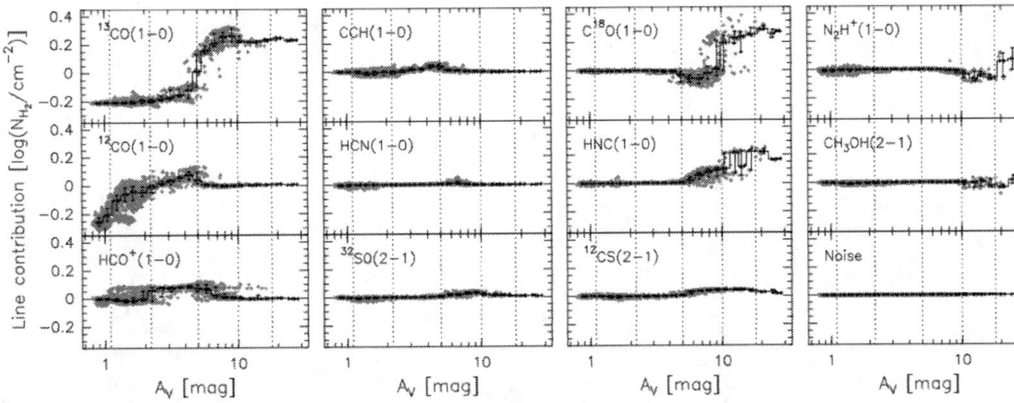


**Figure 2.** Contribution of the studied lines to the logarithm of the $H_2$ column density around the mean column density as a function of the visual extinction [3]. The black points show the median values of all data points falling in a regularly sampled interval of the logarithm of the visual extinction. The black error bars show the range of values where 50% of the points in the current bin are located. The vertical dotted lines show visual extinctions of 1.1, 2.2., 5.0, 10.0, and 18.0. The lines are sorted from top to bottom and then from left to right by increasing values of the minimum $A_v$ at which they start to contribute.

emission of molecular lines is sensitive to the column and volume gas densities on one hand and the gas temperature on the other hand. In a subsequent study [3], we quantitatively characterized the correlations between the line integrated intensities and the dust-derived column density using a machine learning regression tool called random forest. We first showed that this correlation can be used to quantitatively infer the molecular column density from the molecular line emissions, even on data that were not used to train the regression. We also showed that the most important contributors to this new kind of column density estimators are in decreasing order the $^{13}CO$, $^{12}CO$, $C^{18}O$, and $HCO^+$ $(1-0)$ lines for the bulk of the molecular cloud, while other lines as $N_2H^+$ $(1-0)$ prove to be important in the densest environments (dense cores). More quantitatively, the logarithm of the gas column density for each line of sight can be to first order analyzed as the sum

$$\log N = \log N_0 + \Sigma_l \log N_l, \tag{1}$$

where $\log N_0$ is the mean column density of the training data set and $\log N_l$ is the contribution of each considered line to the final estimation. Figure 2 shows the evolution of the contribution of each line as a function of the visual extinction. Except for the $^{13}CO$ $(1-0)$ line that contributes at all $A_v$ and the noise sample that does not contribute at any $A_v$, the other lines contribute inside a given $A_v$ range. Once generalized this new tool should allow us to accurately estimate the mass of the different (potentially velocity separated) components of a giant molecular cloud, in particular the mass of its filamentary structure [6].

These results hints that the relationships between line intensities and environment in Giant Molecular Clouds are more complicated than often assumed. Sensitivity, excitation, and, chemistry certainly contribute to the observed line intensity distributions. In order to develop the next generation of molecular line diagnostics of mass, density, temperature, and radiation field, we need to understand these different contributions quantitatively. We started this effort by studying the abundances and excitation temperature of the three main CO isotopologues within the framework of the local thermodynamic equilibrium (LTE) for the excitation and radiative transfer [7]. Using the $(1-0)$ and $(2-1)$ transitions of the $^{13}CO$ and $C^{18}O$ species and the $(1-0)$ line of $^{12}CO$, estimates and associated confidence intervals of the column density, excitation temperature and velocity dispersion of these species could be obtained

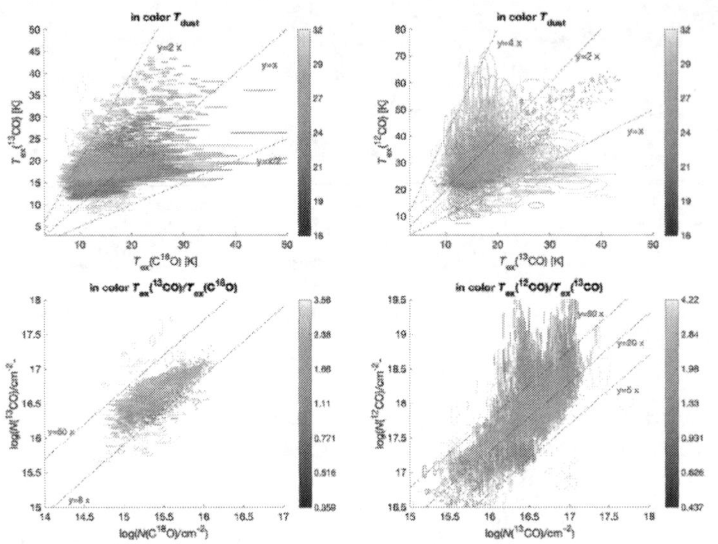


**Figure 3.** Scatter plots between the column densities **(top)** and excitation temperatures **(bottom)** of the main CO isotopologues [7]. The ellipses represent the interval of confidence for each estimation. The color scale encodes the dust temperature $T_{\text{dust}}$ **(top)** and the ratio of the excitation temperatures of the considered species **(bottom)**. Dashed red lines show the loci of ratios: 1/2, 1, 2, and 4 **(top)**, 8 and 50 **(bottom left)**, and 5, 20, and 60 **(bottom right)**.

towards the Horsehead nebula. While the excitation temperature of the $^{13}$CO and C$^{18}$O lines are on average similar, the excitation temperature of $^{12}$CO is typically twice as large as the $^{13}$CO and C$^{18}$O ones (top row of Fig. 3). This suggests that the CO isotopologue lines can be mostly excited in more or less dense or warm regions along the same line of sight. Moreover, the ratio of inferred column densities of the CO isotopologues are significantly different from the values given by the ratio of the carbon and oxygen elemental abundances in the solar neighborhood (bottom row of Fig. 3). Assuming they follow the elemental abundances, the [$^{12}$CO]/[$^{13}$CO] and [$^{13}$CO]/[C$^{18}$O] ratios should be in the ranges 57 – 67 and 7.5 – 9.8, respectively, while we found $5 \leq {}^{12}\text{CO}/{}^{13}\text{CO} \leq 70$ and $8 \leq {}^{13}\text{CO}/\text{C}^{18}\text{O} \leq 50$ with a clear correlation between these values and the dust-traced far UV illumination. These results can be qualitatively understood as the result of the fractionation reaction of C$^+$ that enriches diffuse and translucent gas in $^{13}$CO as compared to $^{12}$CO, and of selective photodissociation that destroys C$^{18}$O much faster than $^{13}$CO in the diffuse envelopes of dense cores and filaments.

## 2 Challenges for the future

In order to go one step further to infer the physical parameters without human supervision, we will need to invert not only radiative transfer models, but also physics and chemistry models of photo-dissociation regions (PDRs) like the Meudon PDR code. Several challenges need to be handled to succeed.

First, datasets like the ORION-B one imply large brightness dynamical ranges between the brightest and faintest detected lines (a factor $\geq 2500$ for the lines integrated over the field of view). This implies that an accurate noise model must be used. In particular, it is necessary to include a multiplicative component in addition to the usual additive noise because the calibration uncertainty (of the order of 5 – 10%) has an important impact at high signal-to-noise ratios. Moreover, the notion of censored information needs to be introduced in the fit to take into account potential upper limits due to the limited instrument sensitivity. Second, a sophisticated PDR code must include many physical and chemical processes, which may have subtle but important impacts on the line strengths. The micro-physical parameters as, eg, the chemical rates, are still often uncertain because physicists and chemists can only study a small fraction of the thousands of reactions that occur in molecular clouds. Finally, there

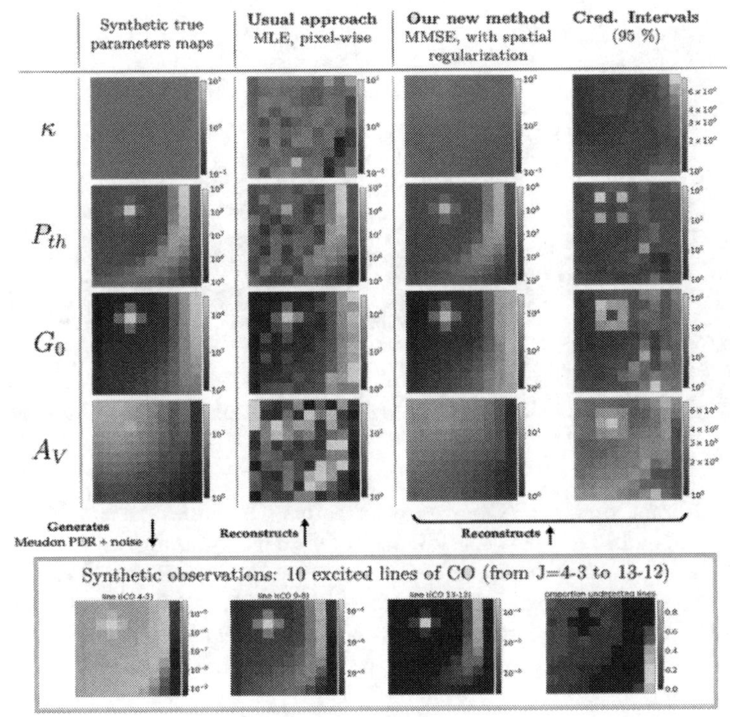


**Figure 4.** Comparison of the fit of several physical parameters (from top to bottom multiplicative geometry factor, thermal pressure, far UV illumination, visual extinction) between a standard maximum likelihood estimator (MLE), and a new Bayesian approach (MMSE) [8]. The first column shows the input parameters used to produce the synthetic observations of ten different $J$-lines of CO (bottom row). The other columns show the MLE fit, MMSE fit and its associated credibility interval.

are several hidden variables in the fit process. The most obvious one is the geometry of the cloud that is probably fractal with sharp changes of properties at several spatial scales as a molecular cloud is composed of a diffuse gas around a filamentary structure that surrounds cold cores where star formation happens. Another hidden information is the mass of the bulk of the molecular clouds as their main components (molecular hydrogen and helium) are mostly invisible to us. We thus need to rely on different minor tracers of the mass as the rotational line emission of species with a permanent dipole moment or the continuum emission of the dust.

To overcome these difficulties, we introduced a new, fast Bayesian framework that takes into account 1) a continuous approximation of large grids of PDR models, 2) additive and multiplicative noise, 3) censored information, and 4) a spatial regularization [8]. Figure 4 compares the performances of a standard maximum likelihood estimator (MLE) to the new proposed Bayesian estimator on a mock dataset. Not only the new solution delivers a much less noisy estimation, but it also includes a credibility interval associated to the best estimated set of parameters.

In less than 1 000 hours, the ORION-B large program imaged 5 square degrees over the 3 mm atmospheric window at a sensitivity of $\sim$ 0.1 K inside a typical resolution element of $27'' \times 0.6\,\mathrm{km\,s^{-1}}$ with a single beam receiver. IRAM currently develops a new receiver that will contain 25 beams, each covering the same instantaneous frequency bandwidth of 16 GHz times two polarizations as the EMIR receiver used in the ORION-B project. This tremendous observing power could be used to observe 125 square degrees of a fair fraction of the Galactic plane at a sensitivity of 0.1 K in 1 000 hours. Alternatively this added power could be used to increase the sensitivity over the same field of view as the ORION-B project to 20 mK. Figure 5 gives an idea of the evolution of the number of detected lines when the sensitivity increases by an order of magnitude. For NOEMA, the first step towards using multi-beam receivers is to develop the On-The-Fly observing mode in the coming years. In

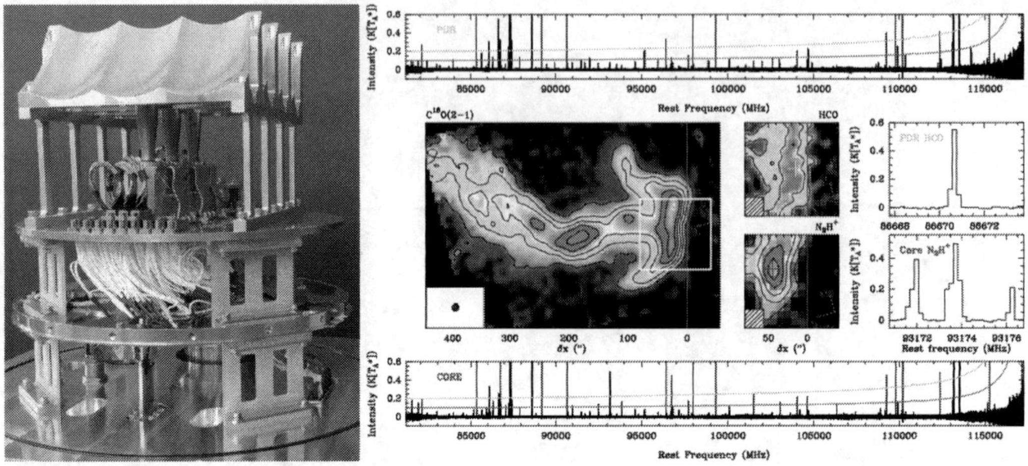


**Figure 5.** **Left:** 3 mm HEMT 3-beam prototype (Courtesy of P. Serres, O. Garnier, & the IRAM front-end group). **Right:** Spectra (smoothed to 195 kHz resolution) of the 3 mm band for two different positions in the Horsehead Nebula, obtained by the WHISPER survey [9–11] at the IRAM-30m. The red and green curves show the 1 and 2$\sigma$ brightness levels achieved in the ORION-B field of view. The photo-dissociation region (PDR) is marked by the green cross on the HCO emission map (middle, top image). The dense core is marked by the blue cross on the $N_2H^+(1-0)$ emission map (middle, bottom image). These fields are reported as a white square on the $C^{18}O$ $(2-1)$ integrated emission map displayed on the left panel. The middle right panels display two lines zoomed from the surveys.

other words, the results obtained by the ORION-B project is just a glimpse into what will be routine science of the interstellar medium in the coming decade.

**Acknowledgements**
This work is based on observations carried out under project numbers 019-13, 022-14, 145-14, 122-15, 018-16, and the large program 124-16 with the IRAM 30m telescope. IRAM is supported by INSU/CNRS (France), MPG (Germany) and IGN (Spain). This work was supported by the French Agence Nationale de la Recherche through the DAOISM grant ANR-21-CE31-0010. It has received financial support from the CNRS through the MITI interdisciplinary programs, and from the Programme National "Physique et Chimie du Milieu Interstellaire" (PCMI) of CNRS/INSU with INC/INP, co-funded by CEA and CNES. J.R.G. and M.G.S.M. thank the Spanish MCINN for funding support under grant PID2019-106110GB-I00.

## References

[1] M. Lombardi et al., A&A**566**, A45 (2014), 1404.0032
[2] J. Pety et al., A&A**599**, A98 (2017), 1611.04037
[3] P. Gratier et al., A&A**645**, A27 (2021), 2008.13417
[4] E. Bron et al., A&A**610**, A12 (2018), 1710.07288
[5] P. Gratier et al., A&A**599**, A100 (2017), 1701.04205
[6] J.H. Orkisz et al., A&A**624**, A113 (2019), 1902.02077
[7] A. Roueff et al., A&A**645**, A26 (2021), 2005.08317
[8] P. Palud et al., Subm. (2022)
[9] J. Pety et al., A&A**548**, A68 (2012), 1210.8178
[10] P. Gratier et al., A&A**557**, A101 (2013), 1305.2371
[11] V.V. Guzmán et al., Faraday Discussions **168**, 103 (2014), 1404.7798

# AUTHOR INDEX

# AUTHOR INDEX

**EDP Sciences – Web of Conferences**
17, Avenue du Hoggar
Parc d'Activité de Courtabouf
BP 112
F-91944 Les Ulis Cedex A
France

CURRAN ASSOCIATES INC.
proceedings.com

ISBN 978-1-7138-6059-4

9 781713 860594